NUCLEAR MAGNETIC RESONANCE

NUCLEAR MAGNETIC RESONANCE

GENERAL CONCEPTS AND APPLICATIONS

William W. Paudler
Department of Chemistry
Portland State University
Portland, Oregon

A WILEY-INTERSCIENCE PUBLICATION
JOHN WILEY & SONS
NEW YORK CHICHESTER BRISBANE TORONTO SINGAPORE

Library of Congress Cataloging in Publication Data:

Paudler, William W., 1932-
Nuclear magnetic resonance.

"A Wiley-Interscience publication."
Includes index.
1. Nuclear magnetic resonance. I. Title.

QC762.P38 1987 538′.362 86-26549
ISBN 0-471-83979-5

Printed in the United States of America

10 9 8 7 6 5 4 3 2

To my family—
Renee, David, Leslie and Gary

PREFACE

The rapid developments of the nuclear magnetic resonance (NMR) method, which led to my writing the quasi-predecessor of this book, *Nuclear Magnetic Resonance* (1971), have amazingly not abated in the intervening 15 years. Quite to the contrary, incredible advances in equipment and applications have been and continue to be made. It would have been impossible to imagine in 1971 that most spectrometers in the 1980s would be based on superconducting magnet technology, that essentially all magnetically active nuclei could almost routinely be examined, that the resolution of NMR spectra of solid samples would approach that of solution samples, and that NMR spectra of human organs, observed in vivo, would be a routine medical diagnostic tool. Yet, all of this—and more—has transpired.

The theory and many applications of the NMR method are presented in this book without detailed quantum mechanical computations on the assumption that the references provided will direct interested readers to the necessary background material. The book is intended to provide sufficiently detailed information to familiarize the reader with not only the theoretical foundation but the numerous applications of this most powerful spectroscopic technique.

Since one of the major expansions of NMR spectroscopy has dealt with its application to nuclei other than the traditional C, H, and F species, the material covered in this book is broad based and draws on examples from across the periodic table. Thus, it is hoped that chemists from all areas of specialization will find this book of interest and of use in their research.

The development of magnetic imaging technology, which has led to a fundamentally new diagnostic tool in medicine, has opened yet another door in the continuing development of the NMR method and should be of interest to readers with expertise in medicine.

Finally, as an important learning tool, a large number of problems are provided both at the end of each chapter and in the appendix.

William W. Paudler

Portland, Oregon
January 1987

ACKNOWLEDGMENT

As is true for any endeavor that culminates in the publication of a book, a number of individuals have in one way or another participated in its preparation. This has been true in this instance as well.

Mrs. Donna Kiykioglu, in addition to providing secretarial support, has been of help beyond the call of duty in her continued willingness to tolerate additional burdens that the writing of this book has placed on her already full daily schedule. My sincere thank you goes to her.

The scientific background and photographic abilities of my son David have helped significantly in developing the necessary figures, tables, and equations. His support has greatly simplified the preparation of this book. I express my most heartfelt thanks to him.

Those readers who have written lengthy articles or books will know that these projects are always greatly facilitated by strong support at the home front. This is certainly very true in my case. I take this opportunity to thank my wife, Renee, for her patience, continuous encouragement, and support.

W.W.P.

CONTENTS

1

ELEMENTARY THEORY OF NUCLEAR MAGNETIC RESONANCE SPECTROSCOPY

1.1 MAGNETIC PROPERTIES OF NUCLEI

An understanding of some of the magnetic properties of atomic nuclei is necessary before any reasonable discussions of the nuclear magnetic resonance (NMR) method can be undertaken. The magnetic properties of atomic nuclei are dependent on and can be described by means of their spin quantum number (I) and their magnetic moment (μ). The spin quantum number of a nucleus, a value that corresponds to the descriptor s for an electron, is either an integral or half-integral value.

While the intrinsic angular momentum of an atomic nucleus is a quantized value of $I^*\hbar$, its magnetic moment (μ) can be expressed in units of the Bohr magneton (or nuclear magneton)

$$\mu \propto \frac{e\hbar}{2mc} \tag{1.1}$$

The value m corresponds to the mass of a proton, and μ is its magnetic moment ($I = \frac{1}{2}$) when the proton is treated as a spherical spinning object with all of its charge distributed uniformally over its surface. In this relationship, the mass (m) is considered to be equally distributed throughout the sphere. Although this is a considerably oversimplified model, order-of-magnitude calculations are possible with it.

In a more general sense, equation 1.2 expresses the magnetic moment of a nucleus with spin I in terms of nuclear magnetons:

$$\mu = g \frac{e\hbar}{2mc} (I) \tag{1.2}$$

In this expression, e and m are the charge and the mass of a proton, respectively, and g is the nuclear g factor. This equation shows that a nucleus with a spin quantum number of zero has no magnetic moment. In fact, the majority of the different elemental isotopes have spin quantum numbers of zero. Only approximately one third of the different isotopes have nonzero spin quantum numbers! The spin quantum numbers, percentages of natural abundances sensitivities (to be discussed in Chapter 2), and nuclear moments of a number of magnetically active elements are listed in Table 1.1. An examination of Table 1.1 reveals the interesting fact that most of the commonly occurring nuclei (e.g., C^{12}, O^{16}, S^{32}, etc.) do not appear on this list of magnetically active ones; they have spin quantum numbers of zero.

Although there is not yet available a comprehensive theory that accounts for the origin of nuclear angular momenta, some empirical generalizations have been developed.

The following three rules delineate these generalizations:

1. If the mass number of a nucleus (A) is odd, the nuclear spin (I) will be

TABLE 1.1 NMR Magnetic Properties of Some Nuclei

Nucleus	Spin quant. No.	Percent abundance[a]	Magnetic moment[b]	Sensitivity[c]
H-1	$\frac{1}{2}$	99.98	2.7926	62.91
H-2 (D)	1	0.0156	0.8574	0.61
H-3 (T)	$\frac{1}{2}$	—	2.9788	76.34
Li-6	1	7.43		0.0005
Li-7	$\frac{3}{2}$	92.57		18.46
B-10	3	18.83	1.8005	1.22
B-11	$\frac{3}{2}$	81.17	2.6880	10.39
C-13	$\frac{1}{2}$	1.108	0.7022	1
N-14	1	99.635	0.4036	0.063
N-15	$-\frac{1}{2}$	0.365	−0.2830	0.066
O-17	$-\frac{5}{2}$	0.037	−1.8930	1.829
F-19	$\frac{1}{2}$	100.00	2.6273	52.383
Na-23	$\frac{3}{2}$	100.00		5.822
Al-27	$\frac{5}{2}$	100.00		12.986
Si-29	$-\frac{1}{2}$	4.70		0.493
P-31	$\frac{1}{2}$	100.00	1.1305	4.173
S-33	$\frac{3}{2}$	0.74	0.6427	0.142
Cl-35	$\frac{3}{2}$	75.4	0.8209	0.296
Cl-37	$\frac{3}{2}$	24.6	0.6833	0.171
K-39	$\frac{3}{2}$	93.08		0.032
K-41	$\frac{3}{2}$	6.91		0.005
Ca-43	$-\frac{7}{2}$	0.13		0.402
Cr-53	$-\frac{3}{2}$	9.54		0.057
Mn-55	$\frac{5}{2}$	100.00		11.013
Fe-57	$\frac{1}{2}$	2.245		0.002
Co-59	$\frac{7}{2}$	100.00		14.762
Ni-61	$\frac{3}{2}$	1.25		0.224
Cu-63	$\frac{3}{2}$	69.09		5.858
Cu-65	$\frac{3}{2}$	30.91		7.201
Zn-67	$\frac{5}{2}$	4.12		0.180
As-75	$\frac{3}{2}$	100.00		1.580
Se-77	$\frac{1}{2}$	7.50		0.436
Br-79	$-\frac{3}{2}$	50.57	2.0991	4.947
Se-79	$\frac{7}{2}$			0.187
Br-81	$\frac{3}{2}$	49.43	2.2626	6.196
Pd-105	$-\frac{5}{2}$	22.23		0.071
Ag-107	$-\frac{1}{2}$	51.35		0.004
Ag-109	$-\frac{1}{2}$	48.65		0.006
Cd-111	$-\frac{1}{2}$	12.86		0.6
Cd-113	$-\frac{1}{2}$	12.34		0.687
Sn-115	$-\frac{1}{2}$	0.35		2.20
Sn-117	$-\frac{1}{2}$	7.67		2.845
Sn-119	$-\frac{1}{2}$	8.68		3.258
Te-123	$-\frac{1}{2}$	0.89		1.133
Te-125	$-\frac{1}{2}$	7.03		1.984

TABLE 1.1 NMR Magnetic Properties of Some Nuclei (*Continued*)

Nucleus	Spin quant. No.	Percent abundance[a]	Magnetic moment[b]	Sensitivity[c]
I-127	$\frac{5}{2}$	100.00	2.7937	5.879
Ba-135	$\frac{3}{2}$	6.59		0.308
Ba-137	$\frac{3}{2}$	11.32		0.432
W-183	$\frac{1}{2}$	14.28		0.005
Os-187	$\frac{1}{2}$			0.001
Os-189	$\frac{3}{2}$	16.1		0.147
Pt-195	$\frac{1}{2}$	33.7		0.615
Au-197	$\frac{3}{2}$	100.00		0.002
Hg-199	$\frac{1}{2}$	16.86	0.4979	0.356
Hg-201	$-\frac{3}{2}$	13.24	−0.5513	0.089
Pb-207	$\frac{1}{2}$	21.11		0.576

[a] Refers to percent of "natural" abundance of the isotopic species.
[b] Magnetic moments are given in multiples of $e\hbar/4\,mc$, where m = mass of a proton and one nuclear magneton is equal to 5.05×10^{-24} erg/gauss.
[c] Sensitivities quoted refer to the response intensity of an equal number of carbon-13 nuclei.
Sources: Taken in part from Varian Associates, *NMR Table*, 5th Ed., and *Other Nuclei NMR on the XL-200, A Collection of Spectra*, 1979.

half-integral. For example, H^1, C^{13}, and N^{15} have spin quantum numbers of $\frac{1}{2}$, and O^{17} has a value of $\frac{5}{2}$.

2. If the mass (A) and the charge (Z) numbers are both even, the nucleus will have a spin quantum number of zero. For example, C^{12}, O^{16}, and S^{32}.
3. If the mass number (A) is even and the charge number (Z) is odd, the nuclear spin has an integral value. For example, H^2 and N^{14}.

Because the magnetic moment (μ) and the angular momentum behave as parallel vectors, the ratio $\mu/I\hbar$ can be used to describe the magnetic properties of a particular nucleus. This ratio is designated by the symbol γ and is referred to as the "magnetogyric ratio"

$$\gamma = \mu/(I\hbar) \tag{1.3}$$

This expression is equivalent to equation 1.2, where $ge/2mc$ has been replaced by γ. Thus, the magnetogyric ratio γ is related by a constant $2mc$ to the nuclear g factor.

1.2 SPINNING NUCLEI IN A MAGNETIC FIELD

If an isolated nucleus with a spin quantum number of $\frac{1}{2}$ is placed into a static magnetic field, it can align itself either with ($I = +\frac{1}{2}$) or against ($I = -\frac{1}{2}$) it. Since the energy difference between these two states is very small (a few millicalories per mole of nuclei), there exists a thermal equilibrium between these two states, and only a very few nuclei are present in excess in the lower energy state. According to the Boltzman distribution law (exp $(2\mu h\pi/kT)$ this excess is less than 1 nucleus in 10,000! Thus, when a large number of nuclei are placed into a magnetic field (H_0) only a very small number will align themselves in excess with it (an almost equal number of nuclei will align themselves against the field). *It is this small excess of nuclei that gives rise to the NMR signals!*

By Newton's law, the applied magnetic field will produce an angular acceleration causing a nucleus to precess, much like a spinning top, in the direction of the applied magnetic field. This phenomenon is depicted schematically in Figure 1.1.

Strictly mechanical concepts suggest that the greater the applied magnetic field (H_0), the more aligned the nuclei will become with this field and, consequently, the smaller will be the alignment angle θ. Concurrent with these changes, the precessional frequency (ω_0) should also increase. However, these factors are not governed by mechanical principles, but, more importantly, by quantum mechanical considerations as well. The latter limit the angular momentum to assuming only whole multiples of $h/2\pi$ (or $\hbar$) when projected in the direction of H_0. Thus, only a *selected* number of alignments of the nuclei in the applied magnetic field are possible.

The precessional frequency, ω_0, of these aligned nuclei is, in addition to the quantum mechanical limitations, also controlled by the magnetogyric ratio, γ, of the nucleus being studied. Equation 1.4 shows these relationships in mathematical form

$$\omega_0 = -\gamma H_0 \tag{1.4}$$

The precessional frequency, ω_0, is referred to as the "Larmor frequency."

The number of possible alignments of a magnetically active nucleus, when placed into a magnetic field, is $2I + 1$! Thus a nucleus with a spin quantum number (I) of $\frac{1}{2}$ (such as H^1, C^{13}, N^{15}), when placed into a static magnetic field, will be either aligned with ($I = +\frac{1}{2}$) or against ($I = -\frac{1}{2}$) it ($2I + 1 = 2$). The aligned

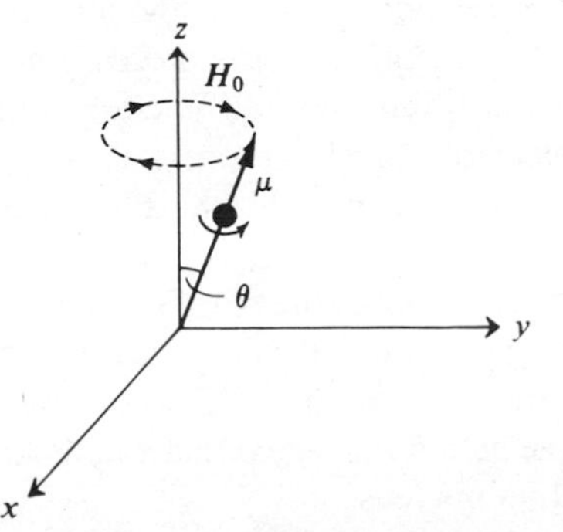

Figure 1.1. Precession of a spinning nucleus in a magnetic field.

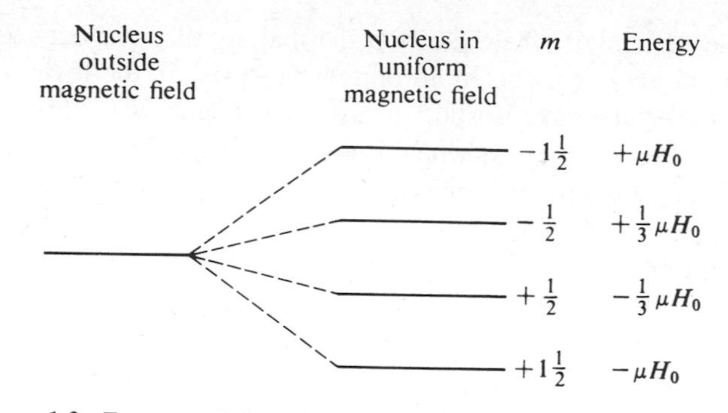

Figure 1.2. Zeeman splitting of nuclear energy levels for an $I = 1\frac{1}{2}$ nucleus.

energy state ($I = +\frac{1}{2}$) corresponds to the lower energy and, consequently, to the more favorable alignment.

This type of "splitting" of energy states is commonly referred to as "nuclear zeeman splitting," in analogy to the magnetic splitting of electronic levels (*Zeeman effect*). The energy of these magnetic states is given by application of equation 1.5.

$$E = -\mu/IH_0m_I \tag{1.5}$$

The magnetic quantum number (m_I) can have values of I, $I - 1$, $I - 2 \ldots - I$. Figure 1.2 is a graphical representation of the four possible nuclear Zeeman levels for a nucleus with a spin quantum number of $1\frac{1}{2}$ (such as B^{11}, S^{33}, or Cl^{35}).

1.3 THE NUCLEAR MAGNETIC RESONANCE EXPERIMENT [1–3]

The fundamental principle governing the NMR technique centers on the induction of transitions between different nuclear Zeeman levels of a particular nucleus. To cause these transitions, a variable radiofrequency (RF) is applied to the nuclei in a magnetic field. The magnetic component of this applied RF, referred to as H_1, acts perpendicular to the applied magnetic field (H_0) which is causing the nuclear alignments. When the frequency of the applied RF is identical to the precessional frequency (ω_0) of the nuclei being observed, a transition between nuclear spin states occurs. Figure 1.3 offers a graphical representation of this process.

The number of intuitively possible transitions is, as is the case for the alignment possibilities, controlled by quantum mechanically imposed limitations. Only those transitions can occur where Δm_I is either $+I$ or $-I$, for example, in the instance shown in Figure 1.2, a transition from the energy level with $E = \mu H_0$ to $E = \frac{1}{3}\mu H_0$, where $\Delta m_I = 1$, is allowed, while the transition to $+\frac{1}{3}\mu H_0$, where $\Delta m_I = 2$, is disallowed. Thus the quantum mechanical selection rules govern the number of observable transitions for any magnetically active nucleus.

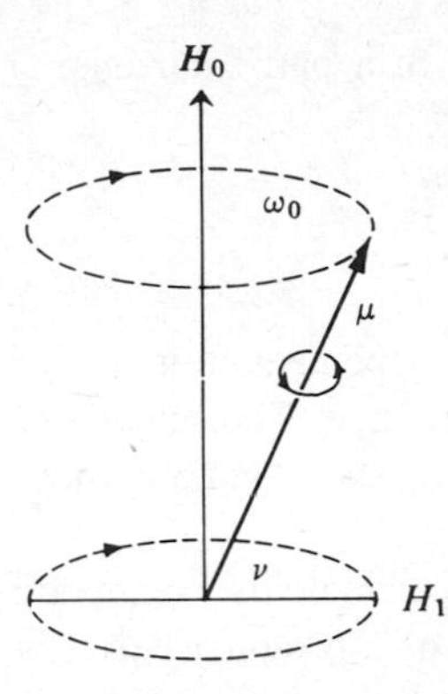

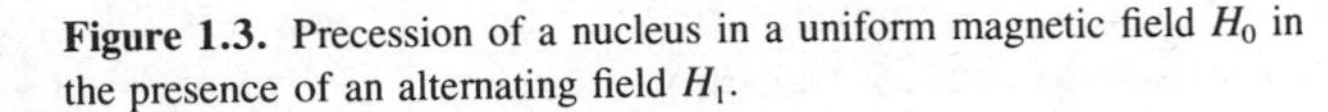
Figure 1.3. Precession of a nucleus in a uniform magnetic field H_0 in the presence of an alternating field H_1.

A nucleus with a spin quantum number of $\frac{1}{2}$ (such as H^1 or C^{13}) and its two allowed magnetic quantum numbers ($+\frac{1}{2}$, $-\frac{1}{2}$) along with the associated energies is represented in Figure 1.4. The energy difference between these two states (ΔE) is given by

$$\Delta E = \frac{\mu H_0(\frac{1}{2})}{(\frac{1}{2})} + \frac{\mu H_0(\frac{1}{2})}{(\frac{1}{2})} = 2\mu H_0 \tag{1.6}$$

The frequency (ω_0) at which these transitions will occur is readily computed by application of the classical Bohr equation 1.7.

$$h\nu = (\mu H_0)/I \tag{1.7}$$

Inclusion of the magnetogyric ratio, γ, into this equation affords equation 1.8.

$$\nu = (H_0 \gamma)/(2\pi) \tag{1.8}$$

Since the energy levels for different nuclei are separated by $\mu H_0/IkT$, the cor-

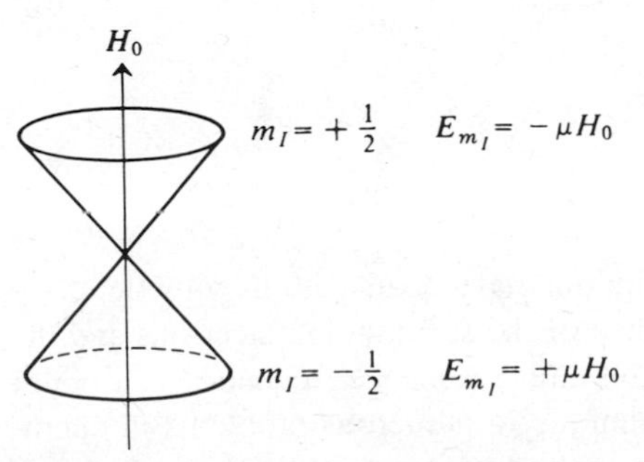

Figure 1.4. The two "cones" created by the spins for an $I = \frac{1}{2}$ nucleus.

responding probability (p) that a selected nucleus will be in a particular energy level is given by equation 1.9.

$$p = [1/(2I + 1)] \times [1 - (m\mu H_0)/IkT] \tag{1.9}$$

Thus, the number of nuclei available for NMR transitions depends, among other factors, on their spin quantum number and their magnetic moment. The sensitivity of particular nucleus toward the NMR technique will vary with these and a number of properties yet to be discussed.

With the principles described so far, it is now possible to calculate the frequency necessary to cause a transition from $m_I = -\frac{1}{2}$ to $m_I = +\frac{1}{2}$ of a proton when it is placed into a particular magnetic field. It is appropriate to recall that magnetic fields are measured in units of gauss. As a reference, the earth's magnetic field is about 0.5 gauss, and the magnetic field of a small (3–4 in. in length) horseshoe magnet is about 15 gauss. Nuclear magnetic resonance spectroscopy deals with very large magnetic fields, with strengths expressed in units of tesla (T), where 1 T = 10,000 gauss.

To calculate the resonance frequency for a *bare* proton, a magnetic field strength of 15,000 gauss (1.5 T) will be used in the computation. The other constant required for this calculation is the nuclear magnetic moment (μ) of a proton (2.79277 Bohr magnetons, where 1 Bohr magneton = 5.0493×10^{-24}:

$$\nu = \frac{2.79277 \times 5.0493 \times 10^{-24}}{6.6256 \times 10^{-27} \times \frac{1}{2}} \times 15{,}000 = 63.9 \text{ MHz} \tag{1.10}$$

Thus, if protons are placed into a magnetic field of 15,000 gauss and a frequency of 63.9 MHz is applied perpendicular to this magnetic field, a transition between allowed spin states will occur. In practice, a frequency in the megahertz range—the FM radiofrequency range—is readily attainable. When this transition occurs, a condition of resonance is said to exist, and the Larmor frequency (ω_0) of the spinning nucleus is identical to the applied frequency. Hence, the term "nuclear magnetic resonance" is applied to this technique.

More than 100 different nuclei have been studied to varying extents by the NMR technique. Table 1.2 lists the NMR field strengths and frequencies of a large number of these nuclei.

1.4 NUCLEAR MAGNETIC RESONANCE OF BONDED ATOMS

In describing the NMR concept, the descriptions have dealt with hypothetical *isolated* nuclei only. The effect that the presence of the surrounding electrons and the bonding of the atoms to other atoms have on their resonance frequency is a major cause for the great applicability of this technique to numerous problems in chemistry, biology, medicine, physics, and many other science-based fields.

TABLE 1.2 NMR Field Strengths and Resonance Frequencies of Some Nuclei

Nucleus	kgauss	Resonance frequency (MHz)
H-1	14.1	60
	23.5	100
	47	200
	70.45	300
	140.9	600
H-2 (D)	18.7	12.21
	47	30.7
H-3 (T)	18.7	84.84
	47	215.9
Li-7	18.7	30.91
	47	77.75
B-11	18.7	25.52
	47	64.185
C-13	18.7	20
	23.5	25
	47	50.3
	93.4	100
N-14	18.71	5.75
	47	14.45
N-15	18.7	8.06
	47	20.27
O-17	18.7	10.78
	47	27.12
F-19	18.7	74.8
	25	100
	47	188.2
Na-23	18.7	21.04
	47	52.92
Al-27	18.7	20.73
	47	52.13
Si-29	18.7	15.80
	47	39.74
P-31	18.7	74.8
	47	188.2
S-33	18.7	6.10
	47	15.34
Cl-35	18.7	7.79
	47	16.32
Cl-37	18.7	6.49
	47	16.315

TABLE 1.2 (*Continued*)

Nucleus	kgauss	Resonance frequency (MHz)
K-39	18.7	3.71
	47	9.34
K-41	18.7	2.04
	47	5.12
Ca-43	18.7	5.35
	47	13.46
Cr-53	18.7	4.50
	47	11.31
Mn-55	18.7	19.62
	47	49.34
Fe-57	18.7	2.57
	47	6.46
Co-59	18.7	17.83
	47	47.24
Ni-61	18.7	7.11
	47	17.88
Cu-63	18.7	21.08
	47	53.03
Cu-65	18.7	22.59
	47	56.80
Zn-67	18.7	4.975
	47	12.512
As-75	18.7	13.62
	47	34.26
Se-77	18.7	15.17
	47	38.145
Br-79	18.7	19.93
	47	50.12
Se-79	18.7	4.15
	47	10.43
Br-81	18.7	21.48
	47	54.03
Pd-105	18.7	3.64
	47	9.16
Ag-107	18.7	3.70
	47	8.10

TABLE 1.2 (*Continued*)

Nucleus	kgauss	Resonance frequency (MHz)
Ag-109	18.7	3.70
	47	9.31
Cd-111	18.7	16.87
	47	42.42
Cd-113	18.7	17.65
	47	44.38
Sn-115	18.7	26.01
	47	65.415
Sn-119	18.7	29.65
	47	74.565
Te-123	18.7	20.85
	47	52.44
Te-125	18.7	25.13
	47	63.20
I-127	18.7	15.915
	47	40.03
Ba-135	18.7	7.90
	47	19.87
Ba-137	18.7	8.84
	47	22.23
W-183	18.7	3.31
	47	8.325
Os-187	18.7	1.83
	47	4.61
Os-189	18.7	1.83
	47	4.61
Pt-195	18.7	17.01
	47	42.78
Au-197	18.7	1.36
	47	3.425
Hg-199	18.7	14.18
	47	35.65
Hg-201	18.7	5.22
	47	13.12
Pb-207	18.7	16.64
	47	41.86

In a sample containing a large number of nuclei, any selected nucleus is subject to a secondary magnetic field generated not only by the induced orbital motions of its own surrounding electrons but also by the diamagnetic moment generated by nearby atoms and molecules. These induced secondary fields will affect the Larmor frequency of the nucleus being examined so that its resonance frequency will be different from that of a bare and isolated nucleus. The strength of these secondary fields is proportional to the applied magnetic field (H_0); that is, the stronger the applied field, the greater the strength of the generated secondary magnetic field.

In effect, the electrons surrounding a nucleus *shield* it from the applied magnetic field so that the field strength ($H_{effective}$) at the location of the nuclear spin is smaller than the applied magnetic field by an amount αH_0:

$$\text{H}_{\text{effective}} = H_0 - \alpha H_0 \tag{1.11}$$

The term α, the *diamagnetic shielding constant*, is independent of the applied magnetic field. It is, however, affected by the electron density around the nucleus and the electronic environment associated with nearby atoms. Thus, the chemical environment of a nucleus will have an effect on the diamagnetic shielding constant. Consequently, the resonance frequencies of various nuclei (with $I > 0$) in a given molecule will be different as long as their chemical environments are different. This brings the NMR technique to the attention of anyone interested in the study of not only the magnetic properties of nuclei but also the chemical environment, with all of its physical and biological implications, of different atoms in molecules.

The values of diamagnetic shielding constants vary from 10^{-5} for protons to about 10^{-2} for the heavier atoms. The difference between the effective field of a proton and a bonded hydrogen atom (a proton shielded by electrons) in a 15,000-gauss (1.50-T) field is approximately $0.00001 \times 15{,}000$, or 0.15 gauss. Since in a 1.5-T magnetic field the proton resonance is expected to be at about 60 MHz, a 0.15-gauss change, caused by environmental or any other effects, corresponds to 600 Hz (1 Hz = 1 cycle per second). Similar although much larger changes, caused by the greater number of surrounding electrons and other factors, are operative in other NMR active nuclei.

These discussions show that only very small changes in magnetic fields, if the RF is kept constant, or in RF, if the magnetic field is kept constant, control the NMR technique. This, along with the earlier described Boltzman distribution situation, points to the fact that very sensitive detection and measurement techniques must be applied for NMR experiments to be successful.

1.4.1 The Original NMR Experiments

The first experimental demonstration of the NMR technique in the condensed phase was described by Bloch and Purcell in 1945. This initial study was soon followed by a major contribution by Arnold, Dharmetti, and Packard [4] who reported the first spectra that showed separate lines for the chemically different protons in a number of alcohols. These spectra were obtained with a 7600-gauss electromagnet

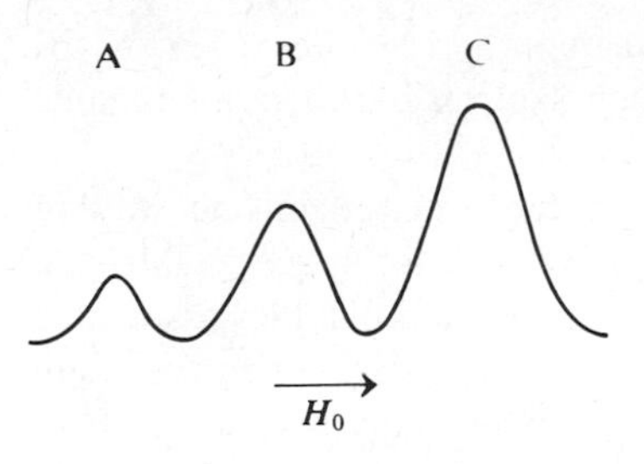

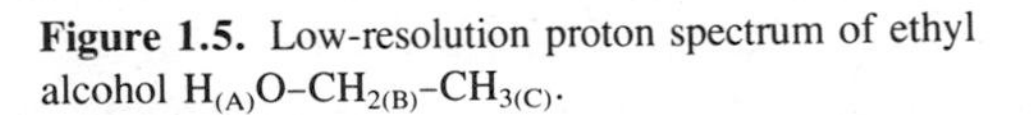

Figure 1.5. Low-resolution proton spectrum of ethyl alcohol $H_{(A)}O–CH_{2(B)}–CH_{3(C)}$.

with 12-in.-diameter pole pieces and a pole gap of 1.75 in. The RF field operated in the 32.4-MHz range. In these early experiments, the magnetic field strength was changed at a sweep rate of approximately 0.05 gauss, covering a total range of about 75 mgauss. Figure 1.5 is a reproduction of the proton spectrum of ethyl alcohol obtained by these scientists.

Since the area ratio of the three peaks is 1:2:3, the respective signals must correspond to the protons on the OH, CH_2, and CH_3 groups, respectively. If the OH peak is assigned a value of 0, the CH_2 and CH_3 peaks are at 16 and 37 mgauss, respectively. The magnetic field, by convention, is always drawn in increasing fashion from left to right (Fig. 1.5).

An analysis of a number of different alcohols reveals that the CH_2 protons, the protons on the carbon bearing the hydroxyl group, absorb at 17 ± 4 mgauss, whereas those on carbons further removed resonate at about 37 mgauss with respect to the hydroxyl proton taken as 0 mgauss. These studies firmly established NMR spectroscopy an ideal tool for the identification of different types of protons by their resonance positions [5–7].

The first NMR studies of C^{13}, the carbon isotope that exists only at 1.08% natural abundance in carbon compounds, were described by Lauterbur and Holm in 1957. The low natural abundance of this isotope coupled with a number of other experimental difficulties severely hampered the early application of C^{13} NMR spectroscopy. Nevertheless, Grant, Lauterbur, Roberts, and Strothers succeeded in obtaining the chemical shifts of the C^{13} nuclei of many different classes of carbon-containing compounds [5, 8].

An early example of a C^{13} NMR spectrum is that of 3,3-dimethyl-1-butene (structure **1**). This compound gives a C^{13} resonance spectrum that shows four different sets of absorption peaks:

$$
\begin{array}{c}
CH_3 \\
| \\
CH_3-C-CH=CH_2 \\
| \\
CH_3
\end{array}
$$

1

On a scale (to be discussed later) where the reference is set at 0, the methyl group carbons resonate at 30.4, the totally substituted carbon at 31.6, the terminal olefinic carbon at 114, and the internal olefinic carbon at 144. Thus, C^{13} NMR spectroscopy lends itself to the study of the carbon skeletal backbone of organic compounds as well [9–11]!

Since most scientists are concerned with molecular structure in one way or another, it is not surprising that studies applying the NMR technique to nitrogen and oxygen were also among the first. The early experimental difficulties associated with C^{13} NMR were equally true for N^{15} NMR studies. These difficulties were magnified by the fact that this nucleus has a negative magnetogyric ratio [5, 10].

Since N^{14} has a spin quantum number of 1, the use of this isotope of nitrogen caused yet a different set of experimental difficulties. Nevertheless, the nitrogen spectra of compounds such as benzofuroxan (**2**) were studied as early as 1964. These studies not only offered some significant insights into the true nature of these unusual compounds but also demonstrated some additional potential applications of this superb experimental technique:

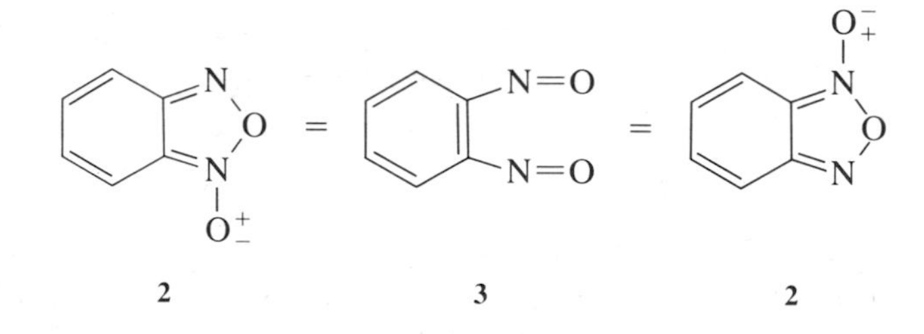

The nitrogen spectrum of compound **2** at 50°C gives only one signal, but at −10°C two signals are observed. The single signal, reported with respect to an appropriate reference, occurs at a value of 369, whereas the low-temperature signals appear at 362.2 and 375.7, respectively. This information establishes the existence of an equilibrium between compounds **2** and **3** at 50°C and a nonexistence of this equilibrium at −10.

The O^{17} spectra of benzofuroxan at −10 and +50°C behave similarly. That is, at −10°C, two different oxygen signals are observed, whereas at +50°C these signals collapse into a singlet [1, 5].

The P^{31} spectrum of adenosine triphosphate (**4**), a molecule of great biological

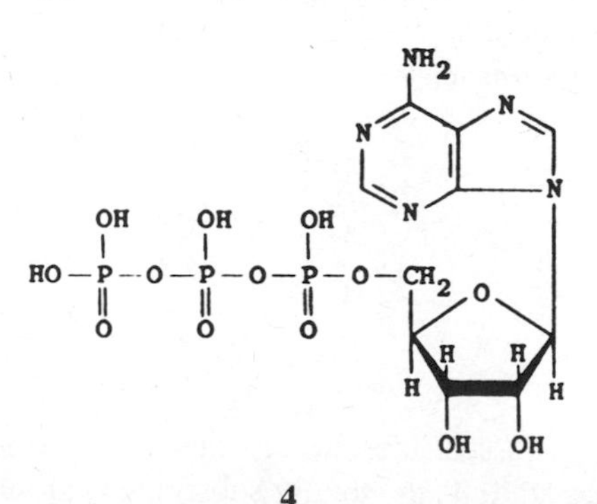

importance, shows three different groupings of phosphorus absorptions, identified as the gamma, alpha, and beta nuclei, respectively, in going from low to high field. Clearly, the general applicability of the NMR technique to molecular structural problems was established 40 years ago!

1.4.2 Fundamental Principles of High-Resolution NMR Spectroscopy

As the experimental techniques applied to NMR spectroscopy improved, it became quickly obvious that the broad peaks obtained during the initial studies were, in fact, composed of many, often overlapping, relatively narrow peaks. For example, the proton spectrum of 1,2-dichloro-1-fluoroethene (structure **4**) showed two peaks even though only one type of proton is present in this molecule. Experimentation also established that the separation of these two peaks is independent of the strength of the applied magnetic field. The relative positions of these peaks with respect to an internal standard are, however, dependent on the strength of the applied magnetic field. Figure 1.6 demonstrates this effect graphically:

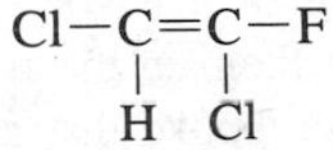

```
Cl—C=C—F
   |  |
   H  Cl
```

4

The nonvariant separation (A) of the two lines is referred to as the "spin–spin coupling constant" or, more commonly, the "coupling constant." This very important value is assigned the symbol J. Chapter 3 is devoted to a discussion of the concept of spin–spin coupling constants.

The actual position of these peaks (1 and 2, 1′ and 2′, respectively, in Fig. 1.6) are given with reference to a standard, generally arbitrarily, set at 0. Since the relative position of the nonvariably separated doublet depends on the applied magnetic field strength, when the RF is kept constant, or the reverse, any description of the resonance position of the doublet of lines must take into account either the strength of the applied magnetic field or that of the applied radiofrequency.

A convenient way to accomplish this is through the following referencing method:

$$\frac{(\text{Position of peak } A) - (\text{position of reference peak})}{H_0} = \delta \tag{1.12}$$

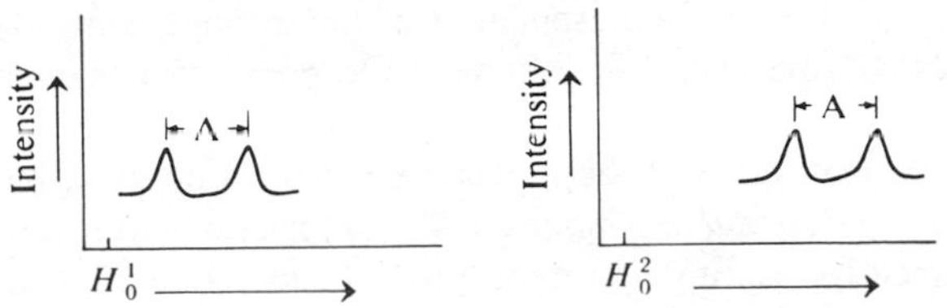

Intensity
A
H_0^1
Intensity
A
H_0^2

Figure 1.6. The effect of H_0 changes on the resonance position of one fluorine doublet in FCCl=CBrF ($H_0^1 > H_0^2$).

If the units of the numbers in the numerator are given in milligauss, H_0 is also given in milligauss, or if the units in the numerator are in Hz (cycles per second), H_0 is given in Hz as well.

The difference of the numerator values are very small (100–1000 mgauss) when compared to the value of H_0 (e.g., 25,000 gauss), and consequently the ratio of the expression in 1.13 is extremely small,

$$\frac{100 \text{ mgauss} - 0 \text{ mgauss}}{25{,}000{,}000 \text{ mgauss}} = 4 \times 10^{-6} = 4 \text{ ppm} \tag{1.13}$$

Since all of the resonance frequencies are of this order of magnitude, the δ values are always given in parts per million (ppm) and are designated by δ(ppm).

1.5 INSTRUMENTATION

Numerous excellent books dealing with NMR instrumentation are available, and the interested reader is referred to them in the reference section at the end of this chapter. Within the context of this book, instrumentation will be described in a general way only [12–14].

The major components of an NMR instrument are:

1. A laboratory magnet to supply the necessary magnetic field (H_0) (three different types of magnets are currently in use: permanent, electro-, and superconducting)
2. An RF transmitter to generate the applied field H_1 with appropriate capabilities to adjust frequencies, intensities, and timing intervals
3. A receiver to determine when the condition of resonance has been reached
4. A device allowing the introduction of the sample between the pole pieces of the magnet
5. A means to cause controlled minor variations of the magnetic (generally accomplished with sweep coils) or the radiofrequency field (the type is determined by the particular instrumentation in use)
6. Computer capability to facilitate analyses, where sample type and/or instrumentation require it.

Figure 1.7 is a schematic representation of the arrangement of the basic components of an NMR spectrometer. Although the permanent magnets are the least expensive, they are also the least flexible, and most modern instruments are equipped with either an electromagnet or the rather expensive to operate, but superbly versatile and stable *superconducting magnets*. The magnetic field in a superconducting magnet is obtained by passing a current through supercooled coils of wire wound around a cylinder that will, ultimately, be the cavity to contain the sample to be

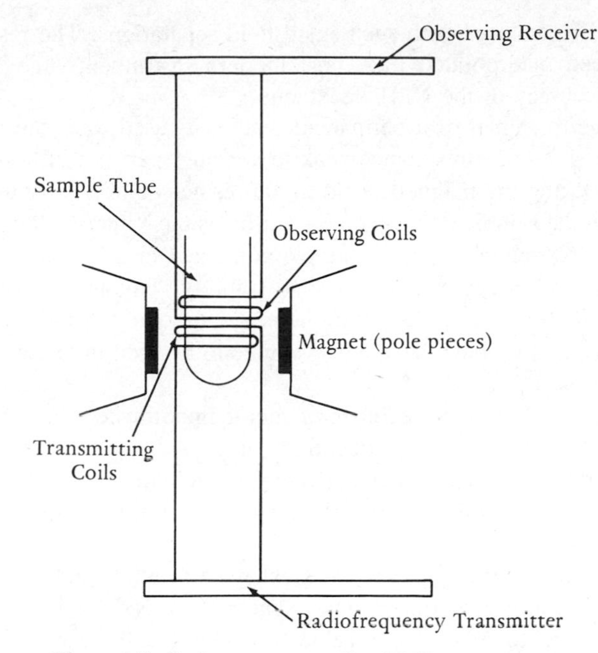

Figure 1.7. Basic components of an NMR spectrometer.

analyzed. When the coils are cooled to near absolute zero (liquid helium is the normal coolant), the resistance to the electric current in the coils is removed and the current density circulates without the requirement to input constant electrical power. The resulting magnetic field is extremely stable. With current technology, fields of up to 11.75 T (a proton resonance frequency of 500 MHz) and higher have been obtained.

Since a particular isotope can resonate over wide ranges of frequencies, depending on its structural environment, in many of the commercial instruments the RF is held constant for any particular type of nucleus while the magnetic field is slowly changed. It is this ''field sweep'' method that was most widely used during the beginnings of NMR spectroscopy. The alternate possibility, an instrument by which the magnetic field is kept constant and the RF is scanned, is important for highly accurate work and has, in fact, become the method of choice for most modern instruments.

1.5.1 Continuous-Wave Spectroscopy

The commercially available continuous-wave (CW) spectrometers operate either in the frequency or field sweep mode. Since, for example, the entire proton resonances are contained within a window of 15 ppm, in a field of 15,000 gauss (1.5 T), much time and effort has been spent to produce fields that are sufficiently stable to permit identification of resonances with such small field separations. The results of these efforts have been incorporated into most modern instruments and have greatly improved the accuracy of the NMR technique.

When an electromagnet or a permanent magnet is used, one can attain stabilization of the position of a resonance peak to within 1 part per billion per hour by electronically locking the magnetic field to the resonance of a substance contained as a reference in the sample being examined. This is the "internal lock" technique. In some of the instruments a water sample contained in a vial separate from the sample container is employed to assure field stability. This is the "external lock" technique, which affords stabilization in the neighborhood of 1 part per 10 million. Both of these techniques have been and continue to be used in many of the routine NMR experiments.

The problem of sensitivity becomes of major importance when the compound being studied is available in small quantities only or is not soluble in appropriate solvents, when the NMR active nuclei are not present in high enough concentrations in naturally occurring atomic isotope mixtures, or when the nuclei being examined exhibit low NMR sensitivity.

Among the troublesome elements of great general utility are the magnetically active isotopes of carbon, nitrogen, and oxygen—C^{13}, N^{15}, N^{14}, O^{17}—along with most of the other isotopes listed in Table 1.1. None of these nuclei can be conveniently and routinely studied by continuous-wave NMR.

1.5.2 Time-Averaging Technique

One of the means employed to overcome sensitivity problems in NMR spectroscopy is based on the average transient computational approach. An NMR spectrum could, in principle, be stored in a computer after the analog signal obtained from the NMR experiment is converted into a digital one. The spectrum might be divided into 4000 different signals with each one stored at a different computer storage location. If the spectrum is repeatedly swept and each set of signals is added to the appropriate storage location, the signals will tend to become reinforced while the instrument noise, being a random process, will tend to cancel itself. The theory of "random processes" states that if n sweeps are carried out and stored, the signal amplitude is proportional to n, and the instrument noise is proportional to the square root of n. Consequently, if 100 sweeps are summed, the signal-to-noise ratio ($S/N = 100/10$) will be improved by a factor of 10. This process is often referred to as "time averaging." An example of it is given in Figure 1.8.

1.5.3 Fourier Transform Spectroscopy

Continuous-wave NMR instrumentation, whether operating in the field frequency or RF mode, even when the time-averaging technique is used, involves a considerable waste of time. For example, in C^{13} NMR spectroscopy at 23.5 kgauss, where the normal range covering all types of C^{13} nuclei is about 5000 Hz, a 1-Hz-

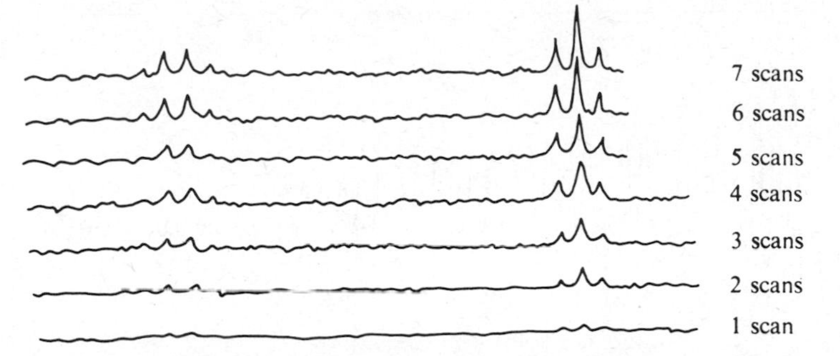

Figure 1.8. Computer-averaged spectra. [Spectra courtesy of Varian Associates, Inc.]

wide resonance line is observed only 1/5000th of the time that the scanning is taking place; the rest of the time is spent observing either other peaks or, more frequently, the baseline only. Even if one increases the scanning speed, the absorption peaks still continue to be accumulated only 1/5000th of the time. While this inefficient time use is not a particular problem when H^1 or F^{19} nuclei are being observed (these nuclei are highly NMR-sensitive and are not diluted by large percentages of NMR inactive isotopes), it becomes a major one when C^{13}, N^{15}, or other low-concentration and/or low-sensitivity nuclei are being investigated [11–14].

A solution to the inefficiency of single-frequency observation is to excite all of the nuclei in a sample simultaneously and to observe the total response of the sample. In principle, this could be done by exciting the sample with x different transmitters tuned to frequencies differing by 1 Hz (if only 1-Hz resolution were required) and observing these x different signals with an equal number of receivers, again tuned to frequencies differing by 1 Hz. To cover 1000 Hz, 1000 different transmitters and receivers would be required!

Fortunately, another technique for excitation and detection is available. A short RF pulse not only produces excitation at one frequency but excites a finite band width of frequencies. For example, a short pulse of 50 μs generates a band width of excitation in excess of 5000 Hz, a range more than adequate for observation of all different C^{13} nuclei. To excite the various nuclei in a sample by using this broad-band, short-time frequency, very high RF power is, however, required. The sample responds to these frequencies by absorbing those of appropriate value for each of the different nuclei present (subject, of course, to the presence of the proper magnetic field, whatever the magnet!). The absorbed and, ultimately, analyzed frequencies are the same absorption frequencies observed with a CW instrument.

The detector, in this broad-band technique, observes a pattern called a "free-induction decay" (FID). It is an exponentially decaying sine wave! The frequency of the sine wave is the difference between the center frequency of the RF pulse and the RF (the Larmor frequency, ω_0) of the nucleus being excited. Figure 1.9 is a graphical representation of this process.

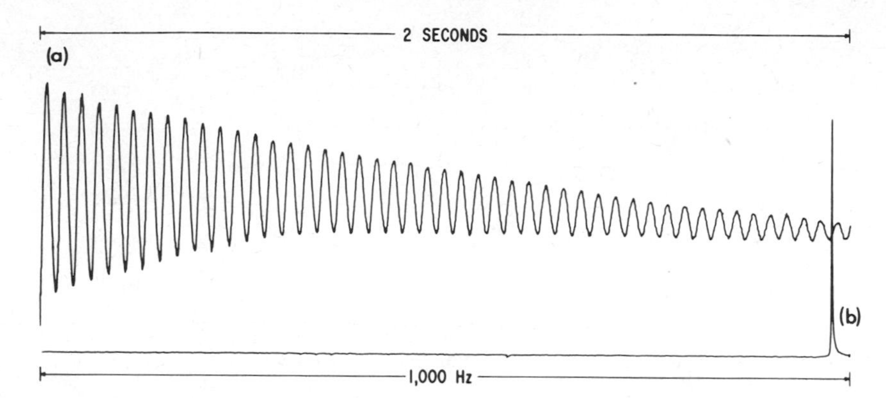

Figure 1.9. Free induction decay spectrum of one nucleus. [G. C. Levy and G. L. Nelson, *Carbon-13 Nuclear Magnetic Resonance for Organic Chemists*, New York, John Wiley, 1972. Reproduced with permission.]

The sine wave in Figure 1.9 contains 46 cycles and was recorded over a 2-s period. Consequently, the precessional frequency of this particular nucleus is 46/2, or 23 Hz (cycles per second). The Larmor frequency of the nucleus, in this instance, is 23 Hz from the center frequency of the RF excitation source [6].

When more than one nucleus is involved, the FID becomes very complicated, since each type of nucleus has its own sine wave. Thus, visual analyses are no longer possible, and a Fourier transformation (FT) must be performed. This mathematical treatment abstracts all of the individual frequency components from the complex sine wave form. A graphical representation of a two-nucleus FID is given in Figure 1.10. It is understood that slowly decaying FIDs correspond to sharp resonance peaks whereas rapidly decaying ones are associated with broad peaks.

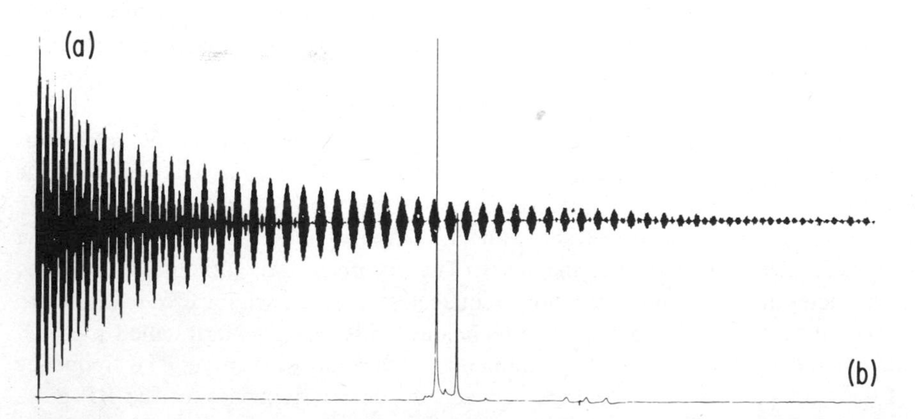

Figure 1.10. Free induction decay spectrum of two nuclei. [G. C. Levy and G. L. Nelson, *Carbon-13 Nuclear Magnetic Resonance for Organic Chemists*, New York, John Wiley, 1972. Reproduced with permission.]

This very cursory description of the FT nmr technique is sufficient to demonstrate that there is no difference in the results obtained from either continuous-wave or Fourier transform NMR. The two techniques give the same spectral information. The advantage of the FT method lies in its considerably improved sensitivity and speed. The advantage of the FT method over the CW technique in a single pulse is proportional to the square root of F/Δ (the total chemical shift range is F, and Δ corresponds to the line width of the narrowest signal). For example, in C^{13} FT NMR spectroscopy, the theoretical increase in sensitivity is the square root of 5000/0.5, or approximately 100 (in practice, factors of 30–50 are normally obtained).

In FT NMR as many as 1 million pulses can be used ranging from 0.1s to several seconds apart. The sensitivity improvement in FT NMR of >10 affords a time improvement in excess of 100. The combination of time averaging and the FT NMR technique is a superb improvement and has enabled a considerable broadening of the applications of the NMR method.

1.5.4 Superconducting Magnets

As discussed earlier, an increase in the strength of the magnetic field brings with it an increase in resolution (i.e., sensitivity) of the NMR technique. However, the maximum magnetic field strength that can be obtained with a magnet, whose core is composed of iron, is 2.0–2.3 T. Consequently, to operate at magnetic fields greater than 2.3 T—that is, at field strengths beyond the upper saturation limit of the iron core magnets—fields generated by other means must be employed. Although the concept of obtaining magnetic fields through superconducting technology has been known for some time, its commercial development did not take place until the need for stable ultrahigh magnetic fields in NMR spectroscopy became obvious. The development of "supercon" application technology was relatively slow, since exhorbitant costs were associated with maintaining the low temperatures (at the boiling point of liquid helium!) required for generating the supercon-maintained fields [15–17].

The new generation of superconducting magnets developed during the 1970s have much superior insulation for the maintenance of the liquid helium coolant. This improvement has decreased the time interval between refills from once every 3 days to once every 2–3 months.

With a typical refill volume of 30 L of liquid helium and an average price of $5.00/L, the annual cost has *decreased* from $18,000 per year in 1967 to $600 per year in 1985. This tremendous improvement has brought the use of supercon NMR spectrometers within the realm of the financial capability of many laboratories. By 1980 the proton spectra of many proteins were obtained at fields of 600 MHz, a frequency that lies somewhere midway between radiofrequencies and microwaves. It is most impressive to recall that the magnetic field strengths used for NMR spectroscopic analyses changed from 30 MHz in the early 1950s to 600 MHz in the 1980s!

The use of these superbly stable and intense magnetic fields in conjunction with

FT analysis has immensely facilitated the study of low-sensitivity and low-concentration nuclides, whether metalic or nonmetalic [16, 17]. The field stability of the supercons also permits the use of increased bore sizes, which, in turn, enable the use of greater-diameter sample tubes (20-mm-diameter tubes are easily employed), thus again increasing the sensitivity of the experiments. In fact, within certain modifications, limitations, and reliance on computer imaging technology, the magnet cavity can be made large enough to hold a human body!

The materials, which are suitable for use in carrying the electric current resistance-free, have one thing in common: they do so only at very low temperatures. The wire materials commonly used in superconductors are an NbTi alloy operating at about 10 K and a somewhat more powerful Nb^3 alloy which is superconductive below 18 K.

1.6 RELAXATION PHENOMENA

The induced and allowed transitions from one spin state to another occur with equal probability in either direction between any pair of energy states as long as Δm = 1. Consequently, to be able to observe a resonance signal, there must be a change in the total magnetization of the sample. This is possible only if the number of transitions from the lower to the upper energy states is not the same as the reverse. Thus, the change in magnetization is due to the inequality of the population distribution among the different nuclear Zeeman levels. This inequality, as already mentioned, is extremely small, and because more transitions will occur from the lower to the upper states, the resonance signal will disappear very rapidly at the resonance condition, and the magnetization will vanish as the spin states become evenly distributed again. The reestablishment of the Boltzman law determined thermal equilibrium distribution is, consequently, required for a continuation of the NMR experiment. The various mechanisms that facilitate the equilibrium reestablishment are referred to as "relaxation phenomena."

1.6.1 Spin–Lattice Relaxation [5, 6, 13]

The energy exchange between the normal thermal motions of a particular molecule or atom and its surrounding molecules or atoms, referred to as the "lattice," is one of the factors that cause the maintenance, or are responsible for the reestablishment, after excitation, of the thermal equilibrium in a spin system. This process is referred to as "spin–lattice relaxation," and it occurs as a result of the interactions of the magnetic moments of nuclei with random fluctuating magnetic fields which, themselves, are caused by the thermal motion of the nuclei in the molecule. The lifetime of a particular set of nuclides in either the upper or lower energy state is described as T_1 the spin–lattice relaxation time. When the nuclei are irradiated, T_1 represents the time required for the nuclear spins to return, exponentially, to the original Boltzman distribution. This concept is mathematically expressed by equation 1.14:

$$T_1 = \frac{1}{2W}, \tag{1.14}$$

In this equation, W stands for the mean probabilities of an upward ($W+$) and a downward ($W-$) transition and is related to the population of nuclei, n, in different spin states through equation 1.15:

$$\frac{dn}{dt} = -2W(n - n_{eq}) \tag{1.15}$$

The difference $n - n_{eq}$ is temperature-dependent. Consequently, when temperature is explicitly considered, equation 1.15 becomes expression 1.16:

$$\frac{dn}{dt} = -2W(n - n_{eq})_0 \times e^{-t/T_1} \tag{1.16}$$

Thus, the difference $(n - n_{eq})$ is decreased by a factor e after time T_1. After 1 T_1 has elapsed $(1/e)$ 36.8% of the nuclei have returned to the original distribution ratio, and more than 99% will be at that distribution after $5T_1$ values. The magnitude of T_1 varies considerably, depending not only on the type of nucleus but also on its environment. Liquids usually have T_1 values of 0.01–100 s, and solids have much longer relaxation times (often on the order of days!).

There exist several different mechanisms that contribute to the value of the spin–lattice relaxation time of magnetically active nuclei.

1.6.1.1 Dipole–Dipole Relaxation

Spin–lattice relaxation can result from fluctuating fields caused by dipole–dipole interaction between neighboring magnetic nuclei or between one nucleus and neighboring unpaired electrons. In most organic molecules, the C^{13}/H^1 dipole–dipole relaxation mechanism dominates the C^{13} spin–lattice relaxation. The two factors that determine the efficiency of this relaxation mechanism are (1) the magnetogyric ratio of the nucleus causing relaxation, and (2) the proximity of that nucleus to the nucleus being relaxed.

In H^1 and C^{13} NMR spectroscopy, protons tend to dominate this relaxation mechanism because of their high magnetogyric ratio (dipole–dipole relaxation depends on the square of the magnetogyric ratio). The proximity factor is extremely important, since it is proportional to the reciprocal of the internuclear distance (r) taken to the sixth power. For example, C^{13} atoms bonded to protons relax very rapidly, whereas those not bonded to one (a trisubstituted carbon bonded to another carbon) do not. The adjacency of an atom with a lone pair of electrons (such as on a nitrogen atom) is even more efficient in improving T_1 than is a bonded proton.

1.6.1.2 Spin–Rotation Relaxation

Freely rotating methyl groups and small molecules can be relaxed by a mechanism involving quantum rotational states of the functional group or molecule. Where this process is in operation, it frequently competes in efficiency with the dipole–dipole relaxation mechanism. In fact, it is this spin–rotation relaxation process that is the dominant feature in the relaxation involving carbons not bonded to protons.

1.6.1.3 Chemical Shift Anisotropy

This contributor to the spin–lattice relaxation process is a readily averaged minimal contributor when the molecule being examined can rotate freely in its matrix (solution spectra, for example). However, in solids the contribution can be significant causing fluctuating magnetic fields when the molecular ''tumbling'' is not averaged. This mechanism will be covered in the discussion on ''magic angle'' solid-state NMR spectroscopy.

1.6.1.4 Scalar Relaxation

A particular magnetically active nucleus that is spin–spin coupled to another rapidly spin–lattice relaxing one can itself be relaxed as a result of the fluctuating spin–spin interactions between the two nuclei. This process becomes of some significance if one of the nuclei in a spin system has a spin quantum number greater than $\frac{1}{2}$. For example, protons bonded to N^{14} frequently have very great line widths.

1.6.1.5 Determination of Spin–Lattice Relaxation Times

The experimental techniques employed to determine spin–lattice relaxation times have most recently been based on the application of pulsed FT NMR (see Chapter 9 for details).

In general, the T_1 values for protonated carbons are inversely proportional to the number of attached protons, if the molecules are tumbling freely. Thus, values close to a 2 : 1 ratio for CH and CH_2 are observed. The freely rotating CH_3– groups have an additional relaxation time advantage in these molecules.

The T_1 for the carbon atoms in cyclohexane is 17 s, whereas in benzene a value of 28.5 s has been determined. By comparison, the carbons coupling the two benzene rings in biphenyl have a T_1 of 61 s.

The *iso*-octane molecule provides a good example of the variations in T_1 observed for differently protonated carbons. In this molecule the methyl group carbons have a T_1 of 9.3–9.5, the methylene carbon has a value of 13, the methine carbon value is 23, and the all-substituted carbon has a T_1 time of 68 s.

1.6.2 Spin–Spin Relaxation

When a sample is placed into an appropriate magnetic field and the proper RF is applied for a sufficient length of time that the individual magnetic moments are drawn into phase and the relationship $\omega_0 = -\gamma H_0$ is fulfilled, and if the RF power is then removed, the nuclei will continue to precess about the magnetic field for an amount of time dependent on the spin–lattice relaxation time, T_1. The decay rate may, however, be faster than predicted from T_1 alone. This rate increase occurs if there are neighboring magnetic moments in the sample whose moments do not move about rapidly and make only small contributions to the magnetic field. This will cause some inhomogeneity of the magnetic field on a microscopic scale, and the individual nuclei will consequently precess in slightly different magnetic fields and will lose phase quickly because of the differences in their precessional frequencies (some applications of this phenomenon will be described in Chapter 9). The spin–spin relaxation time, T_2, the time factor involved in this relaxation process, is defined by equation 1.17:

$$T_2 = \frac{1}{\Delta\omega_0} = \frac{-1}{\gamma \Delta H_0}, \tag{1.17}$$

where ΔH_0 and $\Delta\omega_0$ correspond to the change of the precessional frequency (ω_0) and magnetic field (H_0) caused by the spin–spin relaxation process.

1.7 SATURATION PHENOMENA

Under conditions where the energy level populations equalize, no absorption of the applied RF can occur, and the sample is described as being "saturated." This phenomenon can occur if the magnitude or the time of application of the oscillating field in an NMR experiment becomes too great. This causes a reduction of the excess nuclear population and will decrease the probability of further absorption. The experimentally observable result of the effect, *saturation*, is a decay, in time to zero intensity, of the resonance signal. An average value of about 0.0001 gauss for H_1 in some typical liquids has been calculated as a field intensity that will cause appreciable saturation. This occurrence also manifests itself by the appearance of spurious absorption peaks in the recorded spectrum. These peaks, which occur with the absorption of two or more quanta, can be of potential use in certain structural problems [5, 11, 12, 14].

1.8 PROBLEMS

1. Which of the isotopes of carbon, hydrogen, and nitrogen are suitable for NMR investigations?
2. Estimate the spin quantum numbers of the following elements: Cl^{36}, Cl^{37}, H^2, B^{11}, C^{12}.
3. What are the controlling factors in determining the number of possible transitions of a nucleus with a spin quantum number of 1? How many transitions are allowed?
4. Differentiate between CW and FT NMR spectroscopy.
5. Calculate the resonance frequency for C^{13} in a 25-kgauss field.
6. Give some of the factors that determine the size of the diamagnetic shielding constant.
7. How many different absorption signals would one expect for C^{13} and H^1 in 2-ethyl-butadiene?

8. Why is the fluorine absorption in 1,2-dichloro-1-fluoroethene a doublet?

9. If two signals are separated by 150 mgauss in a 25-kgauss field, what is Δ(ppm) if one of the signals is set at 0 mgauss?

10. In an FT NMR experiment, the FID of one particular nucleus lasts 3 s and contains 68 cycles. What is the precessional frequency of this nucleus?

11. Differentiate between T_1 and T_2 and discuss the major reasons that cause changes in these relaxation times.

12. What is meant by the term ''saturation''?

REFERENCES

1. P. Diehl, H. A. Christ, and F. B. Mallory, *Helv. Chim. Acta.* **45,** 504 (1962).
2. R. A. Friedel and H. L. Retcovsky, *J. Am. Chem. Soc.* **85,** 1300 (1963).
3. H. S. Gutowsky, D. W. McCall, and C. P. Slichter, *J. Chem. Phys.* **21,** 279 (1953).
4. J. T. Arnold, S. S. Dharmetti, and M. E. Packard, *J. Chem. Phys.* **19,** 507 (1951).
5. J. A. Pople, W. G. Schneider, and H. J. Bernstein, *High Resolution Nuclear Magnetic Resonance*, McGraw-Hill, New York, 1959.
6. F. Bloch, W. W. Hansen, and M. E. Packard, *Physiol. Rev.* **9,** 127 (1946).
7. E. M. Purcell, H. C. Torrey, and R. V. Pound, *Physiol. Rev.* **69,** 37 (1946).
8. P. C. Lauterbuhr, *J. Chem. Phys.* **26,** 217 (1957).
9. A. Allerhand, D. Doddrell, and R. Komorski, *J. Chem. Phys.* **55,** 189 (1971).
10. T. C. Farrar and E. D. Becker, *Pulse and Fourier Transform NMR*, Academic, New York, 1971.
11. G. C. Levy and G. L. Nelson, *Carbon-13 Nuclear Magnetic Resonance for Organic Chemists*, Wiley-Interscience, New York, 1972.
12. D. Shaw, *Fourier Transform NMR Spectroscopy*, Elsevier, New York, 1976.
13. J. B. Lambert, H. F. Shurvell, L. Verbit, R. G. Cooks, and G. H. Stout, *Organic Structural Analysis*, Macmillan, New York, 1976.
14. W. Paudler, *Nuclear Magnetic Resonance*, Allyn and Bacon, Boston, 1974.
15. W. Paudler and M. V. Jovanovic, *Org. Magn. Reson.* **9,** 192 (1982).
16. F. W. Wehrli and S. L. Wehrli, *J. Magn. Reson.* 44, (1981).
17. F. W. Wehrli, G. C. Levy, ed., in *NMR Spectroscopy: New Methods and Applications*, ch. 2 ACS Symp. Series 191, Washington, DC, 1982.

2

CHEMICAL SHIFT—GENERAL CONSIDERATIONS

2.1 CONVENTIONS AND TERMINOLOGY

Equation 1.12 is the basic expression that permits a standardized expression of the resonance values independent of the applied magnetic field and the resonance frequency. For example, the value δ, computed from a spectrum obtained at either 60 or 200 MHz, will be numerically the same. Since the resonance frequencies of a particular atom will differ depending on the chemical environment it exists in, the difference is referred to as the "chemical shift" and is given in units of parts per million (ppm).

As a particular nucleus becomes increasingly more *shielded*, because of the surrounding electrons or other factors, it will take a stronger applied magnetic field to cause resonance. These changes are graphically presented in Figure 2.1, with the magnetic field (or the RF) *increasing* from left to right.

Any absorption that is to the *right* of another one, represents a more shielded nucleus with respect to the one that resonates to the *left* of it (peak B vs. peak A in Fig. 2.1). Conversely, peak A is more shielded than peak B. Increasing shielding is associated with *diamagnetic* effects, and increasing *deshielding* results from *paramagnetic* contributions.

2.2 CHEMICAL SHIFTS OF SELECTED NUCLEI

Numerous examples of magnetic field strengths and RFs required to cause resonance in a large number of different nuclei are given in Table 1.2. A graphical representation of a number of the nuclei listed in Table 1.1, indicating where some nuclei would resonate if placed into a magnetic field of 14,092 gauss, is presented in Figure 2.2. The relative intensities (Y axis) refer to a comparison of signal intensities obtained for an equal number of the different nuclei with hydrogen used as the 100% reference.

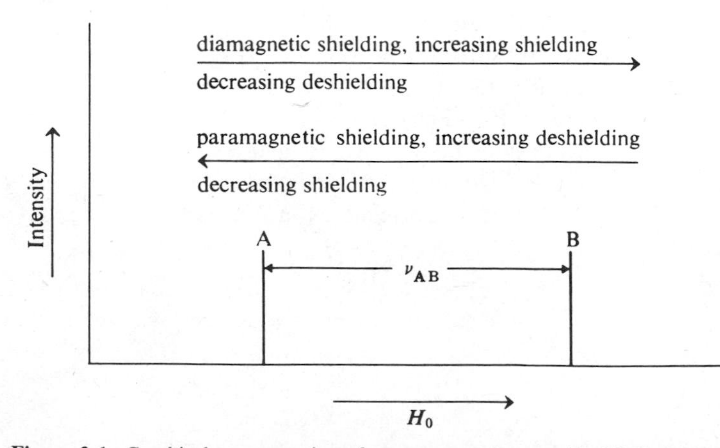

Figure 2.1. Graphical representation of some conventions used in NMR spectroscopy.

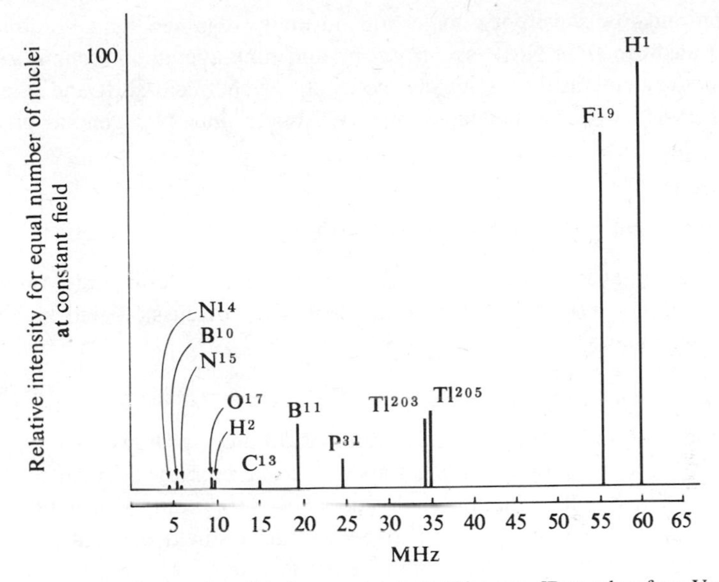

Figure 2.2. Resonance frequencies of various nuclei at 14.092 gauss. [Data taken from Varian Assoc. NMR Table, 5th edition, and used with permission of Varian Associates.]

Clearly, among all of these magnetically active nuclei, the H^1 and F^{19} species are the most sensitive, and C^{13}, N^{14}, and P^{31} are among the less sensitive nuclei toward detection by the NMR method. These very significant sensitivity differences are, among other things, a consequence of the size of the magnetic moment, as shown by the following data:

NUCLEUS	SENSITIVITY	MAGNETIC MOMENT
H^1	100	2.7927
F^{19}	84	2.6273
Tl^{205}	19	1.6114
P^{31}	7	1.1305
C^{13}	2	0.7022
Se^{77}	0.7	0.5333
N^{15}	0.01	−0.2830

From the instrumentation point of view, the sensitivity problem becomes even more enhanced when it is recalled that most of the magnetically useful isotopes of many elements are present at very low concentrations in naturally occurring samples of these elements. For example, C^{13} and N^{15} nuclei are present only to the extent of 1.08 and 0.48%, respectively. This very cursory sensitivity discussion makes it obvious that it may be considerably more difficult to obtain routine NMR spectra of the elements of carbon and nitrogen and most other magnetically active nuclei than those of fluorine and hydrogen. In fact, spectra of these and other low-concentration and low-sensitivity nuclei are normally obtained by a combination of Fourier transform (FT) NMR spectroscopy and time-averaging techniques.

It is convenient, in discussing the concepts of chemical shift and related phenomena, to separate the magnetically active nuclei into two general groups, the nonmetal and metal nuclides.

2.2.1 Magnetically Active Nonmetal Nuclides

Although the number of nonmetal magnetically active nuclei is large, only a select few find common use in NMR spectroscopy. This discussion will be limited to these few nuclides.

2.2.1. Proton Chemical Shifts

To be able to describe the chemical shifts of different protons in a standardized manner, it is useful to agree on one standard, to be employed by all investigators in the field. The requirements for such a standard are that it resonate either upfield (more shielded) or downfield (less shielded, more deshielded) from most protons.

The generally selected internal (and external) standard for H NMR is tetramethylsilane (TMS), $(CH_3)_4Si$. The 12 protons of this compound are chemically identical and resonate at a frequency that places them at a very shielded position. In fact, very few organically bound protons resonate at more shielded positions than these. The TMS protons are designated as resonating at $\delta = 0.0$ ppm. With this compound as reference, the vast majority of protons resonate at lower magnetic field strengths, downfield, from the TMS protons. (It has been shown that the chemical shift of TMS in different aromatic solvent varies as much as 0.15 ppm [1]. Thus, a note of caution is appropriate!)

Since TMS is, effectively, a silicon analog of a hydrocarbon, it is not significantly soluble in very polar solvents. Consequently, a TMS derivative with greater solubility in polar solvents (such as water) has been developed for use as a reference. This material, $(CH_3)_3$–Si–$(CH_2)_3$–CO_2^- Na^+, is the reference standard of choice for aqueous solution studies. The protons of these methyl groups are generally accepted as resonating at $\delta = 0.00$ ppm, although shifts of 0.2 ppm downfield from TMS have been observed in a number of instances.

Since the TMS proton chemical shifts are all upfield from the majority of protons in organic compounds, the δ values obtained are actually numerically negative and *increase numerically* in absolute values on the conventional NMR presentations from *right* to *left*. Traditionally, δ values are given as *positive* when they represent protons *downfield* from TMS and as *negative* when they represent protons *upfield* from it.

This mathematically confusing situation was addressed by G. V. Tiers [2], who proposed a new scale, τ which is related to δ by equation 2.1:

$$\tau = 10.00 - \delta(\text{ppm}) \tag{2.1}$$

The selection of 10.00 is based on the observation that very few protons resonate at δ(ppm) values greater than 10. Thus, this scale gives chemical shift values that increase from left to right on the conventional NMR graphs. Although τ was in use for a number of years, and many NMR data found in the literature of that time (the late 1960s to early 1970s) are given in this unit, it is no longer and the original δ(ppm) convention is employed again.

Since the two units are not mathematically expressed identities but rather represent a scale, the symbol used to designate them must precede the number to which it refers. For example, δ4.55(ppm) is correct, but 4.55δ(ppm) is not.

During the initial developmental phases of NMR it quickly became obvious that there are typical resonance frequencies associated with chemically similar protons. A compilation of a number of the NMR resonance frequencies associated with different types of protons is shown in Figure 2.3. A glance at this figure shows that as the proton frequencies shift upfield (become more shielded), the hybridization of the carbons bonded to the protons changes from sp^2 to sp^3. Consequently, protons bonded to olefinic and aromatic carbons resonate at more deshielded positions (5–11(ppm)) than protons bonded to saturated carbons (0–5(ppm)). The reason for the highly shielded resonance frequencies of protons bonded to acetylenic carbons will be dealt with in Chapter 5.

The chemical shift range of protons bonded to sp^3 carbons can be further subdivided. The protons become more deshielded as the following pattern develops: CH_3-X < X-CH_2-Y < CH-XYZ. Thus, protons bonded to terminal carbons are more shielded than those bonded to internal carbons. Additionally, the nature of the substituents also effect the chemical shift of the attached proton. For example, among the methyl halides, the methyl protons become more deshielded in the

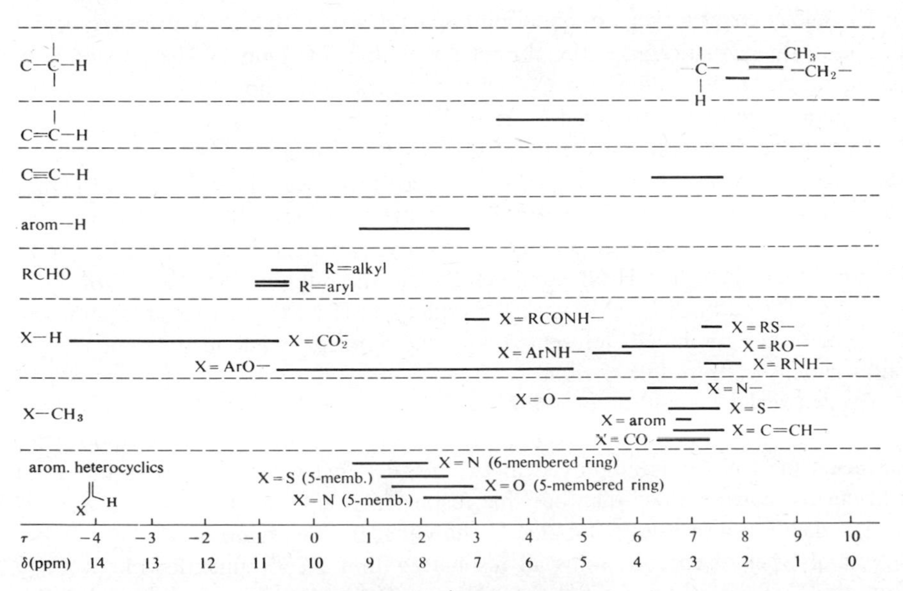

Figure 2.3. H[1] chemical shifts.

sequence CH_3–I < CH_3–Br < CH_3–Cl. Clearly, as the electronegativity of a substituent increases, so does the deshielding of the proton bonded to the same carbon as the substituent.

The chemical shifts of protons bonded to olefinic carbons can also be further subdivided. The data in Figure 2.3 demonstrate that protons bonded to olefinic carbons resonate at more shielded (δ5–7(ppm)) positions than those that are part of an aromatic system (δ7–9(ppm)). As is true for sp^3-bonded protons, "electron-withdrawing" substituents cause additional deshielding of any of these protons. Additional conjugation, such as in 1,3-dienes, for example, deshields these protons even further.

Another very strong deshielding effect occurs when an aromatic carbon is replaced by a nitrogen (benzene vs. pyridine). In this instance the proton bonded to the adjacent carbon (the α carbon) experiences additional drastic deshielding (often in excess of 1 ppm!).

As a group, aldehydic protons resonate at the most deshielded positions (δ10–11(ppm)) of any of the typical protons.

2.2.1.2 Carbon Chemical Shifts

The sole magnetically active carbon isotope of general utility is C^{13}, with a spin quantum number of $\frac{1}{2}$. The low natural abundance of this isotope (1.08%) along with long relaxation times (of the order of 2 min in some organic compounds!) necessitates the use of FT NMR techniques coupled with time-averaging procedures. Since FT NMR instrumentation has been developed to a stage where C^{13} NMR spectra are now almost as readily obtained as proton and fluorine spectra, these limitations are no longer a significant problem.

While the chemical shifts of most protons fall within a range of 10 ppm, those of C^{13} nuclei cover a range of about 600 ppm. (Carbons that are part of molecules that have diamagnetic properties do not fall within this range!) The greater resonance range of the different C^{13} nuclei provides even more detailed structural information than those provided by H NMR.

During the developmental stage of C^{13} NMR, a number of different reference standards were used. Among these, C^{13}-enriched CS_2 was the reference of choice for some time. Consequently, the older C^{13} literature describes the C^{13} chemical shifts with reference to this compound. Currently, and most logically, the C^{13} atoms in tetramethylsilane, the H NMR reference, are the reference for C^{13} NMR spectroscopy. The carbons in TMS, like the protons in H NMR, are set at 0 ppm, and the spectra are graphically presented with the applied magnetic field (or the RF field) increasing from left to right.

On this scale, the carbon in iodoform (CHI_3) resonates at −139(ppm), and the carbonyl carbon in acetone at 205(ppm). The same terminology as developed for and used in H NMR is employed in C^{13} NMR. That is, a C^{13} that resonates at 50(ppm) is more shielded than one that resonates at 200(ppm).

The data given in Figure 2.4 clearly show that the vast majority of the different structural types of carbon atoms all resonate within a 250(ppm) downfield range from the TMS reference. As is true for H nmr, C^{13} NMR resonances are to a large extent dependent on the hybridization of the carbon. Saturated carbons (sp^3) res-

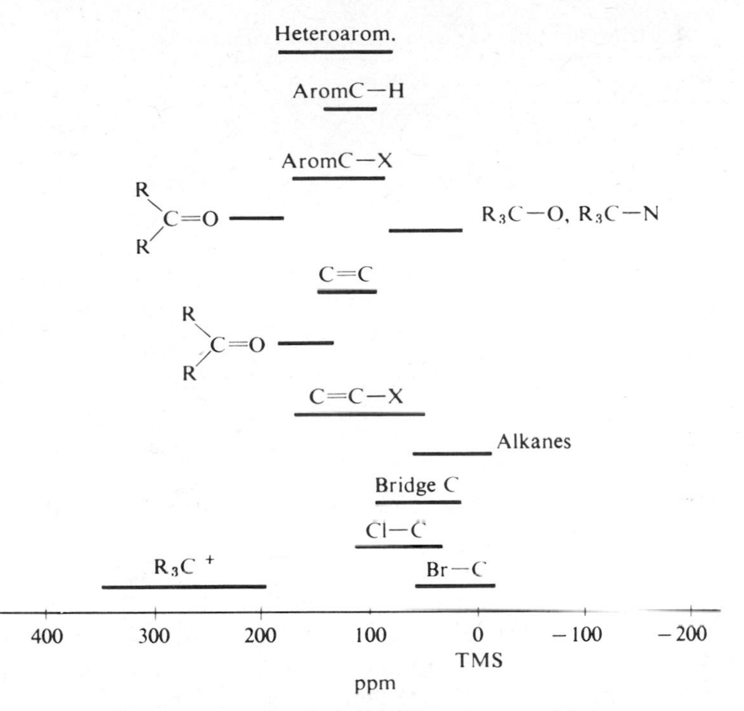

Figure 2.4. Carbon chemical shifts of some structural features.

onate in the range −50 to +100(ppm), sp^2 hybridized carbons from 150 to 300(ppm), and acetylenic carbons (sp) between δ25 and 180(ppm). Again, as in H NMR, acetylenic carbons are abnormal with respect to their resonance frequencies. The theoretical cause for this apparent abnormality will be discussed in Chapter 5.

One of the significant differences between the H NMR and C^{13} NMR resonance patterns lies in the observation that aromatic protons are more deshielded than olefinic ones, whereas aromatic and olefinic carbons resonate within the same absorption range. Another difference of some interest is that the protons in methane (CH_4), for example, resonate at more shielded positions than do those in halomethanes (CH_3-X). The reverse is true for the carbon resonances. The same pattern prevails among aromatic carbons, where the substituted ones resonate at more deshielded positions than do the nonsubstituted.

Of the generally studied stable organic compounds, carbonyl carbons are the most deshielded (δ_C150–240(ppm)). The transient tertiary carbonium ions resonate at 200–280(ppm) and as such represent the most deshielded of carbon atoms so far observed [3–5].

2.2.1.3 Nitrogen Chemical Shifts

Nitrogen has two major isotopes. N^{14}, with a spin quantum number of 1, is the predominant one in naturally occurring samples (<99.5%). The minor isotope is N^{15}, with a spin quantum number of $\frac{1}{2}$ (present to the extent of 0.48% only).

TABLE 2.1 Chemical Shifts of Some Nitrogen-15 Standards

Compound	N^{15} Shift	Comments
NH_3	0.0	25°C, liquid
NH_4NO_3	21.60, 375.59	Shielded: NH_4; deshielded: NO_3
$(CH_3)_4N^+CH^-$	43.54	Saturated aqueous solution
$(CH_3)_2N$–CHO	103.81	Neat liquid
HNO_3	375.80	1 M aqueous solution
CH_3NO_3	380.23	Neat liquid

Unfortunately, the major isotope (N^{14}) has a quadrupole moment (see Chapter 5) which causes significant line broadening and makes the accurate determination of chemical shifts difficult [6].

The availability of FT NMR instrumentation, superconducting magnets, and, as a result, the facility to employ large-size samples (up to 22-mm-diameter sample tubes can be routinely used!) has brought N^{15} NMR measurements to the forefront of recent NMR studies of the nonmetallic nuclides [7]. Although nitrogen NMR has a number of intrinsic problems associated with it (solvent effects, relaxation time, etc.), sufficient information is now available to take these problems into account in routine, although often time-consuming, nitrogen NMR studies.

As was the case for hydrogen and carbon NMR, the development of a suitable reference standard for nitrogen NMR was of primary importance as well. A number of different standards have been used (see Table 2.1) over the past 20 years, and

TABLE 2.2 Nitrogen-15 Chemical Shifts of Some Functional Groups

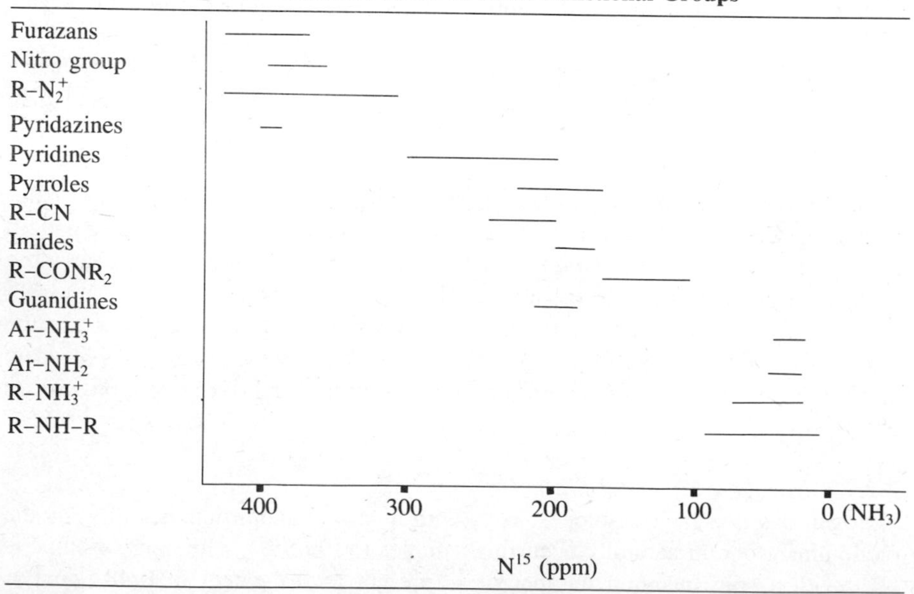

only recently has liquid ammonia, at 25°C, become the accepted primary standard. Since this is not a very convenient standard for general use, nitromethane (CH_3NO_2) is now the generally accepted secondary *external* or *internal* reference standard. The chemical shift of the N^{15} in this compound is 380.23 ppm (in a neat sample) with respect to the primary standard (liquid ammonia). Most of the organically bonded nitrogens resonate within 900 ppm. downfield from ammonia (see Table 2.2).

The chemical shift range can, again, be subdivided into *singly* bonded (δ_N0–150(ppm)), *doubly* bonded (δ_N200–350(ppm)), and *triply* bonded (δ_N175–250(ppm)) nitrogens. As is true in proton and carbon NMR, the chemical shift of the triply bonded atoms falls between the singly and doubly bonded species. For example, the nitrogen in benzonitrile ($C_6H_5{-}CN$) resonates at 258.7(ppm), and that in the imine $C_6H_5{-}CH{=}N{-}CH_3$ resonates at 318.1(ppm).

Another qualitative comparison that can be made is one between a *pyrrole*like nitrogen (**1**) and a *pyridine*like one (**2**). The former is more shielded (δ_N145.1(ppm)) than the latter (δ_N317.3(ppm)) [8].

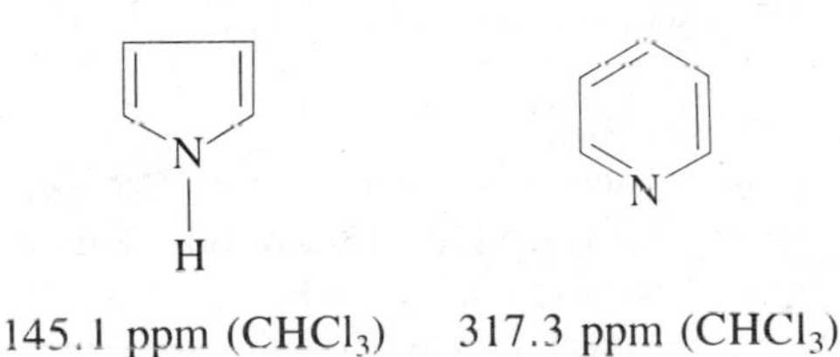

The most deshielded nitrogen observed so far is that in nitrosobenzene (**3**). It resonates at 913 ppm! In comparison, nitrobenzene (**4**) resonates at 370.7 ppm.

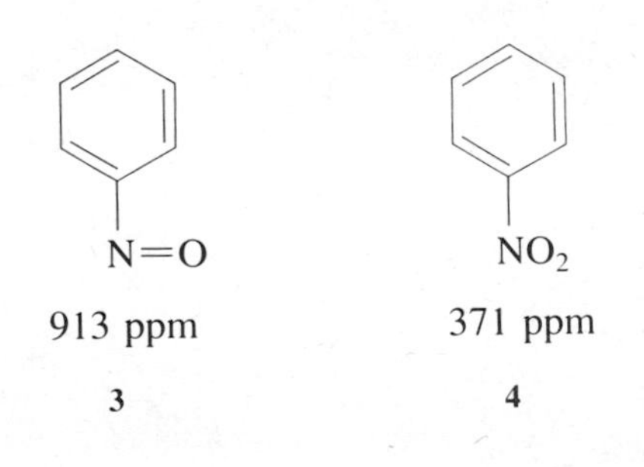

2.2.1.4 Phosphorus Chemical Shifts

The only stable isotope of phosphorus is P^{31}, with a spin quantum number of $\frac{1}{2}$. These properties, along with very favorable resonance spectroscopic properties, similar to those of protons and fluorines, have made this a very extensively studied nucleus. Unfortunately, the magnetic moment of this nucleus is only approximately 7% that of H^1 and F^{19}. Thus, its NMR signal intensity is much lower than that of either of the other two "popular" nuclei.

Phosphoric acid has been the primary reference compound of choice in phosphorus NMR spectroscopy for some time. The compound is used either as an *internal* or, more often, as an *external* standard. Although solutions of phosphoric acid in methylene chloride have been occasionally used, an 85% aqueous solution is the standard of choice [5–7].

The symbol δ_{P31} is employed to indicate the chemical shift with respect to phosphoric acid. These values, for differently bonded phosphorus atoms, lie within ±250 ppm of the standard. The data given in Figure 2.5 present the relative resonance frequency ranges for different classes of phosphorus compounds.

Next to the very highly shielded elemental phosphorus itself, phosphine and other trivalent phosphorus species are the most shielded. The chemical shift differences between phosphine (PH_3, −241) on the one hand and phosphorus tribromide (PBr_3, +227) on the other are dramatic. The pentacovalent phosphorus compounds fall into an intermediate range between phosphine itself and the trialkylphosphines.

The differences of the resonance position in the different types of phosphorus compounds have, generally, been discussed by considering the ionic and double-bond characters of the bonds to the phosphorus nuclei [8–10].

2.2.1.5 Fluorine Chemical Shifts

The spin quantum number of the only stable fluorine isotope, F^{19}, is $\frac{1}{2}$. This spin quantum number, along with favorable relaxation times and high NMR sensitivity akin to that of protons, makes this a readily studied nucleus. In fact, the abundance of published F NMR spectra is second only to those for proton spectra. The fluorine

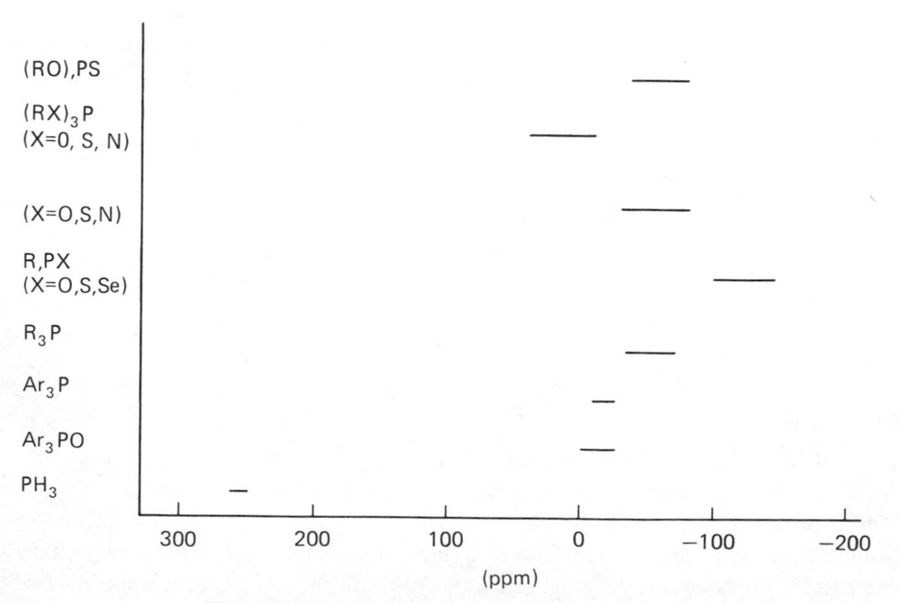

Figure 2.5. Selected phosphorus-31 chemical shifts.

signal of trifluoroacetic acid has found frequent use as a secondary reference standard in F NMR. Not only does the fluorine resonance in trichlorofluoro methane (CCl_3F) serve as a primary reference standard, but the liquid is also a superb solvent for fluorine containing compounds. The fluorine chemical shifts, when referenced to this compound, are indicated by δ_F and include a superscript (*) when no solvent dilution effects have been included. On this scale, the variously bonded fluorines resonate ±200(ppm). Figure 2.6 is a representation of some of the more important fluorine resonance frequencies.

As is true for all of the nonmetal nuclides, fluorines bonded to saturated carbons are among the most shielded (−200 ± 20 ppm) ones, whereas those bonded to olefinic carbons are more deshielded (−130 ± 20 ppm). A fluorine bonded to an aromatic ring is even more deshielded (−110 ± 20 ppm). In analogy with aldehydic protons, a fluorine bonded to a carbonyl group (R–CO–F) is among the most deshielded (15 ± 10 ppm) of fluorine nuclei. This deshielding effect is even more enhanced in R–SO_2–F compounds (70 ± 10 ppm) and finds its ultimate in NF_3, where the fluorine resonates at 147 ppm! A fluorine bonded to an acetylenic carbon appears at a shielded position, similar to that of an acetylenic proton.

2.2.2 Magnetically Active Metal Nuclides

NMR spectroscopy of metal nuclides is significantly different from the spectroscopy dealing with covalently bonded species for a number of reasons, not the least of

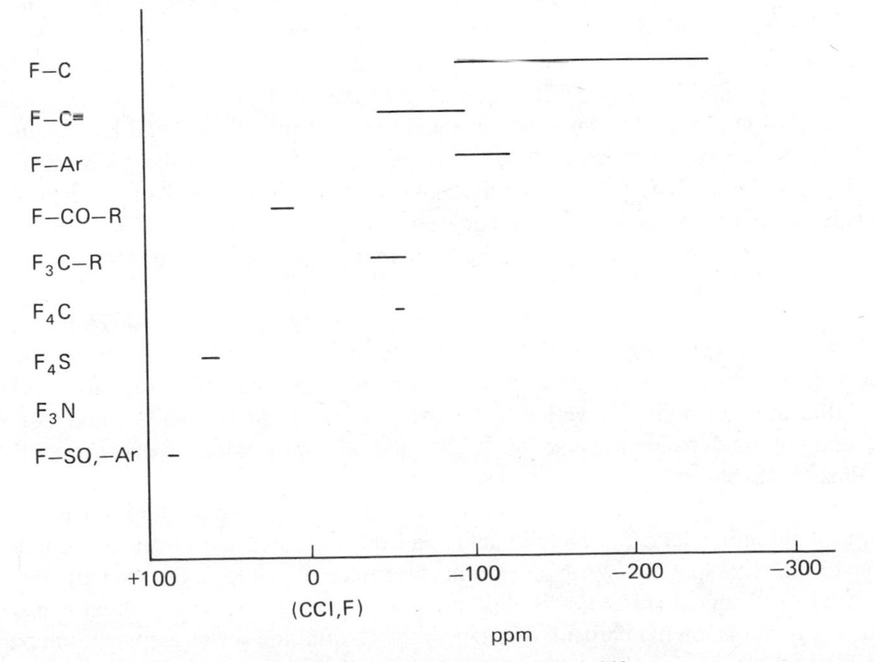

Figure 2.6. Selected fluorine chemical shifts.

which is the existence of very large ranges of chemical shifts observed in the many different species [10–13]. While C^{13} spectra usually fall within 200 ppm, the solvent-dependent differences of Tl^{205} species cover a range of close to 3000 ppm. The apparently largest range so far observed is found in Co^{59}, which has a ligand-depending shift of 12,000 ppm [11].

Within any one row of elements in the periodic table, the chemical shift ranges for a series of compounds containing these elements tend to increase. For example, the Na^{23} chemical shift range for a number of different compounds is about 50 ppm, whereas for Al^{27} it is 300 ppm and for Cl^{35} it is 1000 ppm [14, 15].

There is also a reasonable relationship within any one group going down the periodic table. The solvent-dependent alkali–metal ion chemical shift ranges for Li^{7}, Na^{23}, K^{39}, and Cs^{133} are 5, 17, 30, and 130 ppm, respectively [16, 17].

Among the magnetically active main group elements there appears to exist a fairly *linear* relationship between the local symmetry of the nuclide being studied and the chemical shift. For example, the aluminum nuclides in tetrahedral aluminum compounds resonate about 100 ppm downfield from the corresponding octahedral aluminum derivatives. Similarly, tetrahedral lead compounds resonate at higher fields when solvent coordinated [18–20].

The modified periodic table (Table 2.3) provides some of the important NMR spectral data for most of the magnetically active nuclides, both nonmetalic and metalic.

*2.2.2.1 **Light Metals.*** *Group IA and IIA*

The resonance frequencies of the isolated—that is, structurally not modified—magnetically active nuclides of the group IA and IIa elements in a magnetic field of 4.7 T resonate between 9.34 and 77.75 MHz and between 12.24 and 28.12 MHz, respectively. The chemical shift dependence of the alkali metal ions on the nature of the solvent is well known. The chemical shifts of the alkali metal and alkaline earth ions become more deshielded as the anion is changed in the following sequence: $I^- > Br^- > Cl^-$. When the anion is either NO_3^- or ClO_3^-, all of the alkali metal and alkaline earth nuclides become more shielded. This has been rationalized by arguing that the cations interact less strongly with the nitro and perchlorate ions than they do with the water molecules [21].

Solvent polarity also effects the chemical shifts of the group IA and IIA cations. The chemical shift range among solvents increases in going *down* in either group. This range, between acetonitrile on the one hand and pyridine on the other, is 6, 20, 30, and 130 ppm, respectively, for the cations of the Li, Na, K, and Cs. A similar, anion-dependent range of 1, 20, and 40 ppm exists in the group IIA sequence of cations, Mg, Ca, Sr, respectively [11].

The organometallic compounds of lithium fall into two types, the inorganically bonded lithium salts of carboxylic acids and the like, and the covalently bonded alkyllithium compounds. With an external reference of 1.0 M LiCl, the carbanion-dependent chemical shifts for the lithium salts of organic acids are about 8 ppm. In the cyclopentadienyl carbanion instance, the lithium ion forms a contact ion pair in THF (chemical shift difference = −8.67 ppm) and is essentially solvent solvated in HMPA.

TABLE 2.3 NMR Parameters Superimposed on the Periodic Table

Light Metals | Transition Heavy Metals | Non-Metals | Inert Gases

		III B	IV B	V B	VI B	VII B	VIII	VIII	VIII	I B	II B	III A	IV A	V A	VI A	VII A	Inert Gases
1 1/2 **H** * 200.02																	3 1/2 **He** 152.42
6 1 **Li** * 29.4 3.6	79 3/2 **Be** 28.1 79											10 3 **B** * 21.5	13 1/2 **C** 50.3 1	14 1 **N** * 14.5	17 5/2 **O** 27.1	19 1/2 **F** 188.3	21 **Ne**
23 3/2 **Na** 52.9 525	25 5/2 **Mg** 12.2 1.5											27 5/2 **Al** 52.1 1170	29 1/2 **Si** 39.8	31 1/2 **P** 81	33 3/2 **S** * 15.4	35 3/2 **Cl** * 19.6	**Ar**
39 3/2 **K** 9.3 2.7	43 7/2 **Ca** 13.5 .1	45 7/2 **Sc** 48.6 1710	47 5/2 **Ti** * 11.3 .9	51 7/2 **V** 52.6 2160	53 3/2 **Cr** 11.3 .5	55 5/2 **Mn** 49.3 994	57 1/2 **Fe** 6.5 .004	59 7/2 **Co** 47.2 1570	61 3/2 **Ni** 17.9 .2	63 3/2 **Cu** 53 365	67 5/2 **Zn** 12.5 .7	71 3/2 **Ga** 61.0 319	73 9/2 **Ge** 7.0 .62	75 3/2 **As** 34.3	77 1/2 **Se** * 38.2	79 3/2 **Br** * 50.2	83 9/2 **Kr** 7.7
87 3/2 **Rb** 49.1 277	87 9/2 **Sr** 8.7 1.1	89 1/2 **Y** 9.8 .67	91 5/2 **Zr** 18.7 6.0	93 9/2 **Nb** 48.9 2740	95 5/2 **Mo** 13.0 2.9	99 9/2 **Tc** 45.0 2140	99 5/2 **Ru** 9.2 .8	103 1/2 **Rh** 6.3 .2	105 5/2 **Pd** 9.2 1.4	109 1/2 **Ag** 9.3 .3	113 1/2 **Cd** 44.4 7.6	115 9/2 **In** 43.8 1890	119 1/2 **Sn** 74.6 25.2	121 5/2 **Sb** 47.9 520	123 1/2 **Te** * 54.5	127 5/2 **I** * 40.04	129 1/2 **Xe** * 55.4
133 7/2 **Cs** 26.3 269	133 3/2 **Ba** 19.9 1.8		177 7/2 **Hf** 8.0 .9	181 7/2 **Ta** 23.9 204	183 1/2 **W** 8.3 .1	187 5/2 **Re** 45.5 490	189 3/2 **Os** 15.5 2.1	193 3/2 **Ir** 3.7 .1	195 1/2 **Pt** 43 19.1	197 3/2 **Au** 3.4 .1	199 1/2 **Hg** 35.7 5.4	205 1/2 **Tl** 115.5 769	207 1/2 **Pb** 41.9 11.8	209 9/2 **Bi** 32.2 770	**Po**	**At**	**Rn**
Fr	**Ra**																

*other isotopes:

	I	II	III	IV
B	11	3/2	64.2	
Br	81	3/2	54.05	
Cl	36	2	22.98	
	37	3/2	16.31	
H	2	1	30.74	
	3	1/2	213.43	
I	129	7/2	26.65	

	I	II	III	IV
Li	7	3/2	77.8	1540
N	15	1/2	20.3	
Ti	49	7/2	11.3	1.2
S	35	3/2	23.87	
Se	79	7/2	10.39	
Te	125	1/2	63.22	
Ti	49	7/2	11.3	1.2
Xe	131	3/2	16.40	

Code

I	II
III	IV

I = isotope weight
II = spin quantum #
III = resonance frequency (Mhz) at 4.7T
IV = receptivity

The lithium ions in alkyl lithium compounds resonate at 1.53–0.65 ppm as the branching of the alkyl group increases from an ethyl to a *tert.*-butyl function.

The crown ethers are well known for their ability to form complexes of variable strengths with, among others, the group IA and IIA cations. The reader is referred to the article by J. Dechter for excellent discussions of this and related studies [11].

2.2.2.2 *Some Heavy Metals. Groups IIIA, IVA, and VA*

The metal nuclides of these periodic table groups include Al^{27}, Ga^{71}, In^{115}, and Tl^{205} in group IIIA, and Ge^{73}, Sn^{119}, Pb^{207}, and Bi^{209} in the other groups. Among all of these, only Tl^{205}, Sn^{119}, and Pb^{207} have spin quantum numbers of $\frac{1}{2}$; all of the other nuclides in these groups have larger ones.

With Al $(H_2O)_6{}^{+3}$ as the zero reference for the chemical shift of a number of aluminum compounds, the aluminum ion in the *octahedral* acetonitrile complex of Al^3 has a chemical shift of -46(ppm), whereas in the dimethylsulfoxide complex it has a shift of $+3.4$(ppm). In tetrahedrally bonded aluminum compounds, such as $AlX_4{}^-$, the aluminum chemical shifts vary with the nature of X. For example, for X = I, Br, Cl, these chemical shift are -28, 80 and 102(ppm), respectively [21]. The chemical shift of Al^3 in the much used $LiAlH_4$ is 98(ppm) [22].

The octahedral complexes of Ga^{71} and In^{115} resonate at higher fields than the corresponding *tetrahedral* ones. As expected, the chemical shift ranges of the compounds of these two nuclides are larger than those observed for Al^{+3}. The octahedral $Ga(H_2O)_6{}^{+3}$ ion is used as the zero reference compound in gallium NMR spectroscopy [23]. Relative to this reference, the $GaX_4{}^-$ species with X = I, Br, Cl have chemical shifts of -450, 69, and 257(ppm), respectively. When X = H or OH, the chemical shift for the gallium ion is 682 and 192(ppm), respectively [24].

The same chemical shift pattern, although covering a wider range, exists for indium compounds. For example, with $In(H_2O)_6{}^{+3}$ as the zero point chemical shift reference, the indium nuclide in $InCl_4{}^-$ has a chemical shift of 420(ppm) downfield from the reference.

The solvent effect on the chemical shift of thallium compounds is remarkable, indeed, since it covers a range of at least 2600 ppm! For example, Tl^{+3} dissolved in pyrrolidine has a chemical shift of 1757 ppm (with respect to the usual zero point reference of thallium ion at *infinite* dilution), whereas when dissolved in acetone it has one of -240 ppm [25].

The chemical shift differences between compounds containing $+1$ or $+3$ thallium ions is vast: the $+3$ ions resonate about 3000 ppm downfield from the $+1$ thallium reference.

The coordination sphere effect observed in the other nuclides of this group is also operative in thallium compounds, where the high field shift of 640 ppm in going from $TlCl_6{}^{-3}$ to $TlCl_4{}^-$ is remarkable [26].

One of the more extensively studied classes of thallium compounds are the organometallic ones. The resonance positions of these compounds are separated from the Tl^{+1} resonances and cover a range of 2000 ppm. The thallium chemical shift in trimethylthallium is 4760(ppm) and in the triphenyl analog it is 3821(ppm) [27].

It is not surprising that many of the relationships that have been developed for silicon and carbon NMR are also applicable to the metallic members of the IVA family, Ge^{73}, Sn^{119}, and Pb^{121}. The low sensitivity of Ge^{73} and its low resonance frequency make this nuclide somewhat difficult to study. The chemical shift changes in the GeX_4 series of compounds are 0, −343, and −1117 ppm, respectively, for X = Cl, Br, and I [28].

Among the magnetically active tin species, the chemical shift differences between the +2 and +4 ions is not nearly as great as, for example, between the two different thallium ions. The chemical shift of the tin in stannic chloride is −150 ppm, whereas it is −236 ppm for stannous chloride [29].

With tetramethyltin at 0 ppm, the SnX_4 species resonate at −1699, −632, and −148 ppm for X = I, Br, and Cl, respectively.

The chemical shifts of Pb^{207} compounds cover a range of 3000 ppm, with the lower oxidation state compounds resonating at higher fields. With tetramethyllead as the zero chemical shift reference compound, lead nitrate resonates at −2961 ppm, and tetraphenyllead at −178 [30].

The large quadrupole moment of antimony makes the NMR study of this nucleus somewhat difficult, since this creates immense line width. For example, the hexachloroantimonate anion has a line width of 300 Hz, whereas antimony pentachloride has one of 8000 Hz. The chemical shift range of Sb^{121} is quite large as exemplified by the shift of −2436 ppm in going from the hexachloro to the hexabromoantimonate anion. The SbS_4^{-3} ion resonates at 1032 ppm with reference to $SbCl_6^-$ [31].

Other than the data given in the summary table, no detailed NMR studies of Bi^{209} seem to have been reported [32].

2.2.2.3 Transition Metals

The NMR spectral analyses of the transition metals have experienced major growth as a result of the availability of FT broad-frequency-band spectrometers operating with wide-bore superconducting magnets [33]. There now exists a fairly extensive literature dealing with the B group and group VIII nuclides. The Co^{59} NMR data have been, and continue to be, used as the central body of information on which the analyses of the other nuclides in these groups are based. It has been proposed that trends in chemical shifts for different nuclides be compared in a quasi-standard way. For certain nuclides, the MO_4^{-n} anions represent the high oxidation state and, consequently, the low end of the chemical shift range. The mononuclear carbonyls, existing as low-oxidation-state species, provide the upper end of the chemical shift range for a particular metal nuclide. The different nuclides have also been classified according to the effect of the halides series on the chemical shifts of the cations, whereby the "normal" halogen dependence is $Cl < Br < I$. A number of nuclides are subject to "inverse halogen dependency." This pattern exists among the early-transition metal nuclides and the copper group ones.

Rather than discuss all of the individual groups and their component nuclides in detail, the reader is referred to the excellent review article on NMR of the transition metals by Dechter [33]. The important parameters for most of the magnetically active nuclides are given in Table 2.3. The information given there should be adequate for the general needs of most NMR spectroscopists.

GROUP IIIB [34]. The chemical shifts of Sc^{45} in the chloride, bromide, and perchlorate salts all extrapolate to the same value at infinite dilution. It is this value that is used as the chemical shift reference standard for this nuclide. As far as solvent effects are concerned, the chemical shift difference between an aqueous solution of $ScCl_3$ and one in tetraheydrofuran is 204 ppm.

The NMR behavior of yttrium (Y^{89}), with a spin quantum number of $\frac{1}{2}$, is similar to that found among the alkali metals. The chemical shift reference for the yttrium nuclide also is the infinite dilution chemical shift of yttrium salts taken as 0 ppm. The chemical shifts for the yttrium perchlorate and nitrate compounds are upfield from this reference.

The La^{139} chemical shift reference is, again, the infinite dilution chemical shift of lanthanide salts. Among the LaX^{+2} species, the chemical shifts for the acetate, chloroacetate, dichloroacetate, and trichloroacetate are, respectively, 100, 50, 22, and 0 ppm. The chemical shift of La^{+3} complexed with EDTA is 570 ppm. The solvent effects for a number of La^{139} salts in acetonitrile, dimethylformamide, hexamethylphosphoramide, and dimethyl sulfoxide are −129, −39, 66, and 400 ppm, respectively. An inverse halide dependence for the hexahalo compounds has been observed.

GROUP IVB [35]. Although very little is known about the NMR behavior of the titanium group elements, a most unique feature has been found. The magnetogyric ratios of the Ti^{47} and Ti^{49} nuclides are so similar that *both* of these isotopes can be observed in the same experiment; they are separated by only 266.2 ppm. The most reasonable reference compound for this nuclide appears to be $TiCl_4$, again set at 0 ppm, with a line width of about 3 Hz. The chemical shift of the titanium in the tetrabromo compound, examined as a melt, is 482 ppm, whereas in the tetraiodo compound, in benzene, it is 1287.3 ppm.

GROUP VB [36]. The preferred reference compound for the V^{51} nuclide is $VOCl_3$, which is used as a neat, external sample. With a chemical shift range of 2500 ppm, it is similar to that of 2400 ppm observed in titanium. With the $VOCl_3$ as 0-ppm reference, the VOF_3 resonates at −757 ppm, and the $VOBr_3$ species at 432 ppm.

The receptivity of Nb^{93} is the highest among all of the metal nuclides! This nuclide has a very large quadrupole moment, causing considerable line broadening. The narrowest of these is found in $NbCl_6^-$ dissolved in acetonitrile, where it is 50 Hz. On the other hand, $NbOCl_3$ has a line width of 1000 Hz. Thus it is not surprising that $NbCl_6^-$ is the accepted NMR reference standard for this nuclide. On this scale, the chemical shift of Nb in NbF_6^- is 1550, whereas it is −735 ppm in the Br analog.

GROUP VIB [38]. The NMR behavior of the relatively few chromium compounds studied by the NMR technique is strongly pH-dependent as a result of the chromate–dichromate equilibrium. The chemical shift range for Cr^{53} species is about 2000 ppm and is similar to that of the Mo nuclide in the Mo^{95} species. The chemical shift reference for the latter species is MoO_4^{-2}. The chemical shifts of the lower-oxidation-state derivatives of molybdenum are at higher fields; the higher-oxidation-state species resonate at lower frequencies. For example, MoS_4^{-2} resonates at 2259 ppm, $Mo(CN)_8^{-4}$ at −1309 ppm, and $Mo(CO)_6$ at −1857 ppm. The W^{183} chemical shift range is over 5000 ppm. Again, it appears that the higher the oxidation state the lower the resonance field, with the proviso that, among the lower oxidation states, the +2 and 0 valent species can have very similar chemical shifts. The preferred reference species is WO_4^{-2}. On this scale, the WCl_6 resonates at 2184 ppm and the corresponding fluorine isomer at −1159 ppm. The $W(CO)_6$ species, with a zero oxidation state, resonates at −3505 ppm, and the W^{+2} cyclopentadienide (Cp) complex $[CpW(CO)_3]_2$ resonates at − 4048 ppm.

GROUP VIIB [38]. It is not surprising that it is the permanganate ion that serves as the reference compound for the Mn^{55} species. The Mn species has a very high receptivity and resonances close to that of the C^{13} nuclides. However, except for the permanganate ion, with a line width of 4 Hz, the line widths of most of the compounds of this quadrupolar nuclide are very high. The line width of 6200 Hz observed in $HMn(PF_3)_5$ serves to illustrate this. The chemical shift of the Mn in $Mn_2(CO)_{10}$ is −2325 ppm, with an amazingly narrow line width, for this nucleus, of 48 Hz. In the $Mn(CO)_5X$ sequence, the chemical shift of the metal nuclide changes from −1485 to −1160 to −1005 ppm in going from I to Br to Cl.

GROUP VIII [39–41]. The Fe^{57} nuclide, with a spin of $\frac{1}{2}$, has one of the lowest receptivities of any of the magnetically active nuclides in the periodic table. Even FT technology is insufficient to permit NMR examination of this nucleus within a reasonable experimental time frame, without prior isotope enrichment! The limited data available are a result of this problem. The convenient reference compound $Fe(CO)_5$ is set at 0 ppm. With this reference, the Fe in cyclobutadiene–$Fe(CO)_3$ in benzene solution resonates at −583 ppm, whereas the iron in ferrocene resonates at 1543 ppm. The $Fe(CN)_6^{-4}$ species has an Fe chemical shift of 2497 ppm, while the iron nuclide in the bipyridyl complex $Fe(bpy)_3^{+2}$ resonates at an amazing 11,269 ppm!

The Ru nuclide reference is $Ru(CN)_6^{-4}$, with the Ru^{99} nuclide having narrower line widths than the other magnetically active species, Ru^{101}. Thus it is the Ru^{99} species that has been preferred for the limited NMR studies reported for this element. The Ru incorporated in the hexamine complex resonates at 7680 ppm, in RuO_4 at 1931 ppm, and in $Ru_3(CO)_{12}$ at −1208 ppm.

The Co^{59} nuclide is quadrupolar, with a large quadrupole moment, and, as expected, many of its compounds have extremely large NMR spectral line widths. Values of up to 15,000 Hz have been observed! The reference compound, with a line width of 3 Hz, is $Co(CN)_6^{-3}$. The chemical shift range of Cr compounds varies from 12,520 ppm for $Co(acac)^3$, for example, to 8,150 ppm in the hexamine derivative, and −304 ppm in $Co(P(OMe)_3)_6^{+3}$. Generally, the chemical shifts of the Co^{59} and the Rh^{103} nuclides, the other member of this elemental group, follow the pattern of *the higher the oxidation state, the lower the chemical shift value*. The known range of chemical shifts for Rh^{103} is about 10,000 ppm and is similar to the observed ranges for Co^{59} and Pt^{159} compounds.

The magnetically active Ni^{61} nuclide with a receptivity of about 0.24, with respect to C^{13}, is quadrupolar and has a large quadrupole moment. The preferred reference compound of this nuclide is $Ni(CO)_4$. The Ni in this compound has a conveniently small line width of 4 Hz. The Ni chemical shift in $Ni(PCl_3)_4$ is 267 ppm, whereas that of the corresponding F analogue $Ni(PF_3)_4$ is -929 ppm. Both of these compounds have relatively reasonable line widths (20 Hz).

With a spin of $\frac{1}{2}$, Pt^{159} has a high NMR sensitivity which actually exceeds that of C^{13} by a factor of 19. With a platinum reference of $PtCl_6{}^{-2}$ set at the usual 0 ppm, the corresponding bromine analog resonates at -1800 ppm. The comparable Pt^{+2} species, $PtX_4{}^{-2}$, resonate at -1600 ppm for X = Cl and -2600 ppm for X = Br.

GROUP IB [42]. The quadrupole moment of Cu^{63} is rather large, and as a result reasonable line widths are observed only in compounds with very nearly perfect tetrahedral local symmetry. The lower receptivity of the Cu^{65} isotope does not aid this line width problem. With the Cu isotope in $Cu(MeCN)_4{}^{+}$ as the reference standard, the metal nuclide shift in $Cu(CN)_4{}^{-3}$ is at 480 ppm. The line width of this resonance peak is only 84 Hz. With a Cu chemical shift of 111 ppm, the $Cu(pyridine)_4{}^{+}$ complex has a line width of 880 Hz.

The very low receptivity of the Ag^{107} and Ag^{109} ions accounts for the paucity of NMR studies. The proposed chemical shift reference for the Ag nuclides is the infinite dilution chemical shift of silver salts. Among all of these, AgF is the least sensitive to concentration changes and will, perhaps, become the reference compound of choice. The chemical shift differences between silver complexes of ethylenediamine, pyridine, and acetonitrile are 554, 350, and 335 ppm, respectively.

GROUP IIB [43]. The anion and concentration dependence of the aqueous Zn^{67} nuclide follow the same trends as the alkali metal ions. With concentrations up to 0.8 M, the perchlorate, sulfate, and nitrate salts of zinc have the same chemical shifts. It is the infinite dilution chemical shifts of these salts that are used as the chemical shift references. Nonaqueous solutions of zinc salts generally have a sizable line broadening not present in the corresponding aqueous solutions.

Although dimethylcadmium is frequently used as the chemical shift reference for the Cd^{113} species, it would be more appropriate to use the infinite dilution chemical shift of cadmium salts as the primary reference. With the infinite dilution shift for cadmium chloride taken as 0 ppm, the dimethylcadmium has a chemical shift of 642.9 ppm in the neat form and 576.3 ppm in tetrahydrofuran solvent. The chemical shift range for Cd^{113} covers about 700 ppm, with the organometallic cadmium derivatives at the low end (about 300 ppm) of this range.

Among the extensive Hg^{199} NMR studies, dimethylmercury is most frequently used as the reference standard for mercury compounds. On the basis of this reference, mercuric perchlorate has a chemical shift of -2387 ppm, mercuric acetate resonates at -2396 ppm, and mercuric chloride resonates at -1498 ppm (in ethanol solution). The chemical shift range for Hg^{199} compounds covers a range in excess of 3600 ppm.

2.3 SOME EMPIRICAL RELATIONSHIPS

The literature abounds with an ever increasing number of tables of *additivity constants* that have been developed from examinations of NMR absorption spectra which allow the estimation of chemical shifts for different types of nuclei in different chemical environs. Some of the more generally useful empirical relationships have been selected for description in this chapter.

2.3.1 Proton Additivity Relationships

2.3.1.1 Shoolery's Constants

Although this set of constants is no longer of great general utility, it serves to illustrate some interesting relationships. These constants, based on the analysis of numerous methane derivatives, permit the estimation of proton chemical shifts of disubstituted methanes (X–CH_2–Y) [44–46]. The average additivity parameters for different substituents, ''Shoolery's effective shielding constants,'' are listed in Table 2.4. The relationship between these numbers and the predicted chemical shift is given in equation 2.2

$$\delta(\text{ppm}) = 0.23 + \sum \delta_{\text{effective}} \tag{2.2}$$

TABLE 2.4 Shoolery's Effective Shielding Constants

Substituent Group	Effective Shielding Constant (ppm)
$-CH_3$	0.47
$-CR{=}CRR$	1.32
$-C{\equiv}C-R$	1.44
$-CONH_2$	1.53[a]
$-NRR$	1.57
$-SR$	1.64
$-CN$	1.70
$-COR$	1.70
$-I$	1.83
$-Br$	2.33
$-OR$	2.36
$-Cl$	2.53
$-NO_2$	4.44[a]

Source: Taken in part from L.M. Jackman, *Nuclear Magnetic Resonance Spectroscopy*, p. 59, 1959, and reproduced with permission of the author and Pergamon Publishing Company.

[a] These values were added by the author and are based on an analysis of five different compounds for each of the new constants.

An example serves to demonstrate the applicability of these constants to the prediction of the methylene proton chemical shifts of methoxyacetonitrile, CH_3OCH_2CN. The CN group constant is 1.70 and that of the methoxyl group protons is 2.36. Thus, the Σ is 4.06 and the predicted chemical shift is δ 4.29(ppm). The experimentally determined value for these protons is δ 4.20(ppm) and consequently compares favorably with the *calculated* value. Although with less accuracy, these constants can be used for the estimation of the chemical shift of the methine protons on trisubstituted methanes (*XY*CH*Z*).

The estimation of the methine proton chemical shift in 2-nitrobutane (structure **5**) illustrates this:

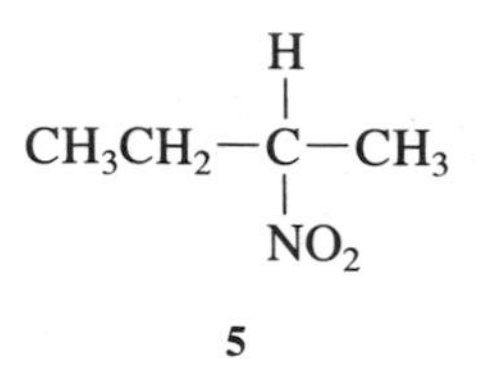

5

The ethyl and methyl groups can be equated for the purpose of this calculation to afford a value of 0.94 (2×0.47). The nitro group has a value of 4.44, and the predicted chemical shift for these methine protons is δ 5.61(ppm). The experimentally determined value is δ 5.48(ppm). Again, the agreement is quite adequate.

2.3.1.2 *Identification of Alcohols and Amines*

The acetylation or benzoylation of alcohols and amines causes significant and, more importantly, constant changes in the chemical shift of the carbon-bonded proton containing the functional group being acetylated or benzoylated [47].

The acetylation of primary alcohols (structure **6**) causes the protons on the hydroxyl group bearing carbon to become more deshielded by 0.50 ppm whereas in secondary alcohols (structure **7**) the deshielding effect is between 1.00 and 1.15 ppm:

RCH_2—OH 0.50 ppm deshielding RCH_2—O—C(=O)—R′

6

R_2CH—OH (R—CH(R)—OH) 1.00–1.15 ppm deshielding RCH—O—C(=O)—R′

7

The acetylation of a primary amine causes a paramagnetic shift (deshielding) of the proton bonded to the amine group bearing carbon by 0.6 ppm.

$$R{-}CH_2NH_2 \quad 0.60 \text{ ppm deshielding} \quad R{-}CH_2NH{-}COCH_3$$

The predictable effects clearly lend themselves to the facile identification of some fundamental structural features of unknown alcohols and amines.

2.3.2 Carbon Additivity Relationships

Since C^{13} NMR is an excellent tool for identification of carbon skeletal structures in molecules, it is not surprising that many empirical relationships relating carbon chemical shifts with structural features have been developed. The reader is referred to the excellent text by Levy and Anet [48] and others listed in the reference section for detailed discussions of a number of these correlations. For the purpose of this text, only a few of the more generally applicable relationships will be discussed.

2.3.2.1 Alcohol Substituent Effects

It is convenient to relate the carbon chemical shifts of a carbon substituted with a methyl group with that of one substituted with a hydroxyl function (structure **8** vs. structure **9**) [49, 50]:

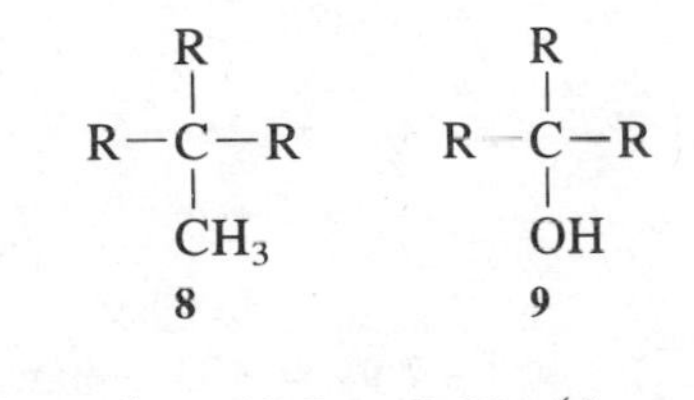

$$\delta_{C\text{-}1}\,(\text{alcohol}) = 43.3 + 0.83 \quad (\delta_{-CH_3})$$

$$\delta_{C\text{-}2}\,(\text{alcohol}) = 0.5 + 1.00 \quad (\delta_{C-CH_3})$$

$$\delta_{C\text{-}3}\,(\text{alcohol}) = -1.7 + 1.00 \quad (\delta_{C-C-CH_3})$$

An example will serve to illustrate the application of these rules. If we wish to compute the chemical shifts of the various carbons in 2-butanol (structure **10**), the hydrocarbon comparator compound is 2-methylbutane (**11**). The experimentally determined chemical shifts of the various carbons in 2-methylbutane are given in structure **11:**

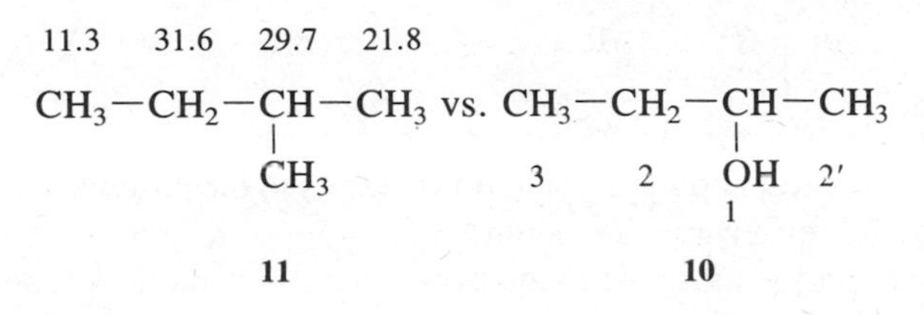

The *transformation* from the hydrocarbon to the alcohol will change the chemical shift of the carbon labeled *1* in structure **10** most significantly since a methyl group is being replaced by a hydroxyl (C—C vs. C—O bond). The effect will be 68.2 ppm (43.3 + 0.83 × 29.7). The other carbon chemical shifts are 22.3 ppm (21.8 + 0.5) for $C_{2'}$, 32.1 (31.6 + 0.5) for C_2, and 10.6 for C_3. The corresponding experimentally determined values are 67.3, 23.6, 35.1, and 13.8 ppm. Clearly, the calculated and observed values agree quite well.

2.3.2.2 Empirical Parameters for Alkenes

Roberts and co-workers [17, 51] have described an empirical method that permits the calculation of the olefinic carbon chemical shifts in acyclic alkenes. These calculations are based on the use of the ethylene carbon chemical shift (123.3 ppm) as a fundamental reference, whereby a saturated carbon bonded to the olefinic one adds 10.6 ppm (α effect) to the reference compound and a carbon bonded to the β carbon, with respect to the ethylenic one, adds 7.2 ppm (β effect). A carbon atom bonded to the α' position (i.e., a carbon bonded as in C—C=**C**—C) will detract from the chemical shift by −7.9, whereas a carbon at the β' position (i.e., a carbon bonded as in C—C—C=**C**—C) will detract by −1.8 ppm.

An example serves to illustrate the application of this empirical relationship. The emphasized carbon chemical shift in 2-pentene (structure **12**) can be obtained as follows:

$$CH_3{-}CH{=}\mathbf{C}H{-}CH_2{-}CH_3$$

12

$$\delta_C = 123.3 + 10.6 + 7.2 + (-7.9) = 133.2 \text{ ppm (found: 133.3 ppm)}$$

Again, the empirically computed value and the experimentally obtained one agree extremely well.

2.3.3 Nitrogen Additivity Relationships [52]

AMINE CORRELATIONS. Duthaler and Roberts [53] have developed an empirical relationship that allows calculation of the nitrogen chemical shifts of primary, secondary, and tertiary amines. The empirical parameters are listed in Table 2.5.

The applicable empirical relationship is expressed by equation 2.3.

$$\delta^j = B^j + n_\beta \beta^j + n_\gamma \gamma^j + n_b(\text{cor } \beta^j + \text{cor } \gamma) \quad (2.3)$$

The superscript j's refer to the parameters for primary (p), secondary (s), or tertiary (t) amines, respectively. The term B^j refers to the base value for a primary, b_p, secondary, B^s, or tertiary, B^t, amine.

The values β^j and γ^j are substitution parameters for carbons two and three bonds removed from nitrogen, respectively. Branching corrections are expressed by the terms cor = β^j for α carbon and cor = γ, which is an additional parameter for α branching if a γ carbon is present. The n_β, n_γ, and n_b constants are, respectively, the numbers of β and γ carbons and the branches at the α carbon. The calculation

TABLE 2.5 Aliphatic Amine Substituent Constants[a]

Parameter	Constant (ppm) in Cyclohexane
B^p	1.4
B^s	8.9
B^t	12.5
p	22.6
s	19.0
t	11.7
p	−3.8
t	−2.2
$\mathrm{cor}^{\alpha'}$	−4.9
$\mathrm{cor}^{\beta'}$	−8.2
$\mathrm{cor}^{\gamma'}$	−1.8

Source: Maciel et al., *J. Phys. Chem.* **81,** 263 (1977).
The standard deviation for all of these constants is 1.4 ppm.
[a] See Ref. [42].

of the N^{15} chemical shift of the tertiary amine **13** illustrates the application of this relationship:

$$\begin{array}{c} CH_2-CH_3 \\ | \\ CH_3-CH_2-C-N(CH_3)_2 \\ | \\ H_3 \end{array}$$

13

1. The amine is tertiary: $B^t = 12.5$
2. Three β carbons: $t = 11.7 \times 3 = 35.1$
3. Two γ carbons: $t = -2.2 \times 2 = -4.4$
4. Two branches at an α carbon: $\mathrm{cor} = \beta' = -8.2 + 2 = -16.4$
5. Two γ carbons on branches: $\mathrm{cor} = \gamma' = -1.8 \times 2 = -3.6$
6. n^b (No. of branches at δ-C's): $n^b = 2$.

When these values are substituted into equation 2.3, the following results:

$$\delta^t = 12.5 + 35.1 + (-4.4) + (-16.4) + (-3.6) = 23.4 \text{ ppm } N^{15}$$

This estimated value agrees with the experimentally observed one of 23.4 ppm.

2.3.4 Empirical Relationships between C, H, and N

2.3.4.1 C^{13} and N^{15} Correlations in Aliphatic Amines

When the carbon chemical shift of a particular carbon is plotted against the chemical shift of a nitrogen, where the carbon is replaced by a nitrogen in a structure, a linear relationship exists [52]. Consequently, comparisons between a particular hydrocarbon and an analogous amine are possible for primary (structures **16** and **17**), secondary (**16** and **18**), and tertiary (**14** and **15**) systems:

$$\underset{\textbf{14}}{CH_3CH_2{-}\overset{\overset{CH_3}{|}}{C}H{-}CH_3} \text{ vs. } \underset{\textbf{15}}{CH_3CH_2{-}\overset{\overset{CH_3}{|}}{N}{-}CH_3}$$

$$\underset{\textbf{16}}{CH_3CH_2CH_3} \text{ vs. } \underset{\textbf{17}}{CH_3CH_2{-}NH_2}$$

$$\underset{\textbf{16}}{CH_3CH_2CH_3} \text{ vs. } \underset{\textbf{18}}{CH_3{-}NH{-}CH_3}$$

The slopes of the correlations are 2.06, 1.96, and 1.39 ppm N^{15} (ppm)/C^{13} (ppm) for aliphatic primary, secondary, and tertiary amines [42], and the complete correlation equations are:

For primary amines: $\delta_C = 2.06\ (\delta_N) - 8$
For secondary amines: $\delta_C = 1.96\ (\delta_N) - 21$
For tertiary amines: $\delta_C = 1.09\ (\delta_N) - 17$

For example, the carbon chemical shift of C_2 in propane is 24.8 ppm. Based on this reference, the chemical shift of its nitrogen analog dimethylamine is 10.1 ppm ($1.96 \times 15.9 - 21$). The experimentally observed N^{15} chemical shift for dimethylamine is 9.1 in cyclohexane and 10.7 ppm in methanol solution.

2.3.4.2 C^{13}, H^1, and N^{15} Correlations in Aromatic Compounds

One of the important uses of NMR spectroscopy centers on its application to structure determination studies of aromatic and heteroaromatic compounds. Since either proton, carbon, or nitrogen NMR can be employed, the use of any one nucleus depends not only on the availability of appropriate instrumentation but also on the specific research needs.

It stands to reason that it would be very useful if knowledge of the chemical shift of any one of these nuclei would allow the computation of the chemical shifts of the other two types of nuclei [54]. For example, the chemical shift of a particular carbon could be estimated from a knowledge of the proton chemical shift of the proton bonded to it (structure **19** vs. **20**):

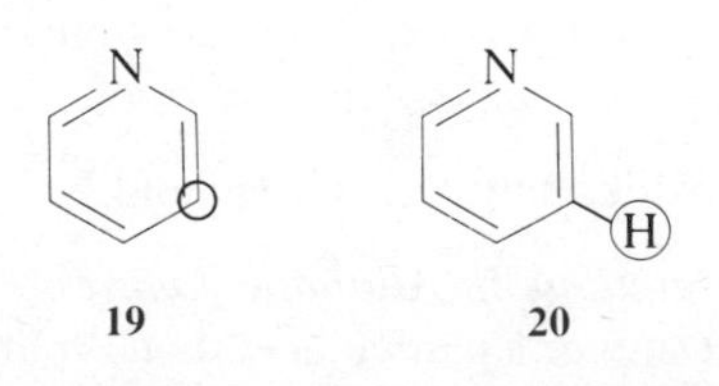

Similarly, it would be useful if the nitrogen chemical shift could be estimated from either the carbon or proton chemical shift of the corresponding carbocyclic analog (**21** vs. **22**):

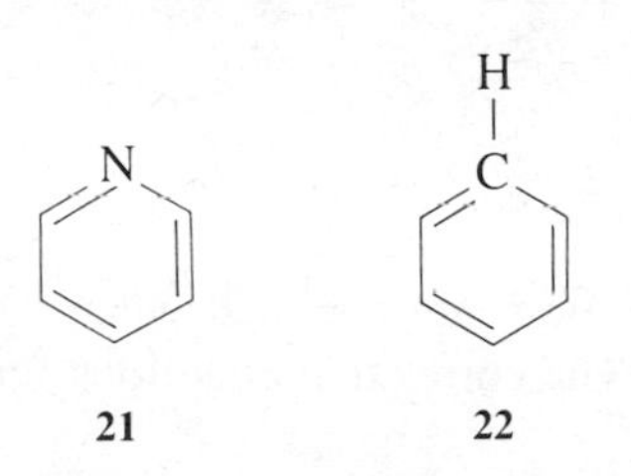

When the C^{13} chemical shifts of a number of benzenoid aromatic compounds and related heteroaromatic analogs are plotted against the H^1 chemical shifts (emphasized carbon in structure **19**) versus the H chemical shift (emphasized hydrogen in **20**), a linear relationship is realized. The correlation equation for this relationship is

$$\delta_{H^1} = 5.6 \times 0.01\ \delta_{C^{13}} + 0.153$$

or, conversely

$$\delta_{C^{13}} = 17.9\ \delta_{H^1} - 2.73$$

Thus, within the limitation that these equations apply only to aromatic and heteroaromatic six-membered rings, these interrelationships become very useful for chemical shift estimates.

These correlations can be extended to include N^{15} chemical shifts, in the following manner: The C^{13} and H chemical shifts in benzene are 128.5 and 7.27 ppm, respectively. If one carbon in benzene is replaced by a nitrogen, pyridine (structure **21**) is obtained. Thus, one might expect a relationship between C^{13} and H chemical shifts of benzene and the N^{15} chemical shift in pyridine (**21**):

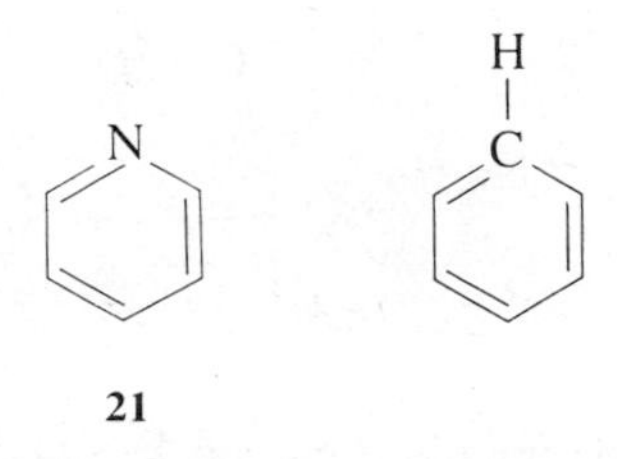

The replacement of a carbon, for example, para to the nitrogen in pyridine (**21**) affords pyrazine (structure **23**). Again, a comparison of the γ carbon and γ hydrogen chemical shifts in pyridine with the N^{15} chemical shift in pyrazine is feasible:

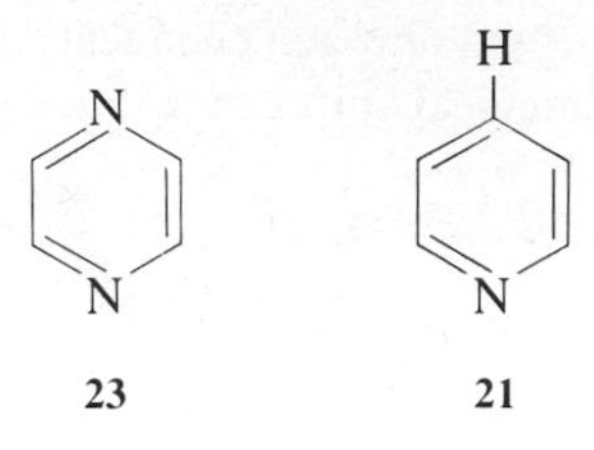

23 21

In fact, excellent correlations are obtained when these plots H versus N^{15} and C^{13} versus N^{15} are made. The correlation equations for these relationships are

$$\delta_{N^{15}} = 65.4\ \delta_{H^1} - 163$$

$$\delta_{N^{15}} = 3.65\ \delta_{C^{13}} - 153$$

Consequently, in aromatic systems, the following relations prevail:

$$H^1/C^{13} = 0.056 \text{ ppm}$$
$$C^{13}/H^1 = 17.9 \text{ ppm*}$$
$$N^{15}/H^1 = 65.5 \text{ ppm}$$
$$H^1/N^{15} = 0.015 \text{ ppm*}$$
$$N^{15}/C^{13} = 3.65 \text{ ppm}$$
$$C^{13}/N^{15} = 0.27 \text{ ppm*}$$

The starred relationships are the reciprocals of the preceding one.

Some examples will illustrate the application of these correlations:

1. The H chemical shift of the β proton in pyridine, the italicized H in structure **24**, is 7.06 ppm. Calculate the chemical shift of the β carbon (emphasized carbon in **24**):

$$C^{13} = 17.9 \times 7.06 - 2.73 = 123.6 \text{ ppm}$$

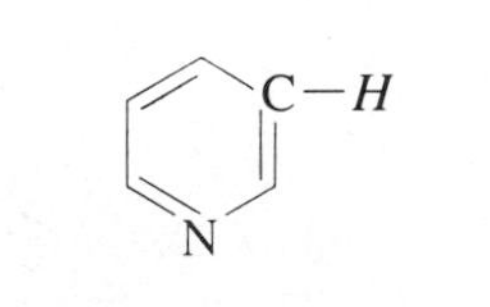

24

The observed chemical shift for the β carbon is 124.5 ppm, a value that agrees quite well with the computed one.

2. The carbon chemical shift of C-4 (the carbon γ to the nitrogen) in pyridine is 136.4 ppm. When this carbon is replaced by a nitrogen, the heterocyclic ring system pyrazine is obtained. Calculate the N^{15} chemical shift of this nitrogen (actually both of the nitrogens are identical).

$$\delta_N = 3.65 \times 136.4 - 135 = 345 \text{ ppm}$$

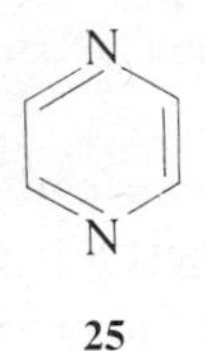

25

The observed value, 338 ppm, again agrees well with this calculated value. Substituted benzenes can be equally readily compared with the corresponding substituted heterocyclic compounds.

It is to be kept in mind that these correlations are directly applicable only to aromatic benzenoid systems and their nitrogen analogs, although reasonable estimates between H^1 and C^{13} chemical shift in heterocyclic compounds of general structure **26** can be made as long as the carbon and hydrogen chemical shift to be calculated are not *ortho* (α position) to the heteroatom. The reason for this limitation is discussed in Chapter 5,

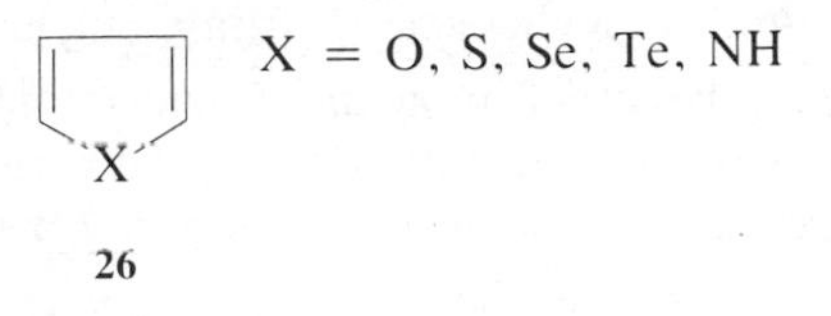

26

2.4 PROBLEMS

1. The difference in chemical shift between two protons, observed in a 15-MHz field, is 230 Hz. Computc the difference if these two protons are observed in a 60-MHz field.
2. A proton in a 60-MHz field resonatcs at +700 cycles with respect to TMS. Translate this chemical shift into δ(ppm).
3. Two protons, A and B, resonate at $\delta_H 6.3$ and 9.5(ppm), respectively. Which one is more shielded? Which one resonates downfield from the other?
4. The chemical shift of a particular carbon in an organic compound is $\delta_C 240$(ppm). What type of compound might this be?
5. The chemical shift of a nitrogen in an organic material is $\delta_N 30$(ppm). Suggest what type of nitrogen this could be.
6. Calculate the proton chemical shifts of the emphasized protons in each of the following compounds: $CH_3CII_2{-}Br$; $CH_3CH_2{-}Cl$; $CH_3CH_2{-}I$. Are there any correlations between these chemical shifts and any other physical properties of these compounds?
7. The proton magnetic resonance spectrum of a compound C_3H_4O shows absorption peaks at $\delta_H 2.50$ (1 H), $\delta_H 2.82$ (1 H), and $\delta_H 4.28$ (2 H). What is the structurc of this compound?
8. The proton NMR spectrum of a compound C_4H_5N has absorption areas at

δ_H6.22 (2 H), δ_H6.68 (2 H), and δ_H8.00 (1 H). The nitrogen resonates at 145(ppm). Deduce the structure of this compound.

9. Predict the approximate chemical shifts of the different protons in 2,6-dimethyl-gamma-pyrone.

10. The hydrocarbon C_5H_{12} shows carbon absorptions at 11, 32, 22, and 300(ppm), respectively. Suggest a possible structure.

11. The proton NMR spectrum of a compound $C_5H_4O_5$ shows the presence of only highly deshielded protons. In order of increasing shielding, their numerical distribution is 1:2:1. The most deshielded proton resonates at δ_H9.92(ppm). What is the structure of this compound?

12. A compound C_8H_6 shows six different carbon signals. Four of these are in the region δ_C120–180(ppm), and the other two are at 83.3 and 77.7(ppm), respectively. What is the structure of this compound?

13. The chemical shifts of the carbons in *n*-butane are δ_C13.0 and 24.8(ppm), respectively. Calculate the carbon chemical shifts of the *n*-propanol carbons.

14. Estimate the carbon chemical shifts of the various carbons in 3-hexene.

15. Calculate the nitrogen-15 chemical shift in *n*-butylamine.

16. The methyl group-bonded carbon in toluene resonates at 125 ppm. The carbon *ortho* to it resonates at 129, whereas the *para*-situated one resonates at 125. Estimate the nitrogen chemical shift in 2-methylpyridine.

17. The proton chemical shift of the β proton in furan is δ_H6.30(ppm). Estimate the carbon chemical shift of the β carbon.

REFERENCES

1. P. Lazlo and A. Speer, *J. Chem. Soc.* **48,** 1732 (1968).
2. G. V. D. Tiers, *J. Phys. Chem.* **62,** 1151 (1958).
3. P. C. Lauterbur, *Ann. N.Y. Acad. Sci.* **70,** 841 (1958).
4. G. V. D. Tiers and G. Filipovitch, *J. Phys. Chem.* **63,** 1701 (1959).
5. R. K. Harris and E. G. Finer, *Bull. Soc. Chim. Fr.* 2805 (1968).
6. S. O. Grim, W. McFarlane, E. F. Davidoff, and T. J. Marks, *J. Phys. Chem.* **70,** 581 (1966).
7. F. Ramirez, S. B. Bhatis, A. V. Patwardham, and R. C. Jones, *J. Org. Chem.* **32,** 2194 (1967).
8. G. H. Hahn and R. Schoeffer, *J. Am. Chem. Soc.* **86,** 1503 (1964).
9. J. R. Van Wazer, C. F. Collins, J. N. Shoolery, and R. C. Jones, *J. Am. Chem. Soc.* **78,** 5715 (1956).
10. M. L. Nielsen, J. V. Pustinger, and T. Strobel, *J. Chem. Eng. Data* **9,** 167 (1964).
11. J. J. Dechter, *Progress in Inorganic Chemistry*, Vol. 29, 1982, pp. 285–385.
12. R. Freeman, G. R. Murray, and R. E. Richards, *Proc. R. Soc. (Lond.)* **A242,** 455 (1957); and J. Mason, *Adv. Inorg. Chem. Radiochem.* **18,** 197 (1976).

13. B. Lindman and S. Forsen, *NMR Basic Princ. Prog.* **12,** 1 (1976).
14. G. Mavel, *Ann. Reports NMR Spectrosc.* **5B,** 1 (1973).
15. J. F. Hinton and K. H. Ladner, *J. Magn. Resonance* **32,** 303 (1978).
16. T. L. Brown, *Acc. Chem. Res.* **1,** 23 (1968).
17. E. Mei, A. I. Popov, and J. L. Dye, *J. Am. Chem. Soc.* **99,** 6532 (1977).
18. J. D. Kennedy, W. McFarlane, and G. S. Pyne, *J. Chem. Soc. Dalton Trans.* 233 (1977).
19. J. D. Kennedy, W. McFarlane, and G. S. Pyne, *Bull. Soc. Chim. Belg.* **84,** 289 (1975).
20. J. Mason, *Adv. Inorg. Chem. Radiochem.* **22,** 199 (1979).
21. F. W. Wehrli, in G. A. Webb ed., *Ann. Reports on NMR Spectroscopy*, Vol. 9, Academic, New York (1979).
22. J. Huett, J. Durand, and Y. Infarnet, *Org. Magn. Resonance* **8,** 382 (1976).
23. S. F. Lincoln, *Aust. J. Chem.* **25,** 2705 (1972).
24. J. W. Akitt, N. N. Greenwood, and A. Storr, *J. Chem. Soc.* 4410 (1965).
25. J. F. Hinton and R. W. Briggs, *J. Magn. Resonance* **25,** 555 (1977).
26. C. Schramm and J. I. Zink, *J. Magn. Resonance* **26,** 513 (1977).
27. H. Koeppel, J. Dallorso, G. Hoffmann, and B. Walther, *Z. Anorg. Allgem. Chem.* 427 (1976).
28. C. R. Lassigne and E. J. Wells, *Can. J. Chem.* **55,** 927 (1977).
29. J. J. Burke and P. C. Lauterbur, *J. Am. Chem. Soc.* **83,** 326 (1961).
30. D. C. Van Beelen, H. O. Van der Kooi, and J. Wolters, *J. Organomet. Chem.* **179,** 37 (1979).
31. G. L. Kok, M. D. Morris, and R. R. Sharp, *Inorg. Chem.* **12,** 1709 (1973).
32. J. P. Manners, K. G. Morallee, and R. J. P. Williams, *J. Chem. Soc. Chem. Commun.* 965 (1970).
33. J. J. Dechter, *Progress in Inorganic Chemistry* **33,** 393–507 (1985).
34. G. A. Melson, D. J. Olszanski, and A. Rahimi, *Spectrochim. Acta* **33A,** 301 (1977); R. M. Adam, G. V. Fayakerly, and D. G. Reid, *J. Magn. Resonance* **33,** 655 (1979); and P. L. Rinaldi, S. A. Khan, G. R. Choppin, and G. C. Levy, *J. Am. Chem. Soc.* **101,** 1350 (1979).
35. N. Hao, B. G. Sayer, G. Denes, D. G. Bickley, C. Detellier, and M. J. McGlinchey, *J. Magn. Resonance* **50,** 50 and references therein to related nuclides (1982).
36. D. Rehder, *Bull. Magn. Resonance* **4,** 33 (1982).
37. J. Y. LeGall, M. M. Kubicki, and F. Y. Petillon, *J. Organomet. Chem.* **221,** 287 (1981); and A. F. Masters, R. T. C. Brownlee, M. J. O'Connor, and A. G. Wedd, *Inorg. Chem.* **20,** 4183 (1981).
38. A. Kececi and D. Rehder, *Z. Naturforsch.* **36b,** 20 (1981).
39. C. Brevard and P. Granger, *Inorg. Chem.* **22,** 532 (1983).
40. F. Wehrli, *Ann. Rep. NMR Spectrosc.* **9,** 126 (1979); and E. Maurer, S. Rieker, M. Schollbach, A. Schwenk, T. Egolf, and W. von Philipsborn, *Helv. Chim. Acta* **65,** 26 (1982).
41. P. S. Pregosin, *Coord. Chem. Rev.* **44,** 247 (1982).
42. G. E. Maciel, L. Simeral, and J. J. H. Ackerman, *J. Phys. Chem.* **81,** 263 (1977).
43. J. D. Kennedy and W. McFarlane, *J. Chem. Soc. Perkin II* 1187 (1977).

44. C. W. Heitsch, *Inorg. Chem.* **4,** 1019 (1965).
45. H. A. Christ, P. Diehl, H. R. Schneider, and H. Dahn, *Helv. Chim. Acta.* **44,** 865 (1961).
46. J. A. Pople, W. G. Schneider, and H. J. Bernstein, *High Resolution Nuclear Magnetic Resonance*, McGraw-Hill, New York, 1959, p. 59.
47. B. P. Bailey and J. N. Shoolery, *J. Am. Chem. Soc.* **77,** 3977 (1955).
48. G. C. Levy and G. L. Nelson, *C-13 Nuclear Magnetic Resonance for Organic Chemists*, Wiley, New York, 1972.
49. J. D. Roberts, F. J. Weigert, J. I. Kroschwitz, and H. J. Reich, *J. Am. Chem. Soc.* **92,** 1338 (1970).
50. L. M. Jackman, D. P. Kelly, *J. Chem. Soc. B* 102 (1970).
51. D. E. Dorman, M. Jautelat, and J. D. Roberts, *J. Org. Chem.* **36,** 2757 (1971).
52. G. C. Levy and R. L. Lichter, *N-15 Nuclear Magnetic Resonance Spectroscopy*, Wiley, New York, 1979, p. 28 ff.
53. R. O. Duthaler and J. D. Roberts, *J. Phys. Chem.* **18,** 2507 (1974).
54. G. R. Newkome, W. W. Paudler, *Contemporary Heterocyclic Chemistry*, Wiley, New York, 1982, p. 397 ff.

3

SPIN–SPIN COUPLING CONCEPTS—GENERAL CONSIDERATIONS

3.1 GENERAL BACKGROUND

The development of high-resolution NMR spectroscopy brought with it the recognition that the peaks originally observed for various nuclei are in fact often composed of a number of relatively narrow lines. A close examination of many of the spacings of the multiplets revealed that in each group of lines, many, if not all, were equally separated from each other. Figure 3.1 shows the high-resolution proton NMR spectrum of ethyl alcohol at 60 and 100 MHz. Clearly, although the relative positions of the groups of lines are different, the separation between the peaks in each group remains the same, regardless of the magnetic field strength employed.

To differentiate between lines that are part of a splitting pattern and those that are not, a given spectrum has to be obtained at two different field strengths, since the former is invariable with field strength and the latter is not. Fortunately, patterns are normally recognizable, and the need for obtaining two spectra at two different field strengths arises relatively rarely. The technique of selective decoupling is another method that allows an analysis of these patterns (see Section 3.3.3).

3.2 MAGNETIC INTERACTIONS

3.2.1 Solid State

Magnetically active nuclei in the solid state interact not only with their surrounding electrons but also with each others' local static fields. These local dipoles are arranged in all possible directions and consequently have different intensities depending on their angular alignment with the direction of the applied field. The contribution of these dipoles on a particular nucleus also varies with changes in the distances from any magnetically active nucleus or molecular grouping.

When this conglomerate of nuclei is placed into a magnetic field, no given nucleus will resonate at a precise value but will resonate over a fairly wide fre-

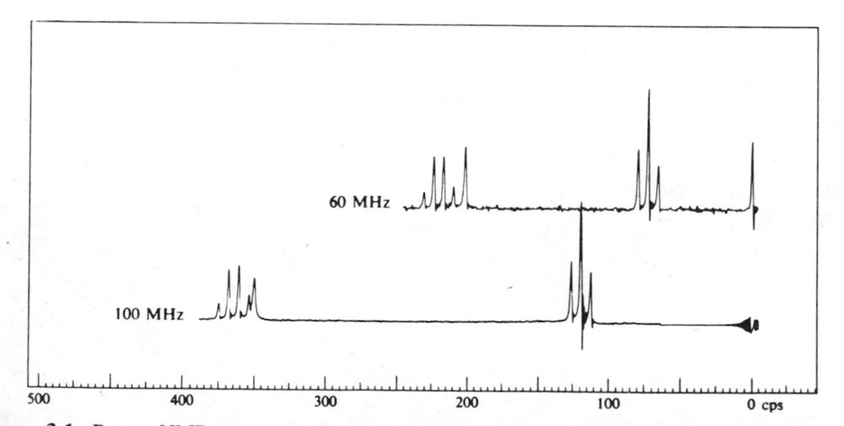

Figure 3.1. Proton NMR spectrum of ethyl alcohol. Upper trace obtained at 60 MHz; lower trace obtained at 100 MHz.

quency range. This line broadening is generally severe enough (at least 400–500 ppm for most C^{13} nuclei, for example) to prevent the observation of NMR fine structure.

The superb technique of "magic angle" NMR, is based on appropriate averaging of the various solid-state interactions, and provides NMR spectra of compounds in the solid state. The process is described in Chapter 4.

3.2.2 Liquid State

When the compound to be examined by NMR is in solution, the local variations in magnetic field become very short-lived, because the neighboring dipoles can now orient themselves with relative ease. This reorientation process causes an effective averaging of the motions, which reduces the local field broadening to almost zero. As a result, the resonance lines become considerably narrowed (0.01 Hz for protons, for example).

The transmission of magnetic interactions between two nuclei occurs through intervening chemical bonds and becomes readily discernible when the compound under study is either in the liquid state or in solution, since the external dipole contributions are averaged and the absorption peaks are narrowed.

The transmission of magnetic information from one nucleus to another occurs by slight polarization of the spins or orbital motions of the bonding electrons and is consequently independent of the motion of the molecule itself. The existence of these interactions adds another dimension to the use of NMR spectroscopy.

3.3 QUALITATIVE ANALYSES OF SPIN–SPIN INTERACTIONS

3.3.1 General Concepts

The analysis of a two-nucleus system, with each nucleus having a spin quantum number of $\frac{1}{2}$, in terms of its overall energy levels serves to introduce and describe the subject of spin–spin coupling. In any system of nuclei that can interact with each other, each of the spin states of any one nucleus can interact with each spin state of any other near nucleus. In a two-nuclei system involving nuclei with spin quantum numbers of $\frac{1}{2}$, there are four possible additive combinations of spin states. Table 3.1 shows these interaction possibilities.

TABLE 3.1 Spin Quantum Number Combinations for a Two-Proton System

Magnetic state	$I_z(H_A)$	$I_z(H_B)$	ΣI_z
1	$+\frac{1}{2}$	$+\frac{1}{2}$	$+1$
2	$+\frac{1}{2}$	$-\frac{1}{2}$	0
3	$-\frac{1}{2}$	$+\frac{1}{2}$	0
4	$-\frac{1}{2}$	$-\frac{1}{2}$	-1

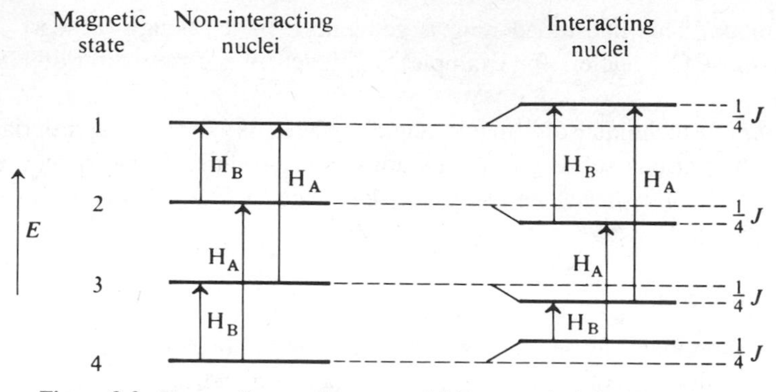

Figure 3.2. Energy diagram for two nuclei in magnetic field ($H_A > H_B$).

If there is no interaction between the four energy levels of this two-nucleus system, the left part of the diagram in Figure 3.2 prevails. Since *Iz* must be 1 (see Chapter 1) for any transition to occur, nucleus A cannot go from magnetic state number 4 to magnetic state number 3 ($-\frac{1}{2}$ to $-\frac{1}{2}$). However, nucleus B can be involved in the transition from magnetic state number 4 to magnetic state number 3 ($-\frac{1}{2}$ to $+\frac{1}{2}$). Within this limitation, each nucleus in a two-nucleus spin $\frac{1}{2}$ system is subject to the following two transitions (see Fig. 3.2):

Nucleus A: States 4 to 2 and 3 to 1

Nucleus B: States 4 to 3 and 2 to 1

The energies of the 4-to-3 and 2-to-1 transitions are identical, as are the 4-to-3 and 2-to-1 transitions. If this were the entire picture, only two resonance lines would be observed for this system, one line for each nucleus.

However, the two nuclei can interact with each other under some specific conditions. If the two nuclei that compose the magnetic state number 4 interact with each other, the resulting energy level will be *less* stable, since the two nuclei have the same spin quantum number. The same is true for the two nuclei that make up the magnetic state number 1.

On the other hand, when the nuclei that form magnetic state numbers 2 and 3 interact, *more* stable energy levels will result, since the interacting nuclei have spin quantum numbers of opposite signs. These respective increases and decreases in the four energy levels will be numerically identical. (The four energy levels resulting from these interactions are depicted on the right side of Figure 3.2.) When the permitted transitions are superimposed on these new energy levels, the two transitions for each nucleus are no longer identical, and four observable resonance lines will result for these two interacting nuclei.

If the change in each energy level is designated by $\frac{1}{4}J$, the energy of the nucleus B transitions from state 4 to 3 changes by $-\frac{1}{2}J$, and the state 2-to-1 transitions change by $+\frac{1}{2}J$. Similarly, the transitions involving nucleus A change by $-\frac{1}{2}J$ and $+\frac{1}{2}J$ as well. Consequently, with reference to a noninteracting system (dotted lines

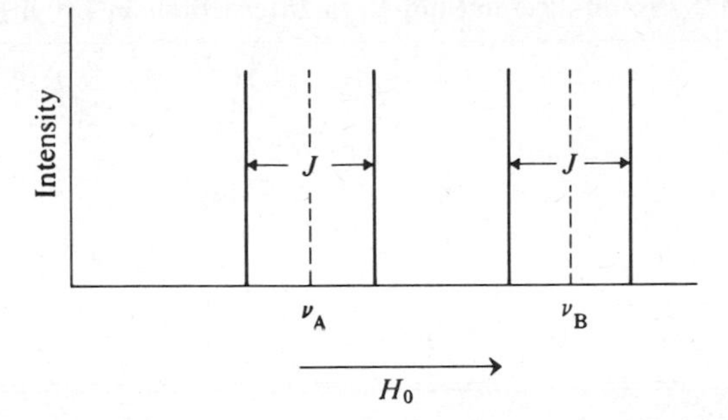

Figure 3.3. Hypothetical splitting pattern for two interacting protons. (The dotted lines represent the line positions for the hypothetical case where the two protons are not interacting.)

in Fig. 3.3), the two absorption frequencies for each nucleus will be greater and less by $\frac{1}{2}J$. The separation between the two resonance lines of each nucleus is, consequently, J. This value is referred to as the "spin–spin coupling constant" or, simply, the "coupling constant."

It is important to recognize that this concept has been developed without consideration of the strength of the applied magnetic field! Consequently, the coupling constant is *independent* of the strength of the applied magnetic field. Clearly, the experimental observation (see Fig. 3.1) that the separation of the groups of lines of the protons in ethanol are independent of the applied field strength is now explained.

The approach developed for the general case of two interacting spin $\frac{1}{2}$ nuclei can readily be expanded to explain the multiplicity pattern observed for the protons in ethanol, where the methyl protons appear as a triplet and the methylene protons as a quartet.

The various possible combinations of the interactions between the methylene protons, H_A and H_B, with the three equivalent methyl protons, are shown in Table 3.2. In this tabulation, an upward-pointing arrow refers to a spin of $+\frac{1}{2}$ and a downward-pointing one to a spin of $-\frac{1}{2}$. Four different magnetic fields result from the methyl group protons (H_A) acting on the methylene protons. There exists one possibility each of obtaining $\Sigma I_z = 1\frac{1}{2}$ and $\Sigma I_z = 1\frac{1}{2}$—three each of obtaining a summation of $+\frac{1}{2}$ and three of $-\frac{1}{2}$. A similar analysis of the effect of the methylene protons (H_B) generates three different magnetic field summations: one possibility each of a summation of $+1$ and -1 and two possibilities of a zero value summation.

Thus, this analysis predicts that the methylene protons of ethanol will appear as a *quartet*, and the methyl group protons as a *triplet*. The separation of the lines in the quartet and triplet will be identical and represent J, the coupling constant (11 Hz, or cps (cycles per second)). A glance at Figure 3.1 confirms this analysis.

Some reflection leads to the conclusion that the multiplicity of the splitting pattern can be arrived at by the simple expedient of counting the number of *identical* spin $\frac{1}{2}$ nuclei interacting with the nucleus being examined and adding 1. For ex-

TABLE 3.2 Proton–Proton Spin–Spin Interactions in Ethyl Alcohol

	H_B	H_B	H_B	H_B	H_B	H_B	H_B	H_B
H_A	↑	↑	↑	↓	↓	↑	↑	↓
H_A	↑	↑	↓	↓	↓	↓	↑	↑
H_A	↑	↓	↓	↑	↓	↑	↓	↓
ΣI_A	$1\frac{1}{2}$	$\frac{1}{2}$	$-\frac{1}{2}$	$-\frac{1}{2}$	$-1\frac{1}{2}$	$\frac{1}{2}$	$\frac{1}{2}$	$-\frac{1}{2}$

	H_A	H_A	H_A	H_A
H_B	↑	↑	↓	↓
H_B	↑	↓	↓	↑
ΣI_B	1	0	−1	0

$CH_3—CH_2—OH$
(H_A) (H_B)

$I = +\frac{1}{2} = \uparrow$

$I = -\frac{1}{2} = \downarrow$

ample, the methine proton (H_C in 2-bromopropane (structure **1**) is adjacent to six *identical* carbons,

H_C
|
$CH_3—C—CH_3$
|
Br
1

Thus, this proton will appear as a *heptaplet* (6 + 1), with the spacing of the lines being equal to J. The methyl group protons are adjacent to a carbon bearing one hydrogen. Consequently, they will appear as a *doublet*, with the line separation J being equal to the heptaplet line separation. In a general sense, the number of lines resulting from spin–spin interactions is $2I + 1$, where I is the spin quantum number of the particular coupling nucleus.

The observant reader will have noted that although the protons are bonded to carbons, 1.08% of which have a spin quantum number of $\frac{1}{2}$, has been ignored in this analysis. The protons are, indeed, coupled to the magnetically active carbons (C^{13}). However, since the concentration of these nuclei is so low, the coupling patterns resulting from these interactions are not readily observed under circumstances that give adequate proton spectra. In fact, these secondary coupling patterns are "buried" under the instrumental "noise." This additional coupling will be described in Section 3.7.

3.3.2 Relative Line Intensities

The relative line intensities of the methylene proton quartet of ethyl alcohol (Fig. 3.1) are approximately 1 : 3 : 3 : 1, while those of the methyl group triplet are 1 : 2 : 1. The data presented in Table 3.2 show that the four different spin quantum number summations for the methylene protons are composed of one $+1\frac{1}{2}$ and one $-1\frac{1}{2}$ combination. The $-\frac{1}{2}$ and $+\frac{1}{2}$ summations are each composed of three possible spin combinations. Consequently, the $-\frac{1}{2}$ and $+\frac{1}{2}$ summations will result in absorptions that have three times the *area* (in a general sense, line *intensities* can be used) of the + and $-1\frac{1}{2}$ summation areas, each. The observed relative intensities of 1 : 3 : 3 : 1 are thus readily explained. Similar arguments show that the triplet line intensities should be 1 : 2 : 1. Clearly, they are!

Examination of other spin $\frac{1}{2}$ systems establishes that the relative line intensities of increasingly complex multiplets are the same as the coefficients of the *binomial* expansion. A convenient mnemonic for these coefficients is Pascals's triangle:

$$\begin{array}{ccccccccc} & & & & 1 & & & & \\ & & & 1 & & 1 & & & \\ & & 1 & & 2 & & 1 & & \\ & 1 & & 3 & & 3 & & 1 & \\ 1 & & 4 & & 6 & & 4 & & 1 \end{array}$$

For example, the relative line intensities for a *pentuplet* are 1 : 4 : 6 : 4 : 1.

3.3.3 Spin-Spin Decoupling

The development of the elementary principles leading to an understanding of chemical shifts and spin-spin coupling constants make it clear that NMR spectra can only be obtained when an unequal distribution of nuclei in the various possible spin states exist. Thus, in the instance of two interacting nuclei, when one of them is not present in an unequal spin distribution, the other one will not be coupled to it and will appear as a singlet. If this phenomenon could be artificially induced, the multiplicity of splitting patterns could be selectively modified. For example, H_B in the three-proton system $CH_A—CH_B—CH_c$ would appear as a "doublet of doublets" (not a quartet) under normal circumstances. If, however, the unequal distribution of the H_A nuclei could be made even, this doublet of doublets would collapse into a doublet. In fact, this can be experimentally accomplished! If during the observation of H_B, H_A is subjected to a strong RF field at the frequency equivalent to its chemical shift, its spin distribution becomes equal and a doublet is observed for H_B.

This process is known as "spin-spin decoupling" and "double resonance." (Chapter 9 deals with this and related techniques in some detail).

3.4 HOMONUCLEAR COUPLING CONSTANTS

The spin–spin interaction between two or more identical types of nuclei (i.e., proton to proton, carbon to carbon, etc.) is referred to as "homonuclear coupling." A number of the more important homonuclear spin systems are the discussion topic for this section.

3.4.1 Proton–Proton Coupling Constants

The qualitative discussion of the spin–spin interaction process implies that the size of the coupling constants should depend on the strength of the interaction between the spin-coupled nuclei. For example, we would envision that nonequivalent hydrogens bonded to the same carbon, $H_A—C—H_B$, will interact with each other much more strongly than two nonequivalent hydrogens bonded to two different carbons—that is, $H_A—C—C—H_B$. In other words *geminal* coupling constants are expected to be larger than *vicinal* ones.

The proton–proton coupling constants for a number of different structural systems are listed in Table 3.3. The fact that some of these constants are negative is the subject of discussion in Chapter 7.

In absolute values, geminal coupling constants vary between 6 and 20 Hz, and vicinal ones between 5.5 and 7.5 Hz. The spin–spin coupling between protons separated by more than two sp^3 hybridized carbon atoms is normally either very small or, more generally, zero.

Spin–spin interactions across sp^2 hybridized carbons persist over a considerable number of bonds. For example, spin–spin coupling across four or more olefinic bonds—that is, $H_A—C=C—C=C—C=C—C=C—H_B$, is routinely observed.

In a general sense, the size of proton–proton coupling constants decreases with increasing distance between the interacting protons and is most effective between protons bonded to carbon atoms that are part of a conjugated system. For example, *meta*-situated protons in benzenoid systems (structure **2**) are coupled to each other to the extent of 0.5 to 4 Hz, whereas those in the nonconjugated allylic systems, such as structure **3**, are on the average smaller (0.2–2.5 Hz).

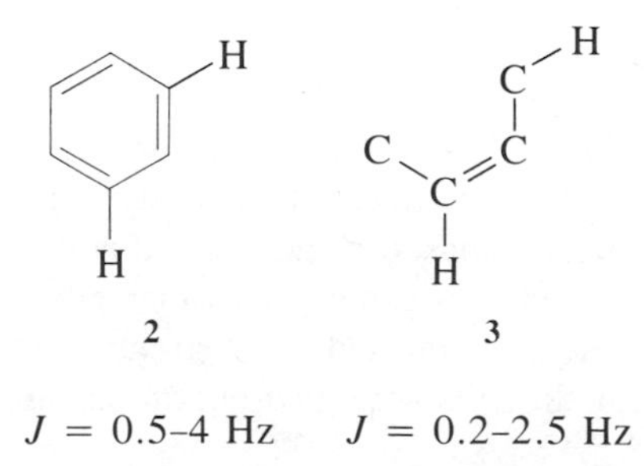

J = 0.5–4 Hz J = 0.2–2.5 Hz

When coupling between protons separated by more than two carbons is discussed, the term "long-range coupling" is often employed.

TABLE 3.3 Some Typical Proton–Proton Spin Coupling Constants

Structural type	*J*(cps)
$H_2C<$ (geminal H–C–H)	−20 to +6
H–C–C–H (−CH−CH−)	5.5 to 7.5
$H_2C{=}C<$	about 2
−C(H)=C(H)− (cis)	7 to 10
−C(H)=C(H)− (trans)	12 to 19
benzene	*o* 6 to 9 *m* 0.5 to 4 *p* 0 to 2.5
pyridine (positions 2–6, N)	J_{23} 5 to 6 J_{24} 1.5 to 2.5 J_{25} 0.5 to 1.[illegible] J_{26} 0 to 0.5 J_{34} 6.5 to 8.5 J_{35} 1 to 2
H_4, H_3, H_5, H_2 five-membered ring (X = S, O, NR)	J_{23} 1.0 to 2.5 J_{24} 0.5 to 1.5 J_{25} 2.5 to 4.0
H–C=C–C–H (H and CH on opposite sides)	0.5 to 2.5
H–C=C–C–H (H and CH on the same side)	0 to 0.3
H–C≡C–C(H)–	1.5 to 3.0
cyclohexane chair (ax., eq.)	$J_{ax,ax}$ 9 to 14 (vicinal) $J_{ax,eq}$ 2 to 4 (vicinal) $J_{eq,eq}$ 2.5 to 4 (vicinal)

The table of proton-proton coupling constants introduces a new symbolism which needs to be briefly described. To enable unambiguous identification of coupling constants with structures, the constant, J, is subscripted by the identification numbers of the protons derived from their location on the structure. The numbers are those of the carbon atoms to which the hydrogen is bonded.

In a more accurate sense, the type of nucleus to which a particular coupling constant refers should, if the possibility of ambiguity exists, be described by including the elemental symbol as well. For example, $J_{C\text{-}2,H\text{-}3}$ refers to the coupling constant between C_2 and the protons on C_3 in a particular structure.

It is instructive to predict the splitting pattern of each of the protons in a substituted benzenoid system, *m*-dinitrobenzene (structure **4**) represents an interesting structure for this purpose, since it incorporates most of the coupling features frequently observed in proton NMR spectroscopy

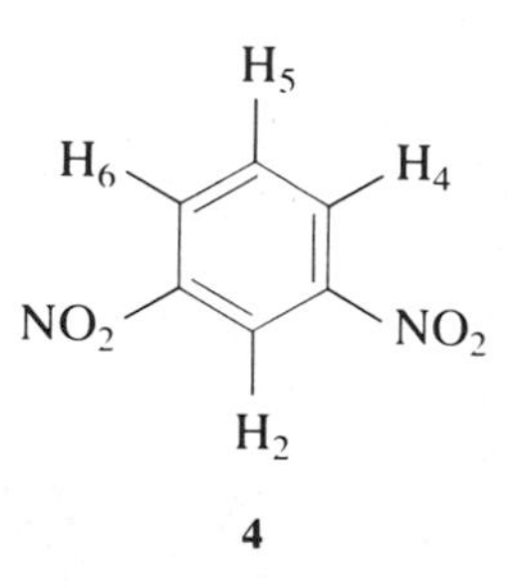

4

The compound has two equivalent protons, H_4 and H_6, and two different ones, H_2 and H_5. In a qualitative sense based on earlier discussions, H_2 should be the most deshielded proton in this compound, since it is situated between two powerful electron-withdrawing groups. The two chemically identical protons, H_4 and H_6, are each adjacent to one nitro group and consequently would be more shielded than H_2; H_5 would be the most shielded proton in this compound.

The H_2 proton will be coupled with two identical protons. H_4 and H_6. Thus, a *triplet* with peak intensities 1 : 3 : 1 will be generated. The lines in the triplet will be separated by the value of the coupling constant $J_{2,4}$. Table 3.3 lists the range of the coupling constant for *meta*-situated protons as being in the range of 0.5–4 Hz. The data in this table also list the range of coupling constants for *para*-situated protons to be in the range of 0–2.5 Hz. Consequently, if *para*-coupling is present in compound **4**, each member of the triplet will be further split into a doublet with its absorption lines being separated by the value of the *para*-coupling constant, $J_{2,5}$.

The various coupling constants can be *extracted* from the experimental pattern in the following manner:

$J_{2,4}$ = separation between lines 1,3; 2,4; 3,5; or 4.6
$J_{2,5}$ = separation between lines 1,2; 3,4; or 5,6

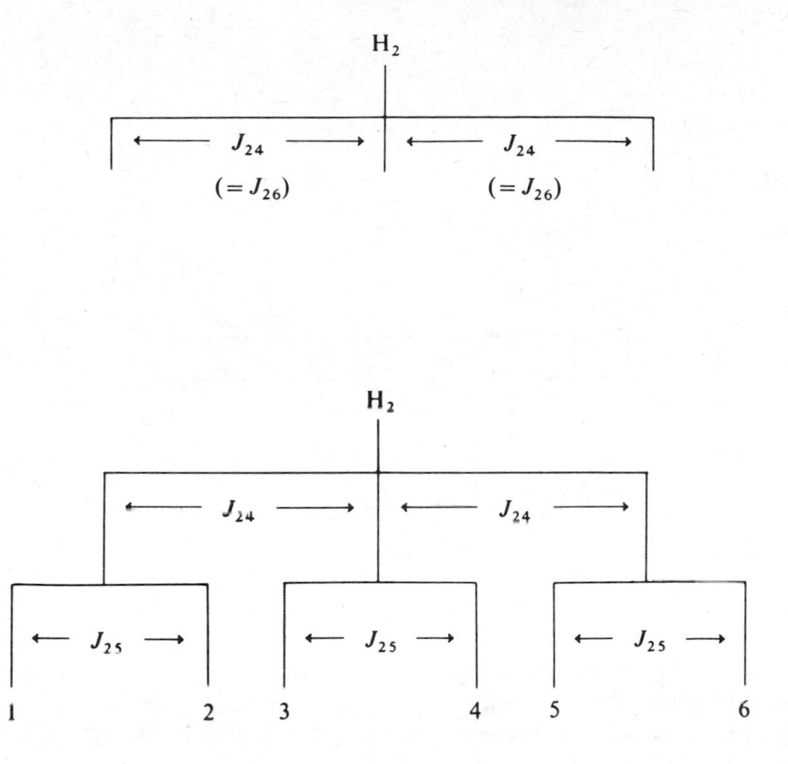

The experimental spectrum of *m*-dinitrobenzene (Fig. 3.4) shows a somewhat line-broadened, very deshielded triplet. The chemical shift of H_2 (904 Hz at 100 MHz = δ_H9.04(ppm)) is equal to the position of the central line of the triplet. The line broadening indicates the existence of some *para*-coupling ($J_{2,5}$); it is too small to be identifiable from this particular spectrum. The separation of the three members of the triplet is 2.0 Hz. Thus, $J_{2,4}$ = 2.0 Hz.

The H_5 proton, which is predicted to be the most shielded one in this compound, is adjacent to two identical protons, H_4 and H_6. Its interaction with these will cause the generation of a triplet, with the members being separated from each other by

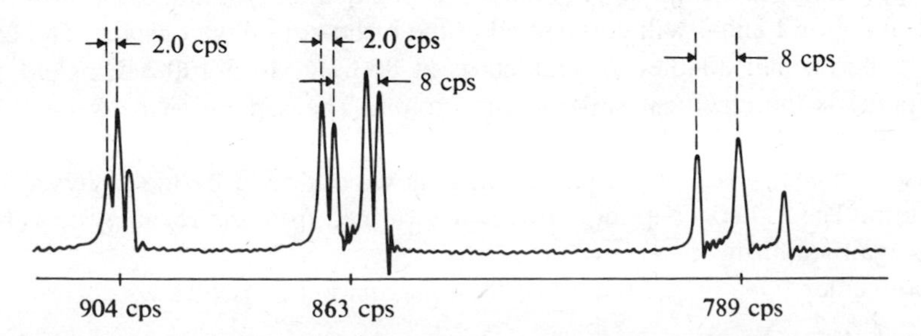

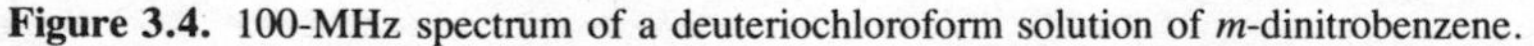

Figure 3.4. 100-MHz spectrum of a deuteriochloroform solution of *m*-dinitrobenzene.

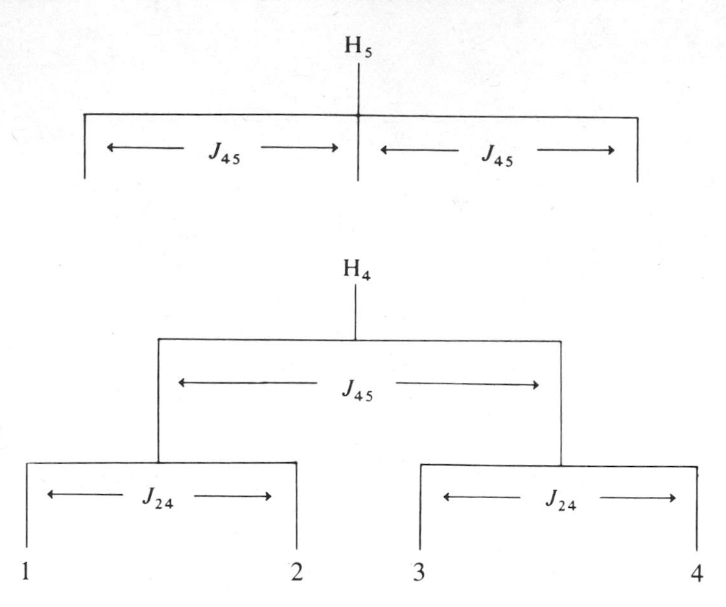

the value of $J_{4,5}$. Although no significant *para*-coupling is observed in this spectrum, it is nevertheless instructive to include it in this analysis.

Again, the separations between lines 1,3; 2,4; 3,5; and 4,6 will give the size of the coupling constant, in this instance $J_{4,5}$, whereas the line separations between 1,2; 3,4; or 5.6 will give $J_{2,5}$. The spectrum reproduced in Figure 3.4 shows that the most shielded triplet is centered at 789 Hz (δ_H7.89(ppm)) and that each member is line-broadened. Thus, again, a small $J_{2,5}$ constant is indicated. The separation of the members of the triplet is 8.0 Hz ($J_{4,5}$).

The H_4 and H_6 protons are identical, the absorption area of the pattern generated by them will have twice the area (or itentity) of the other proton patterns in Figure 3.4. The *ortho*-coupling expected to exist between both of these identical protons and H_5 will cause the generation of a doublet. The separation of these doublet lines corresponds to $J_{4,5}$. In addition, H_4 and H_6 will also interact with H_2, causing each member of the doublet to be split into a yet another doublet. The following diagram represents this pattern analysis.

Consequently, in the doublet of doublets (not a quartet !), the separation between lines 1 and 3 or 2 and 4 will correspond to the ortho-coupling constant. The center of the experimental doublet of doublets is at 863 Hz. In a 100 MHz field, i.e., δ8.63(ppm) is the chemical shift of this proton. The separation between lines 1 and 3 is 8 Hz and the lines 1 to separation is 2.0 Hz. The analysis of the H_5 pattern established that $J_{4,5}$ is 8 Hz, a value that is now confirmed by the analysis of the H_4 pattern. The 2.0 Hz coupling, already determined from the H_2 analysis as being $J_{2,4}$, is again confirmed.

A convenient way to present all of this information is as follows:

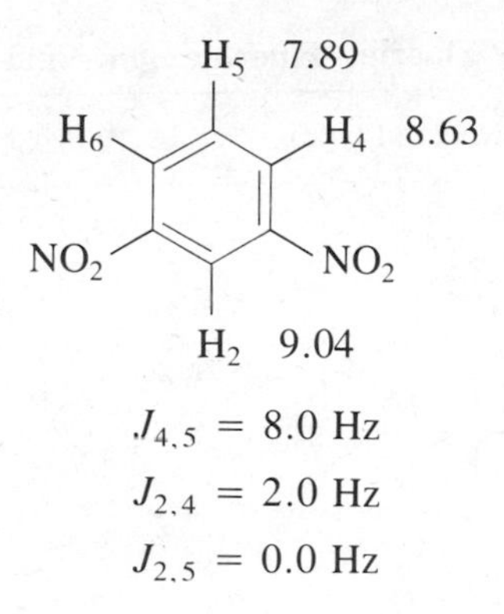

3.4.2 Fluorine–Fluorine Coupling Constants

The spin–spin interaction of fluorine nuclei is, generally, numerically larger than the corresponding proton interactions. For example, the geminal coupling between two fluorines bonded to an sp^2 hybridized carbon (F—C—F) is about 40 Hz, whereas geminal protons bonded to this type of carbon are coupled to the extent of about 2 Hz only. This numerically larger coupling facilitates determination of long-range coupling constants in F-compounds. Interactions between fluorines removed by six or more bonds from each other have been observed! The fluorine–fluorine coupling constants in highly fluorinated organic compounds have occasionally shown some unexpected behavior. For example, some vicinal coupling constants are often nearly zero, whereas interactions between fluorine nuclei separated by as many as 4–5 bonds show spin–spin interactions of 5–15 Hz.

An examination of the fluorine–fluorine coupling constants given in Table 3.4 nevertheless points to the general similarities between these coupling constants and those of related proton-to-proton ones. In both instances couplings between geminally situated nuclei are larger when they are bonded to an sp^3 (150–220 Hz) than when they are bonded to an sp^2 (25–90 Hz) hybridized carbon.

The *cis*- (35–60 Hz) versus *trans*- (115–125 Hz) fluorine–fluorine coupling constants follow the same size pattern as the protons in olefinic compounds.

The proton–proton coupling constant's insensitivity to the nature of substituents in aromatic systems also exists for the fluorinated analogs insofar as the *ortho*-coupling constants are concerned. This insensitivity does not, however, exist in the *meta*- and *para*-fluorine coupling constants. In fact, an additivity relationship exists between the kind and number of substituents on fluorinated aromatics and the size of these coupling constants (see Chapter 6) [1, 2].

3.4.3 Carbon–Carbon Coupling Constants

The low natural abundance (1.08%) of the magnetically active C^{13} nuclei makes the probability of finding two adjacent C^{13} nuclei in a particular structure very low

TABLE 3.4 Some Fluorine–Fluorine Spin–Spin Coupling Constants

Structural type		J_{FF} (cps)
$F_2C<$ (F, F on the same C)		150 to 220 (dependent on ring size)
$-CF-CF-$ (F, F on adjacent C)		0 to 20
$F_2C{=}CXY$	(X not F) (Y not F)	25 to 90
$(F)(X)C{=}C(F)(Y)$, F and F cis	(X not F) (Y not F)	33 to 58[a]
$(F)(X)C{=}C(Y)(F)$, F and F trans	(Y not F) (X not F)	115 to 125[a]
$(F_1)(F_2)C{=}C(F_3)(Y)$		J_{12} 85 to 102 J_{13} 30 to 66 J_{23} 107 to 115
$C_6H_4F_2$ (difluorobenzene)		*ortho* 20 *meta* 2 to 5 *para* 10 to 15
C_6F_5R		*ortho* −18 to −22 *meta* −10 to +10[b] *para* ∓4 to 8[b]

[a] These two coupling constants are generally of opposite sign
[b] Size is very dependent on the nature of R.

indeed. For any specific pair of positions in an organic molecule this probability is 1 in 10,000! Thus, most studies directed toward establishing $J_{C,C}$ values have been done using C^{13} enriched samples. The major types of these coupling constants are given in Table 3.5 [3–9].

The spin–spin interactions between two different adjacent C^{13} nuclei will cause each one of them to appear as doublets, if no other magnetically active nuclei are bonded to these carbons. If other magnetically active nuclei are bonded to these carbons, they will, of course, cause additional splitting of these doublets. The

TABLE 3.5 C^{13}–C^{13} Spin–Spin Coupling Constants

Structural type		$J_{C^{13}C^{13}}$ (cps)
H—C≡C—H		171[1]
$H_2C{=}CH_2$		67[1]
$H_3C{-}CH_3$		35[1]
$CH_3{-}CO_2H$		57.6[4]
$(CH_3)_3{-}C{-}X$	($X{=}CH_3$, NH_2, OH, Cl, Br)	36.9 to 40.2[3]
R—cyclopropyl		13 to 16[2]

dramatic effect of the degree of hybridization on the size of these coupling constants is amply shown in Table 3.5. The sp^3-to-sp^3 bonded carbons are coupled to the extent of 35–40 Hz, sp^2-to-sp^2 bonded ones in the range of 55–70 Hz, and sp-to-sp bonded carbons 170–180 Hz.

An interesting observation is the relatively small (13–16 Hz) carbon-to-carbon coupling constant in cyclopropane derivatives.

In a general sense, coupling constants between differently hybridized adjacent carbons (i.e., in acetic acid) lie between the values established for the corresponding equivalently bonded carbons.

The carbon–carbon coupling constants in aromatic systems are subject to the effect of electron-withdrawing substituents, which can increase the coupling constants by 10–20 Hz.

3.4.4 Nitrogen–Nitrogen Coupling Constants

One of the significant contributing factors to the size of spin–spin coupling constants is the size of the magnetogyric ratio γ of the interacting nuclei. The magnetogyric ratio of N^{15} is very small, and as a result the magnitude of the different $J_{N,N}$ values are also small. Table 3.6 lists a number of the more important N^{15}–N^{15} coupling constants [10–14].

As is the case with all other nuclei, the more double-bond character there is in the nitrogen-to-nitrogen bond, the larger is the value of the coupling constant. The 6–7 Hz observed for hydrazine derivatives (—N—N—) changes to 17–21 Hz in

TABLE 3.6 Representative N–N Coupling Constants

Structural Type	$J_{N,N}$ (Hz)
R—N—N—R	6–7
R—N=N—R	17–21
R—N—N=C	10.7
C=N—N=C	11–13
R—N—N=O	19
$R{-}N{-}NO_2$	9

azo compounds (—N=N—). Again, the coupling constants of *mixed* hybridized nitrogen-to-nitrogen bonds lie between these extremes; that is, —N—N= has a coupling constant of 10.7 Hz [10–18].

Chapter 7 will deal with the more theoretical and quantitative aspects of these coupling constants.

3.4.5 Phosphorus–Phosphorus Coupling Constants

Three excellent examples serve to provide a reference point for the size of P^{31}–P^{31} coupling constants. The diphosphite anion (structure **5**) has a P^{31}–P^{31} coupling constant of 480 Hz. The size of this constant is considerably decreased in phosphorus sesquisulfide (**6**), where it is only 86 Hz. Separation of two phosphorus atoms by an oxygen atom, as in the isohypophosphate anion (structure **7**), causes a further attenuation of the interaction constant, to 17 Hz, [1, 2, 17]

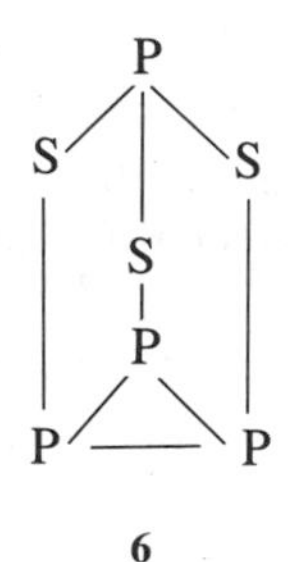

6

```
    O   O   ⌉ -3
    |   |   |
O — P — P — O
    |   |   |
    H   O   ⌋
```

5

```
    O       O ⌉ -3
    |       | |
H — P — O — P |
    |       | |
    O       O ⌋
```

7

3.5 HETERONUCLEAR COUPLING CONSTANTS

"Heteronuclear coupling" refers to the spin–spin interactions between magnetically active nuclei of different atomic species— that is, the coupling between a C^{13} nucleus and a proton bonded to it. This section will deal, *qualitatively*, with a number of the more important heteronuclear coupling systems. The quantitative aspects of these interactions will be dealt with in Chapter 5.

3.5.1 Proton–Carbon Coupling Constants

Since there are as many different types of $J_{C,H}$ constants as there are different carbon structural types, only the more important ones are described in this section [1, 2, 18, 19].

In a compound such as formaldehyde (structure **8**), the C^{13} nucleus will appear as a triplet (any potential interaction with a magnetically active oxygen isotope is not considered in this analysis), since it is coupled to two *identical* protons. On the other hand, the protons will appear as a doublet as a result of coupling with the C^{13} nuclei in formaldehyde,

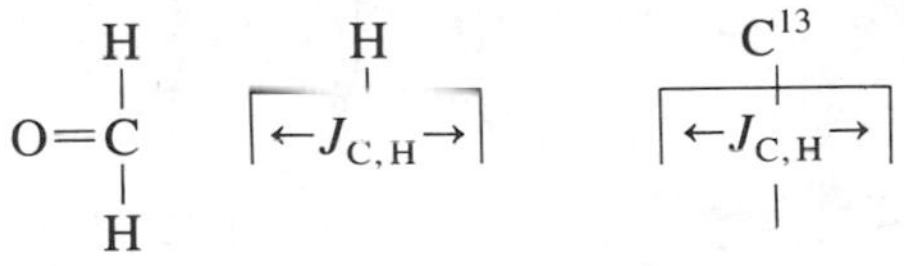

H
|
O=C
|
H

8

H: $\leftarrow J_{C,H} \rightarrow$

C^{13}: $\leftarrow J_{C,H} \rightarrow$

However, since the C^{13} isotope is only present to the extent of 1.08% in a natural sample of formaldehyde, the proton *doublet* will be of very low intensity. The major portion of the protons will appear as a singlet where the doublet component is buried in instrument noise under normal recording circumstances.

If this coupling constant is to be observed as part of a proton NMR examination, either very concentrated samples of formaldehyde and/or FT NMR techniques have to be used, and the recording must be done in a manner in which the proton singlet is far off the recorded spectrum.

The carbon–proton coupling in formaldehyde, or in any other compound, can also be observed by a study of the C^{13} spectrum, where again, because of the low concentration of C^{13} nuclei, enhancement techniques such as FT NMR spectroscopy are required. The separation of the multiplets, either the doublet obtained from the H NMR spectrum or the triplet resulting from the C NMR analysis, is the same.

If the process of *heteronuclear decoupling*, as briefly outlined in Section 3.3, is applied to the carbon triplet in formaldehyde, by irradiation at the proton frequency, a singlet will result. The total area of this singlet will be the sum of the areas of the members of the triplet observed under nondecoupling conditions. Thus, the heteronuclear decoupling process causes a considerable increase in signal intensity! As a matter of fact, most C NMR spectra are observed under conditions of total proton decoupling. The rare instances in which the very time-consuming, nondecoupled carbon spectra are obtained occur only when proton–carbon coupling information is being sought.

Figures 3.5 and 3.6 are the proton and carbon spectra of 2,5-dimethylpyrazine-

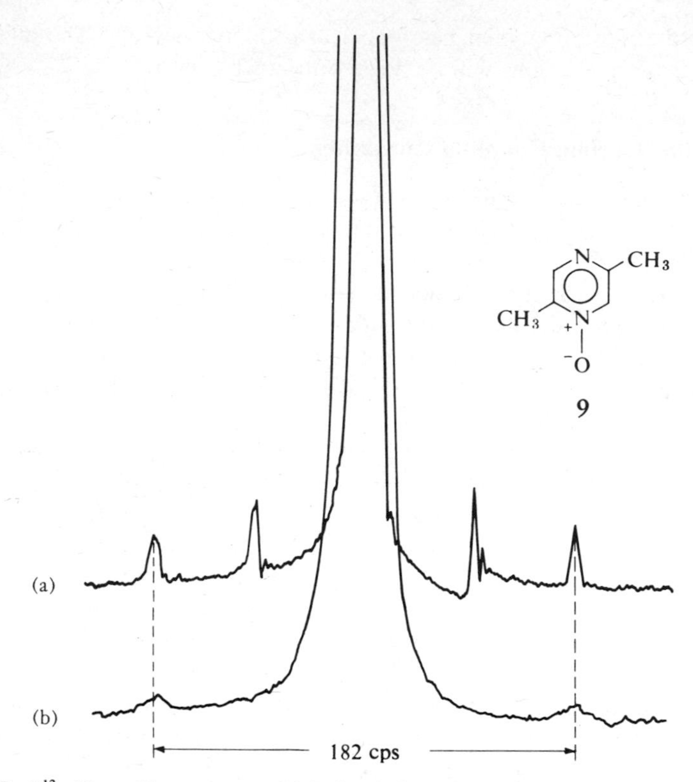

Figure 3.5. C^{13}—H satellite spectrum of 2,5-dimethylpyrazine-N-oxide. (Only the aromatic proton region is shown; spectrum (a) was obtained from a spinning sample, and spinning side bands are visible; spectrum (b) was obtained from the sample without spinning it.)

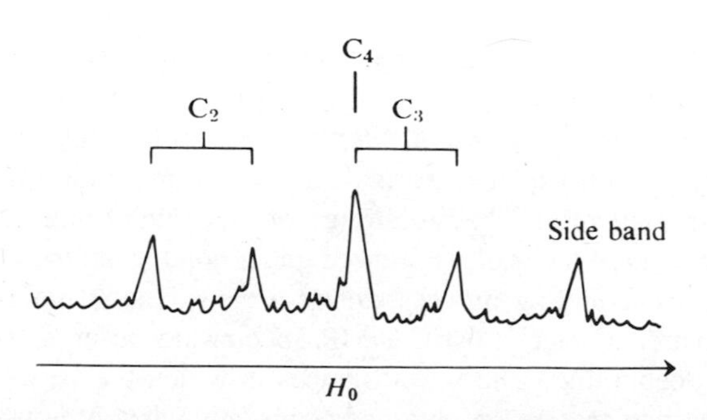

Figure 3.6. C^{13} nuclear magnetic resonance spectrum of 4-bromopyridine (**10**). Reprinted with permission from H. L. Retcofsky and R. A. Friedel, *J. Phys. Chem.*, 3592. Copyright (1967) American Chemical Society.

1-oxide (structure **9**) and 4-bromopyridine (structure **10**), respectively. The former spectrum shows the relative intensity differences between the proton spectrum involving the protons that are coupled to C^{13} nuclei and those that are not. The latter figure shows the proton-decoupled carbon spectrum of 4-bromopyridine.

The proton spectrum of a very concentrated solution of pyrazine (structure **11**), where the sattelite peaks caused by the $H—C^{13}$ interactions are discernible, are given in Figure 3.7. The separation between the centers of the multiplets is 180 Hz and corresponds to the coupling between $H_1—C^{13}$.

A proton bonded to a C^{13} nucleus is magnetically different from a proton bonded to the corresponding C^{12} nucleus! Consequently, in a pyrazine molecule a C^{13} nucleus will interact with two different protons, the one that is bonded to it, and the one that is bonded to an adjacent, but chemically equivalent, C^{12} nucleus. One might also consider that the adjacent carbon could also be a C^{13} isotope. However, the probability of finding two adjacent C^{13} nuclei in a molecule is extremely low. The result of this nonequivalence of protons is that each member of the $C^{13}—H_1$ doublet will be further split into a doublet by H_2. Additionally, the *meta*-coupling between H_1 and H_5 causes further splitting of each multiplet member. The following diagram shows this analysis:

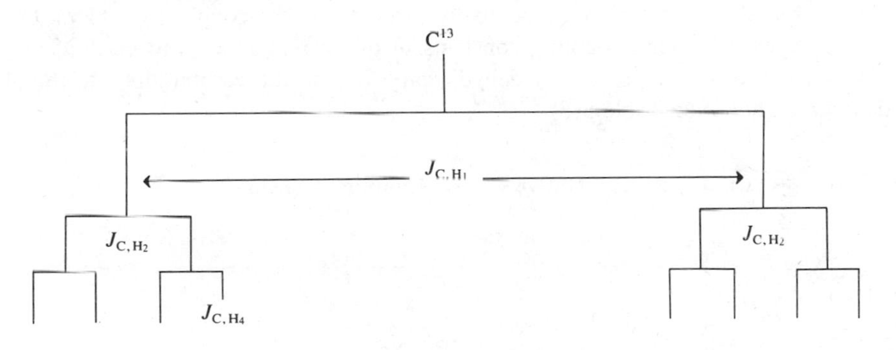

The various coupling constants are $J_{C,Hgeminal} = 180$ Hz, $J_{C,Hmeta} = 2.0$ Hz, $J_{CHortho} = 1.4$ Hz.

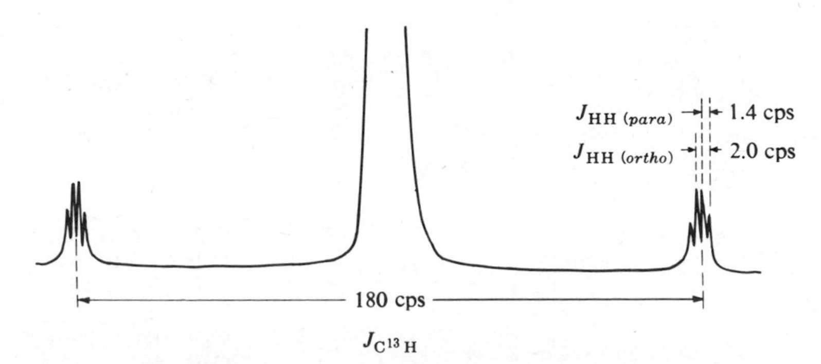

Figure 3.7. Carbon-13 satellite spectrum of pyrazine (**11**).

It is of considerable importance to recognize that the analysis of $C^{13}-H$ satellites affords coupling constants of protons that have identical chemical shifts! These protons appear simply as singlets under normal conditions.

Similar analyses can, of course, be done on the C^{13} spectra of organic compounds. The various coupling constants of this type, presented in Table 3.7, demonstrate the great variance in these values.

The C—H coupling constant range for directly bonded carbons and hydrogens is 120–250 Hz and suggests that there may well be a relationship between the degree of hybridization (% of *s* character) of the carbon and the size of the coupling constant of the proton directly bonded to it.

Sp^3 carbons bonded to a proton have the smallest one-bond coupling constant in this series (about 125 Hz), followed by the sp^2 to proton constant of about 160 Hz. The largest of these one-bond coupling constants is observed in sp-hybridized carbons bonded to a proton (250 Hz). The theoretical interrelationships among these coupling constants are dealt with in Chapter 7.

The long-range proton-coupling constants are considerably smaller (1–50 Hz) than the one-bond coupling values (125–250 Hz).

In aromatic benzenoid and heteroaromatic systems, the *meta*-coupling is generally larger (7.4 and 2.5 Hz, respectively), than the *ortho*-coupling (1.0 Hz). One of the largest two-bond coupling constants in this series appears to be the 25- to 50-Hz value observed between aldehydic protons and the carbon alpha to the aldehydic carbon (see Table 3.8).

TABLE 3.7 Representative $C^{13}-H$ Coupling Constants

Structural type		$J_{C^{13}H}$ (cps)
CH_4		125
CH_3-X	(X=OH, halogen)	141 to 153
CH_3-X	(X=CO_2H, CN, $SOCH_3$, NO_2, $COCH_3$, aromatic, heterocylic)	126 to 138
CH_2X_2	(X=halogen)	173 to 185
CHX_3	(X=halogen)	193 to 208
H(C=C) (vinyl C–H)		155 to 171
H—C≡C—		247 (one compound)
benzene ring–R		153 to 159
pyridine ring (N)–R		C_2H 174 to 180 C_3H 158 to 164 C_4H 150 to 156
five-membered ring (X)–R	(X = NR—, O—, S—)	C_2H 180 to 201 C_3H 165 to 175

TABLE 3.8 Representative Long-Range C^{13}—H Coupling Constants

Structural Types	$J_{C,H}$ (Hz)
(benzene ring)—H; 2, 3, 4	$^{2}J = 1.0$ (*ortho*-) $^{3}J = 7.4$ (*meta*-) $^{4}J = -1.1$ (*para*-)
(pyridine ring, N)—H; 3, 4	$^{3}J = 1.0–1.5$ (*ortho*-) $^{2}J = 2.0–3.0$ (*meta*-)
$C_nH_{2n+1} - X$	$^{2}J = -4.5$
C—C(=O)—H	$J = 25–50$

3.5.2 Proton–Nitrogen Coupling Constants

Since both N^{14} and the much less abundant N^{15} species are magnetically active, any protons bonded to either one of the isotopes will be spin–spin coupled. However, in the N^{14} nuclei, with a spin quantum number of 1, rapid quadrupole-induced relaxation occurs, generally resulting in the elimination of the proton coupling to it. Only in those instances where the field about the nitrogen nucleus is symmetrical, such as in the ammonium ion, can N^{14}—H coupling be observed. The size of the N^{14}—H coupling constant can be related to the corresponding N^{15}—H one by multiplying these constants by the ratio of the magnetogyric ratio ($N^{15}/N^{14} = -1.402$).

Since the magnetic moment of N^{15} is small in comparison to that of C^{13} and H, the heteronuclear coupling constants are smaller. Furthermore, since the magnetic moment is negative, the coupling constants whose signs are opposite to those of related systems involving C^{13} nuclei, are negative as well.

Representative J_{N-H} coupling constants of some of the more usual structural arrangements are given in Table 3.9. In general, the size of these coupling constants are, again, related to the degree of hybridization of the N^{15} species involved in the interaction (details and limitations of this statement are dealt with in Chapter 7).

The coupling constants between a proton bonded to a pyramidal nitrogen are about 75 Hz; those to a trigonal nitrogen, about 90 Hz; and these to a linearly bonded one, about 140 Hz. The two-bond coupling constants, (cf. Table 3.10) as expected, are considerably smaller (-15 to $+15$ Hz) than the one-bond, or geminal ones.

When protons are spatially close to a rigidly held lone pair of electrons on a nitrogen atom, the two-bond coupling constants tend to be larger numerically and

TABLE 3.9 Representative N^{15}—H Coupling Constants

Structural Type	$J_{N,H}$ (Hz)
R_2N—H	−(61.2–67)
R—Ar—NH_2	−(76–93)
$R_2N^+H_2$	−(73.3–76)
Ar—NHR	−87
R—Ar—NHNH_2	−(90–100)[a]
H—CO—NHR	−(88–93)
—R N H	−(96–98)
R—C≡N^+—H	−(134–136)
Ar—C≡N^+—H	−(134–136)

[a] Spin sign has not yet been experimentally determined.

TABLE 3.10 Representative Long-Range (Two-bond) N^{15}—H Coupling Constants

Structural Types	$J_{N,H}$ (Hz)
H—CN	+8.7
R—CH=N^+=N	+0.1–0.2
H—C—N—	−(0.6–1.0)
—H N H	−4.5
—H N	−(10–12)
H—C=N—OH	−(14∓13)[a]
R_2—N—CHO	−15[b]

[a] The size depends on the stereochemical arrangement. If the proton is close to the nitrogen lone pair of electrons, *J* is large and negative.
[b] Temperature-dependent.

negative (111 Hz in quinoline, for example). Removal of the lone pair of electrons by protonation decreases the size of these two-bond coupling constants considerably (−2 Hz in quinolinium salts).

3.5.3 Proton–Phosphorus Coupling Constants

Some typical $P^{31}-H$ coupling constants are listed in Table 3.11. The size of some of these, 707 Hz in H_3PO_3, is large, indeed. The charge on the phosphorus atom in a particular structure has a drastic effect on the P–H coupling constant. There is a significant increase in the sequence: dialkylphosphine (179–210 Hz), dialkyl phosphite, trialkyl phosphonium ion (500–530 Hz).

The size of geminal coupling constants in systems such as H–*X*–P range from 3 to 25 Hz with the nature of *X* controlling the size of the interaction. When *X* is a carbon, there is a smooth dependence between the dihedral angle in H—C—P and the size of the coupling constant. The constant is largest (+25 Hz) when the dihedral angle is zero, and becomes a negative value (−6 Hz) when the angle is 120°.

3.5.4 Proton–Fluorine Coupling Constants

Since the vast majority of organic fluorine-containing compounds also contain covalently bonded protons, spin–spin interactions between these two nuclei are well known. The fluorine spectrum of a 1,2-difluoropropane derivative, where F–F and F–H coupling occurs, is given in Figure 3.8.

It is instructive to analyze the splitting pattern of this compound (structure **9**):

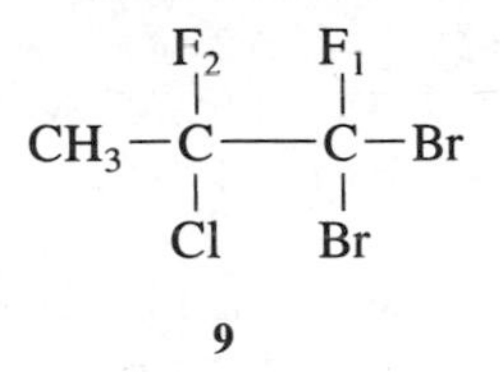

TABLE 3.11 Representative P^{31}–H Coupling Constants

Structural Type	$J_{P,H}$ (Hz)
—P—X—H (*X*=C, NC, OC)	−6 to 25
PH_3	179
R_2P-H	179–210
PH_2-PH_2	186
$(RO)_2\underset{X}{\underset{\mid}{P^+}}-H$ (X = O, S)	515–695
$R_3\overset{+}{P}-H$	approx. 525

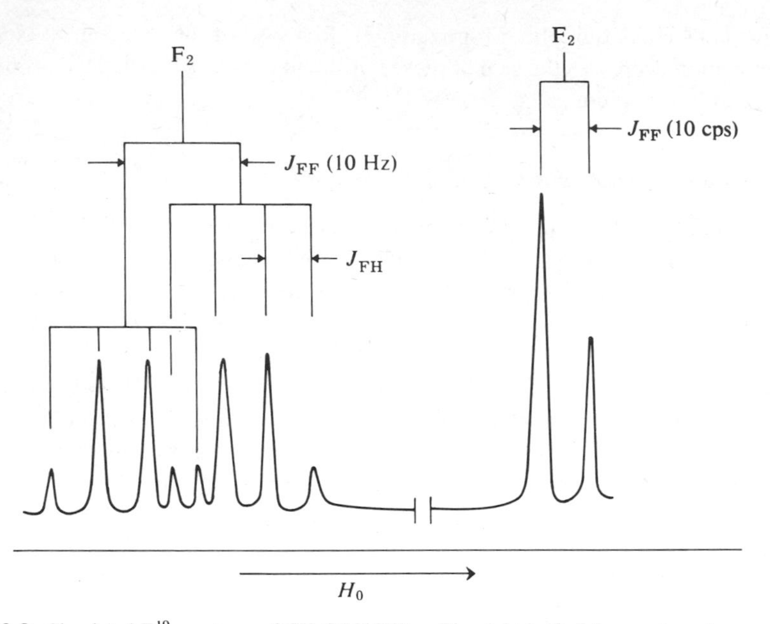

Figure 3.8. Simulated F^{19} spectrum of $CH_3CFClCFBr_2$. The right half of the spectrum is compressed by a factor of 2 as compared with the left-hand side.

The F–F coupling pattern in the fluorine spectrum of this compound is expected to be composed of a doublet for each of the fluorine nuclei. In addition, F_2 will be coupled to the three methyl group protons, causing each doublet member to be further split into quartets ($2nI + 1 = 2 \times 3 \times 1/2 + 1$). This analysis is graphically given in Figure 3.8. The proton NMR spectrum of this compound would be a simple doublet, as a result of the coupling of the equivalent methyl group protons with F_2.

Some of the more commonly encountered F–H coupling constants are presented in Table 3.12. As expected, the sizes of these coupling constants lie between the proton and fluorine constant values observed in the corresponding all-proton and all-fluorine compounds. In contrast to the fluorine–fluorine coupling constants in aromatic compounds, the fluorine–proton constants are not effected by the nature of the substituents on the aromatic ring.

Another unusual feature, observed neither in the proton–proton nor in the fluorine–fluorine coupling constants, is the size of the geminal fluorine–proton constants in olefinic systems (72–90 Hz) as compared to the same constant (45–80 Hz range) in similar saturated environments.

3.5.5 Fluorine–Carbon Coupling Constants

As is the case with H–C coupling constants, the interaction between F and an sp^3 carbon is less than that with an sp^2 hybridized one (155–170 vs. 235–380 Hz). However, while the difference between spin–spin coupling constants in olefinic

TABLE 3.12 Proton–Fluorine Spin–Spin Coupling Constants

Structural type	J_{FH} (cps)
H–C(–F) (H and F on same carbon)	45 to 80
H–C–C–F	5 to 30
H(F)C=	72 to 90
H–C=C–F (cis)	−3 to 20
H–C=C–F (trans)	12 to 52
H and F on benzene ring	*ortho* 6 to 10 *meta* 6 to 8 *para* about 2
R-substituted ring with F_2, H_3, H_4, H_5, F_6	2,3: 9.1 to 10.5 2,4: 5.9 to 6.6 2,5: −0.85 to −1.90

and aromatic systems and protons is not significant, the interaction with fluorine nuclei is (see Tables 3.13 and 3.14). For example, the F–C coupling constant of a fluorine bonded to an aromatic nucleus lies within the range 235–260 Hz, whereas that bonded to an olefinic carbon is within the 285–376 Hz range.

The effect that the electronegativity of a substituent on an olefinic carbon has on the F–C coupling constant is demonstrated by the observation that the coupling in the system F—C= varies from 285–290 Hz, if the olefinic carbon contains an additional fluorine to 300–376 Hz, if another halogen is the substituent.

The existence of significant two-bond coupling (22–61 Hz) is also noteworthy.

3.5.6 Nitrogen–Carbon Coupling Constants

For most organic molecules the carbon–nitrogen coupling constants are negative with their sizes dependent, to a major extent, on the hybridization of the nitrogen atom. Some of the more important C–N coupling constants are presented in Table 3.15 [15].

TABLE 3.13 Some Fluorine-Carbon Spin-Spin Coupling Constants

Structural type	$J_{FC^{13}}$ (cps)
$F-\overset{\vert}{\underset{\vert}{C}}-$	155 to 170
$F-\overset{F}{\underset{\vert}{C}}-$	about 230
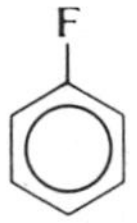 Fluorobenzene (C_6H_5F)	235 to 260
CF_3-X (X = O, N, CF_2, Ar, R)CO_2H)	265 to 290
$F_2C=$	285 to 290
$F(R)C=O$ (R = H, alkyl)	240 to 370
$F(X)C=$ (X = halogen)	300 to 376
$F-\overset{\vert}{\underset{\vert}{C}}-\overset{\vert}{\underset{\vert}{C^{13}}}-$	22 to 61

TABLE 3.14 Two Bond C-N Coupling Constants

Structure Type	J_{C-N} (Hz)
$C-\overset{\vert}{\underset{\vert}{C}}-NH_2$	1.2
$C-C=N$	3.0
$=C-C=N-$(pyridine)	2.53
$=C-C=\overset{+}{N}H-$(pyridinium)	2.01
$=C-C=\overset{+}{N}-\overset{-}{O}$(pyridine N-oxide)	1.43
$=C-C-NH_2$ (aniline)	−2.68
$=C-C-\overset{+}{N}H_3$ (anilinium)	−1.5
$C-C-NO_2$ (pyrrole)	−3.92
$=C-C-NO_2$ (nitrobenzene)	−1.67
$C-C=\overset{N}{\underset{H}{}}-OH$ *cis*-oxime	+1.0 to 11.4
trans-oxime	−7.3 to 5.5
$C=C-\overset{\vert}{N}-H$	5.5
$=C-\underset{\underset{O}{\Vert}}{C}-NH_2$	9.0

TABLE 3.15 Representative N^{15}-C^{13} Coupling Constants

Structural Type	$J_{C,N}$ (Hz)
—C—N—	−2.1−−4.5
—C—N^+—R_3	−4.4−−6.2
O=C—N—	−13.4−−15.1
O=C—$(NR)_2$	−20.2
R—Ar—NH—R	−11−−15
R—Ar—NH—R_2	−8.9−−9.2
C—NO_2	−10.5
Ar—NO_2	−14.5
R—CN	−17.5
RC—NC	−8.9 (NC) −9.8 (RCN)

In a general sense, protonation of a nitrogen will cause an increase in the geminal coupling constants. For example the C–N coupling constant in aniline is −11.43 Hz. Protonation to form the anilinium ion increases this constant to −8.9 Hz. Among the largest of this type of coupling constant is the one observed in compounds with the structural feature —N—C=C—, where the value is 36.2 Hz. Among the smallest of these interactions is the one observed in urea, where it is −20.2 Hz.

A number of qualitative rules have been developed to account for some of the small changes in the coupling constants observed with changes in the electronic and structural nature of the nitrogen nuclei:

1. In the absence of a lone pair of electrons with *s* character, the operative mechanism will, generally, decrease the size of the constant (make it *more* negative).
2. A lone-pair orbital with *s* character tends to make a *positive* contribution to the constant, often bringing the value close to *zero*.
3. A decreased *s* character reduces the positive contribution and consequently makes the coupling constant more negative. This often occurs when minor structural changes have occurred in a particular molecule.
4. Where the possibility of stereoisomerism exists, the more stable isomer will have the more negative coupling constant.

3.6 SPIN SYSTEM NOTATIONS

The concept of chemical shifts, defined the terms shielded and deshielded, and involves terminology that can be conveniently used to simplify the description of the different types of spin–spin coupling systems. The letters *A*, *B*, *C*, . . . are used to designate nonequivalent nuclei of a particular species. For example, the two different types of protons in ethyl iodide (CH_3CH_2—I) are described as H_A and H_B, the more shielded proton being, by convention, designated as the *A* proton.

TABLE 3.16 Examples of Notations for Different Spin Systems

Compound	Description
$F_2C{=}CH_2$	A_2X_2
F_3CCl	A_3 (or X_3)
CH_3CHO	A_3X
Pyridine (H_A, H_B, H_B, H_A, H_A; N)	AB_2X_2
2,4-Dichlorotoluene (Cl, Cl, CH_3)	ABC
3-Methylthiophene (CH_3, H_X, H_A, S, H_B)	ABX

In instances in which there is a significant difference between the chemical shifts of two types of nuclei in a particular structure, the more deshielded nuclei are designated by the letters *X*, *Y*, and *Z*. For example, in toluene, the methyl group protons are described as *A*, whereas the five aromatic protons (3 different types) are defined as the *X*, *Y*, and *Z* protons.

Some typical applications of these conventions to a number of different spin systems are given in Table 3.16.

3.7 PROBLEMS

Problems one through 14 are NMR spectra of different organic compounds. In the proton NMR spectra, the number above each set of peaks corresponds to the number of protons equivalent to each set of *multiplets*. The carbon spectra (Problems 15–19) are proton-decoupled ones (i.e., the proton–carbon coupling patterns are absent!). The peak intensities of the nonproton spectra do not necessarily correspond to the relative number of carbons they represent (relaxation time interferences!).

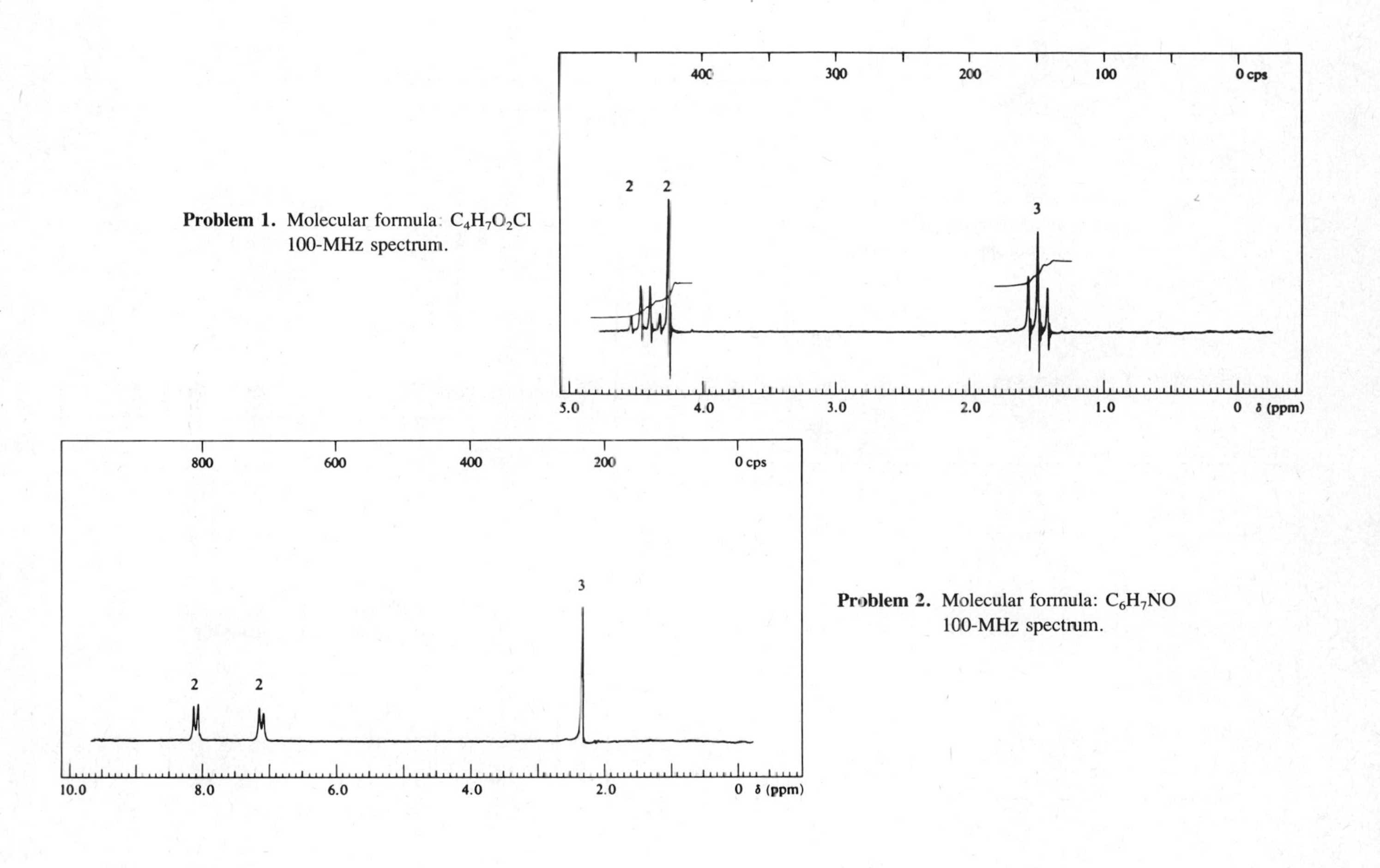

Problem 1. Molecular formula: $C_4H_7O_2Cl$
100-MHz spectrum.

Problem 2. Molecular formula: C_6H_7NO
100-MHz spectrum.

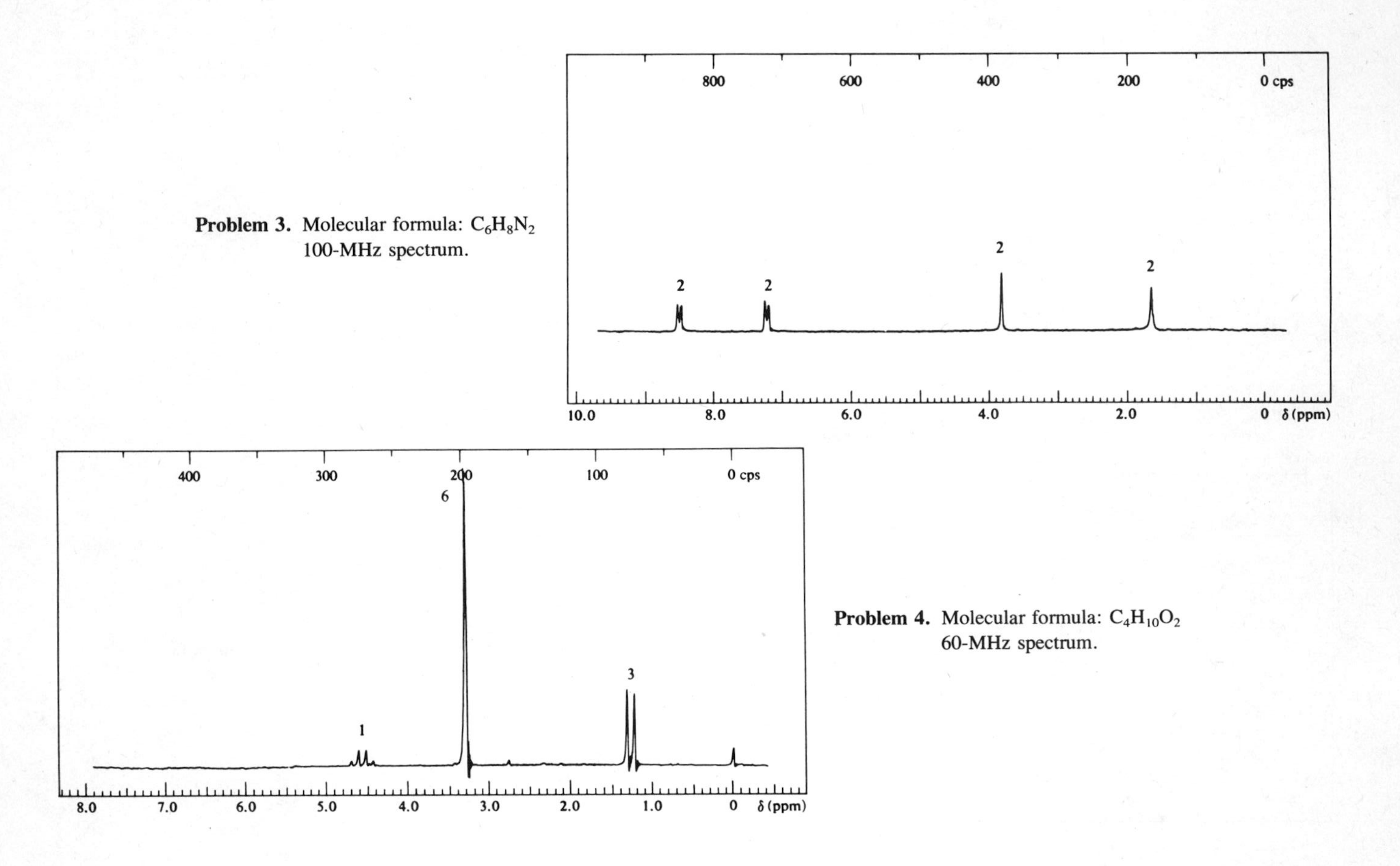

Problem 3. Molecular formula: $C_6H_8N_2$
100-MHz spectrum.

Problem 4. Molecular formula: $C_4H_{10}O_2$
60-MHz spectrum.

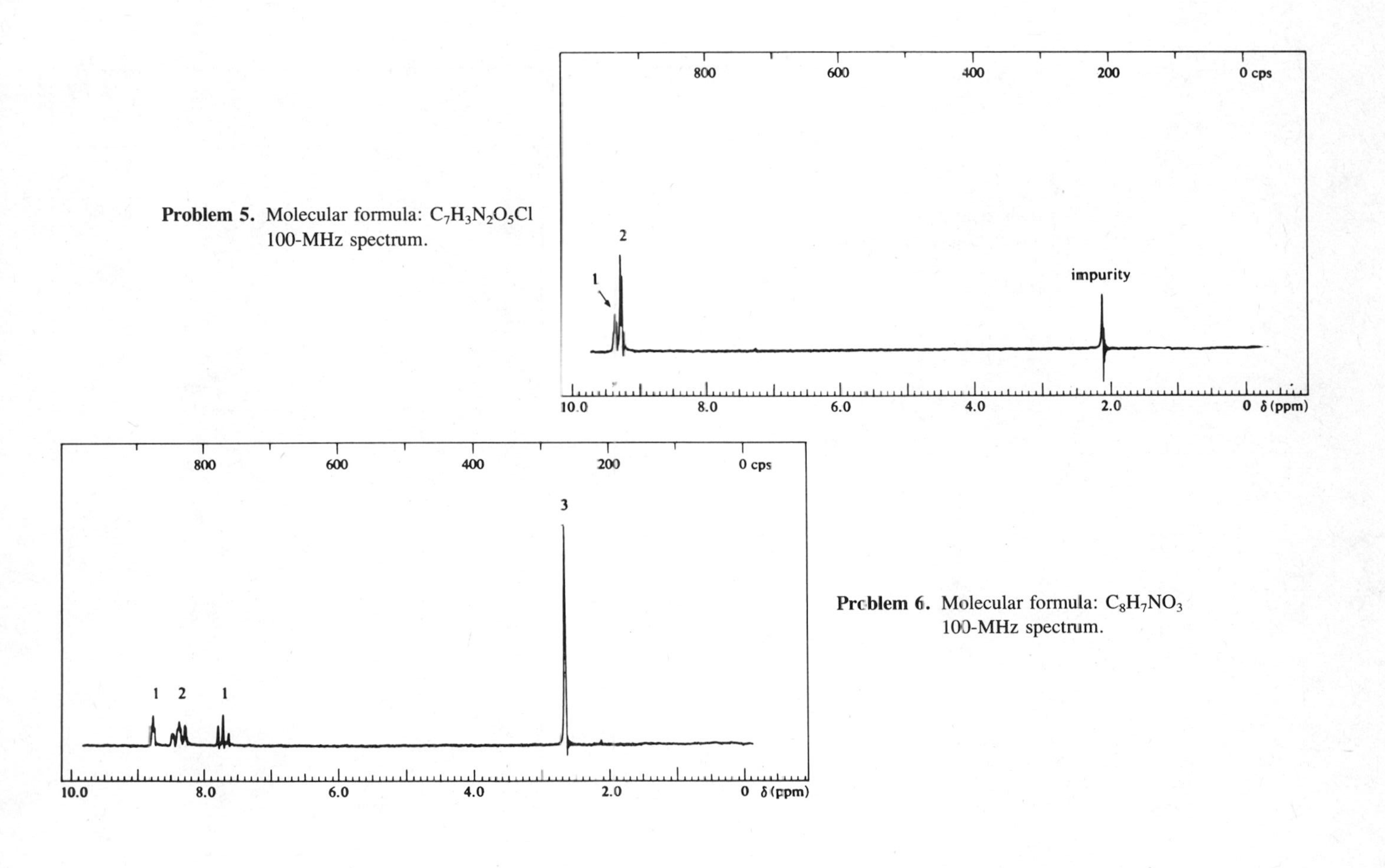

Problem 5. Molecular formula: $C_7H_3N_2O_5Cl$
100-MHz spectrum.

Problem 6. Molecular formula: $C_8H_7NO_3$
100-MHz spectrum.

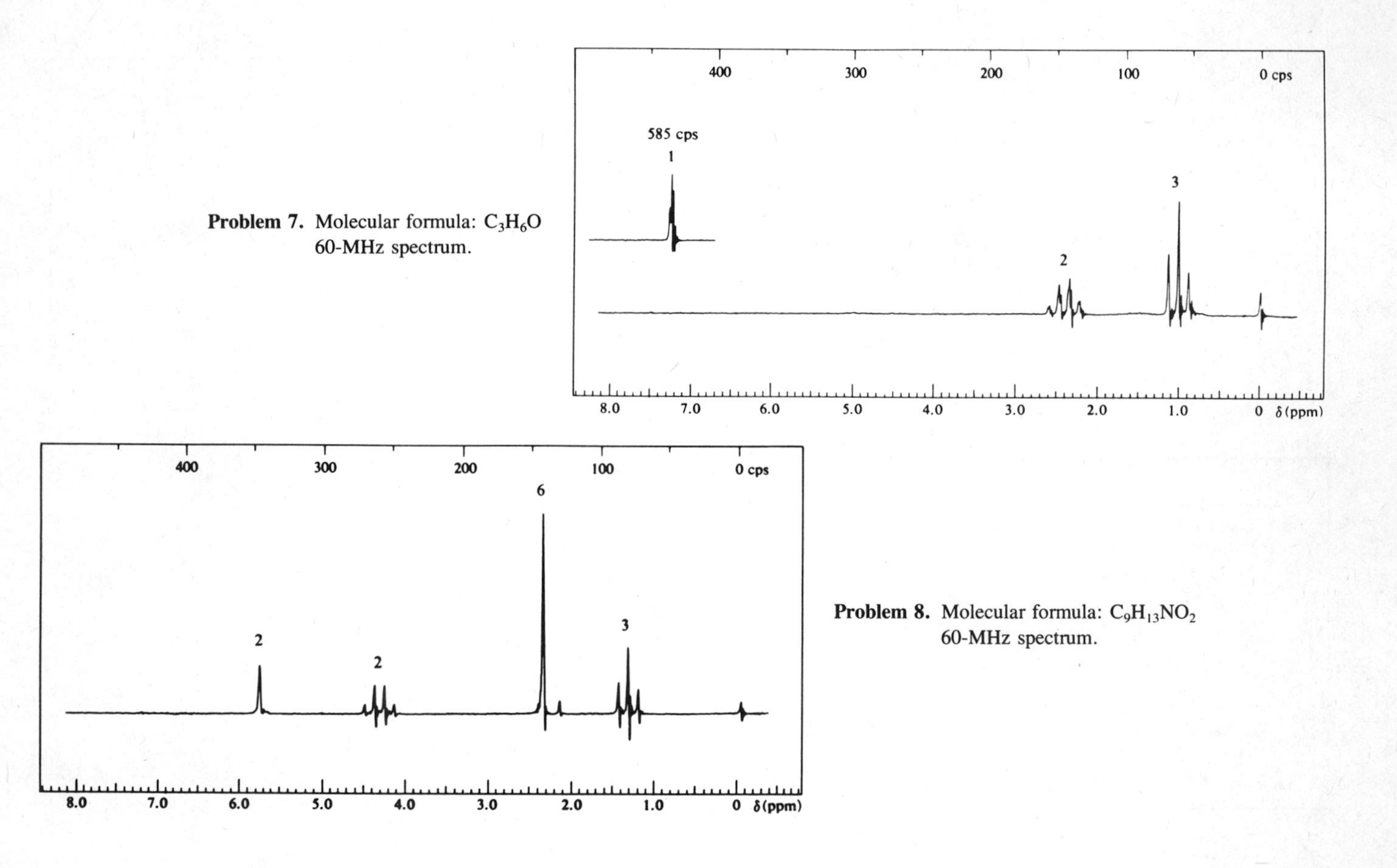

Problem 7. Molecular formula: C_3H_6O
60-MHz spectrum.

Problem 8. Molecular formula: $C_9H_{13}NO_2$
60-MHz spectrum.

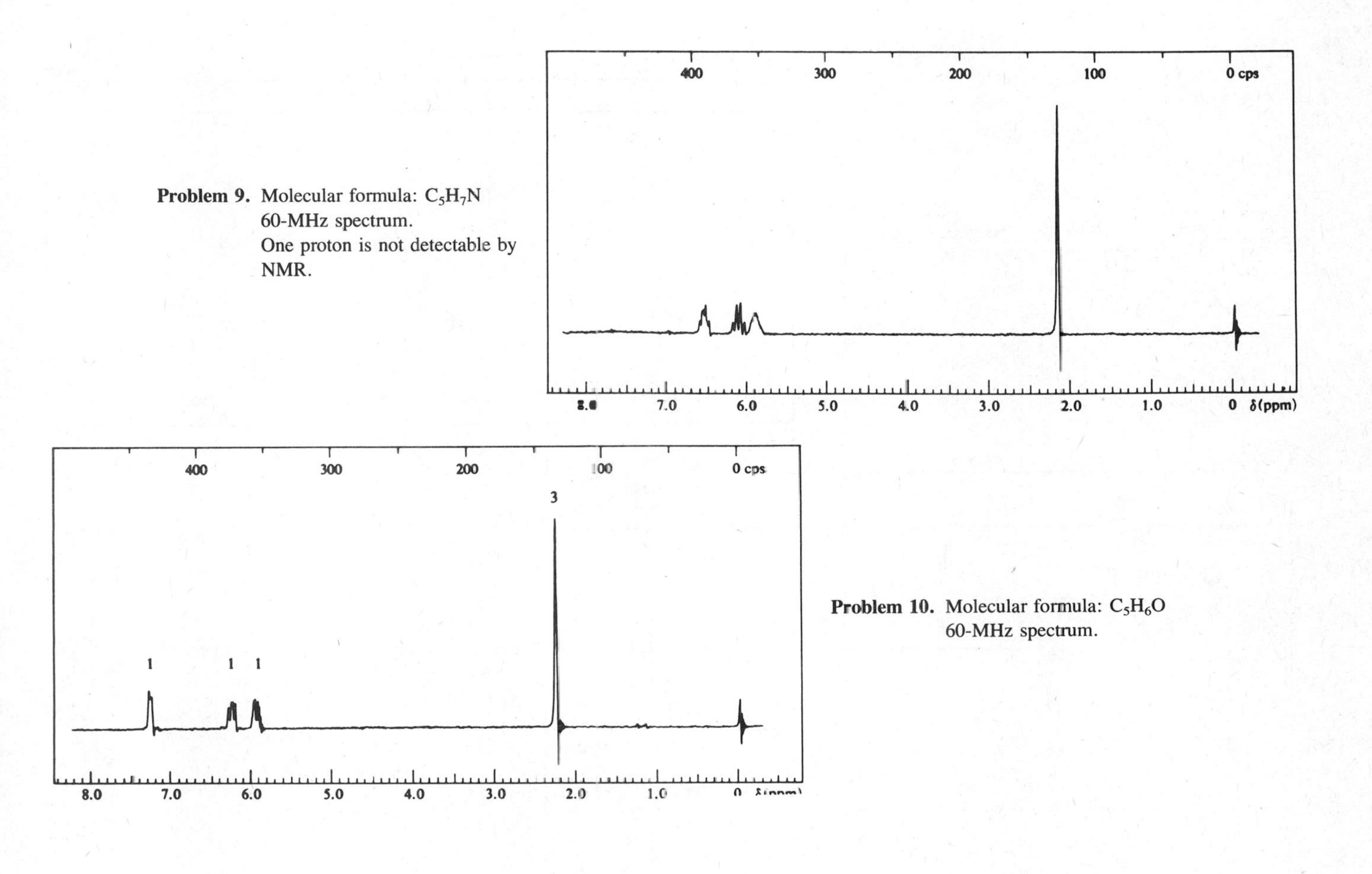

Problem 9. Molecular formula: C_5H_7N
60-MHz spectrum.
One proton is not detectable by NMR.

Problem 10. Molecular formula: C_5H_6O
60-MHz spectrum.

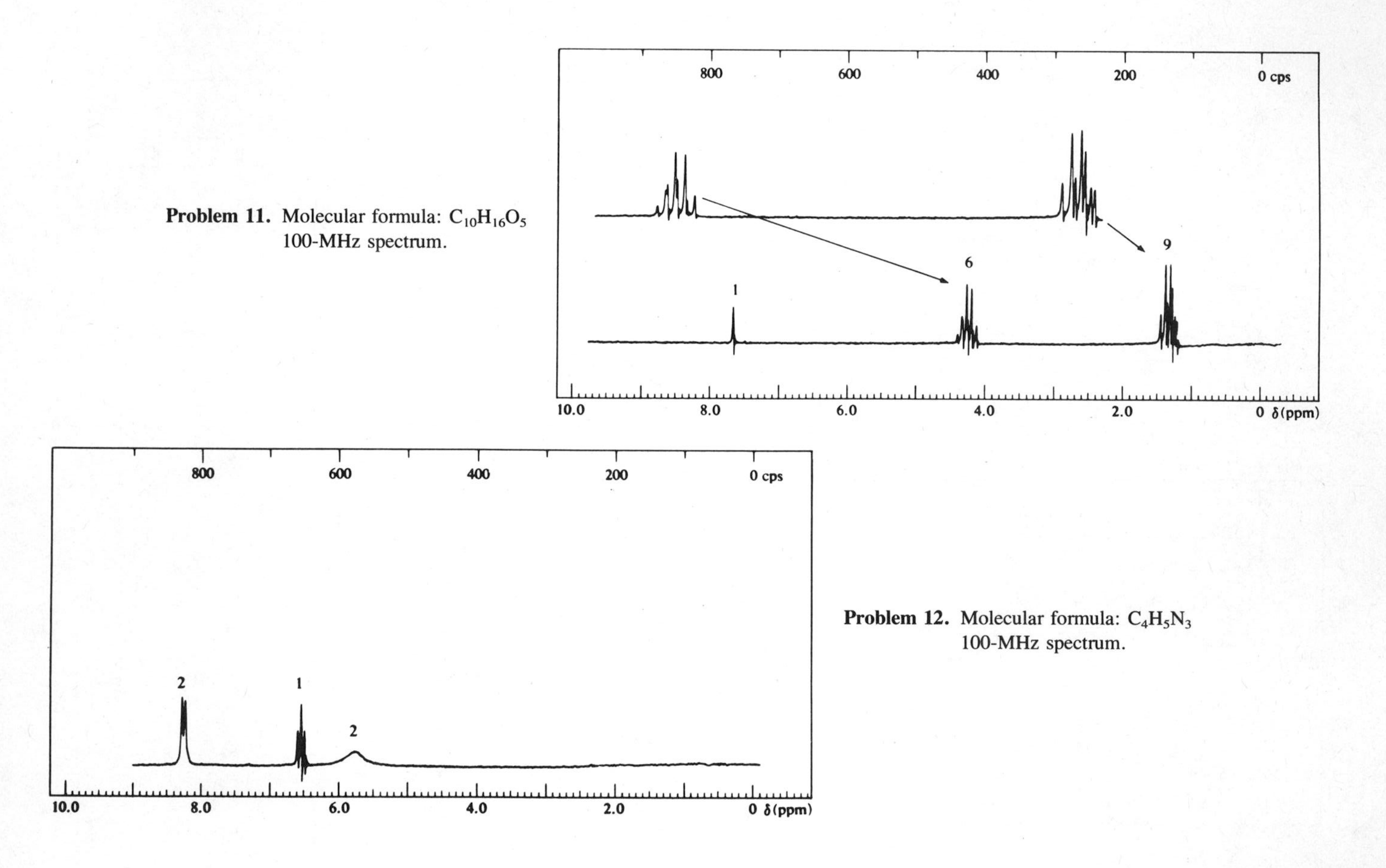

Problem 11. Molecular formula: $C_{10}H_{16}O_5$
100-MHz spectrum.

Problem 12. Molecular formula: $C_4H_5N_3$
100-MHz spectrum.

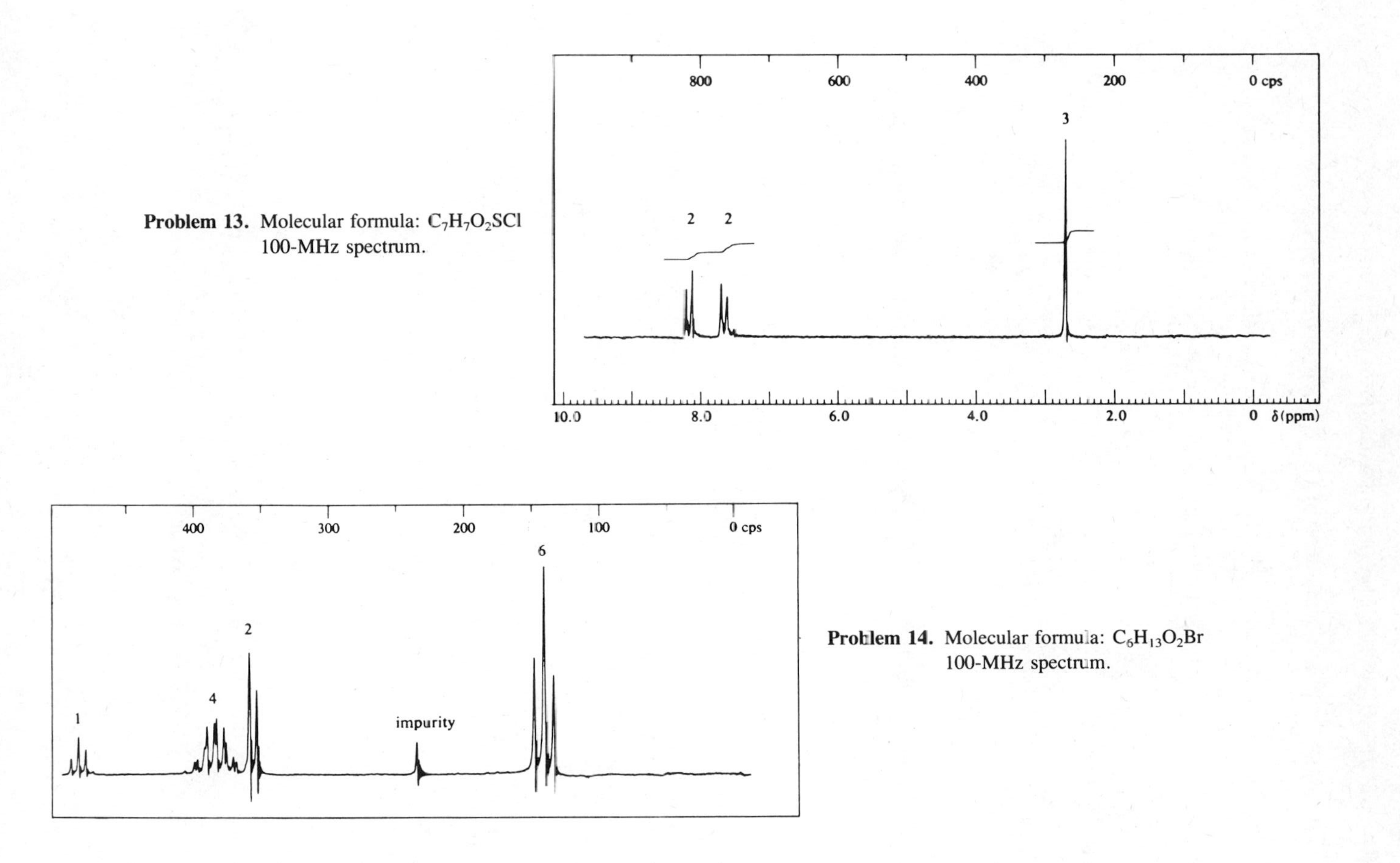

Problem 13. Molecular formula: $C_7H_7O_2SCl$
100-MHz spectrum.

Problem 14. Molecular formula: $C_6H_{13}O_2Br$
100-MHz spectrum.

Problems 15–19 are tabulations of C^{13} NMR spectra (TMS = 0):

15. Molecular formula $C_8H_{12}O_2$. What is the structure of the compound?

Peak Position	Relative Intensity
16.0	0.3
16.2	0.32
20.0	1.0
53.0	0.25
102.0	0.22
146.0	0.22
173.0	0.22

16. Molecular formula $C_4H_6O_2$. What is the structure of the compound?

Peak Position	Relative Intensity
20.0	0.3
98.0	1.00
142.0	0.28
170.0	1.0

17. Molecular formula $C_{10}H_7N$. What is the structure of the compound?

Peak Position	Relative Intensity
150.9	0.78
121.5	0.80
136.0	0.83
128.3	0.68
126.8	0.68
129.7	0.8
130.1	0.8
149.0	0.3
128.7	0.2

18. Molecular formula $C_6H_{15}NO$. What is the structure of the compound?

Peak Position	Relative Intensity
13.0	0.6
21.0	0.9
33.0	1.0
48.0	0.70
52.0	0.65
60.0	0.70

19. Molecular formula $C_9H_{19}NO$. What is the structure of the compound?

Peak Position	Relative Intensity
13.5	0.8
13.7	0.75
20.0	0.8
20.8	1.0
30.0	0.92
32.5	0.90
43.0	0.70
46.0	0.65

20. Predict the proton and fluorine NMR spectra of 2,6-difluoroaniline.

21. The molecular formula of an olefin is $C_3H_4F_2$. The F NMR spectrum of the compound consists of two sets of peaks. One of these is a doublet of doublets, and the other is a doublet of quartets. What is the structure of this compound?

22. Predict the H NMR spectrum of the compound in the previous problem.

23. The proton-decoupled C^{13} spectrum of compound $C_{11}H_{10}$ has the following peaks: δ_C129, 131, 141, 27, 133, and 134(ppm), respectively. The proton spectrum has an A_2B_2 pattern in the aromatic region corresponding to four protons. What is the structure of this compound?

24. The N^{15} NMR of $C_4H_{12}N_2$ has a proton-decoupled singlet at δ_N24.1. The proton-decoupled carbon absorption peaks are at 52 and 36(ppm), respectively. Identify the compound.

25. The molecular formula of a compound is C_7H_5NO, the carbon chemical shifts are at δ_C134, 124, 129, 125(ppm). The N^{15} isotopes in the compound resonate at 46.5. What is the structure of this substance?

26. The C^{13} spectra of two isomeric compounds, C_7H_5N, have almost four identical absorption patterns in the vicinity of 180(ppm). Compound A has an N^{15} absorption peak at δ_C180(ppm), and compound B has one at δ_C259(ppm). Identify the two isomers.

27. The N^{15} spectrum of a compound $C_3H_5N_2$ has a singlet at δ_N209(ppm). When this compound is treated with diazomethane, the resulting material, $C_4H_5N_2$, shows two singlets for the N^{15} nuclei; δ_N158 and 252(ppm). Identify the compounds and explain the nitrogen chemical shift behavior.

28. The N^{15} nuclei in diazomethane, CH_2N_2, resonate at 286 and 397(ppm). Which of the two types of nitrogen nuclei in this compound is the more deshielded? Why?

29. Suggest a reason for the chemical shift differences between the sulfonamide N^{15} nuclei in compounds A and B:

$$C_6H_5SO_2{-}NH{-}C_6H_4{-}R\ (p)$$

$$\text{A: R} = \text{OCH}_3 = 117\,(\text{ppm})$$

$$\text{B: R} = \text{NO}_2 = 127\,(\text{ppm})$$

30. Why are the N–N coupling constants smaller than structurally related C–C constants?

31. The carbon chemical shifts of a compound C_5H_6SO are δ_C98, 160, 120, and 126(ppm). Suggest a possible structure and assign the various chemical shifts.

32. The C^{13} chemical shifts of $C_{10}H_{11}N$ are: δ_C12, 29, 100, 120, 120.5, 129, and 137(ppm). The N^{15} chemical shifts is 129(ppm). Propose a structure!

REFERENCES

1. ref. 5 ch. 1.
2. ref. 13 ch. 1.
3. D. M. Graham and C. E. Holloway, *Can. J. Chem.* **41,** 2114 (1963).
4. G. E. Maciel, J. W. McIver Jr., N. S. Ostlund, and J. A. Pople, *J. Am. Chem. Soc.* **92,** 1, 11 (1970).
5. R. E. Carhart and J. D. Roberts, *Org. Magn. Res.* **3,** 139 (1971).
6. W. M. Lichtman and D. M. Grant, *J. Am. Chem. Soc.* **89,** 5962 (1967).
7. F. S. Wright and J. D. Roberts, *J. Am. Chem. Soc.* **89,** 22967 (1967).
8. *Ibid.* **89,** 5962 (1967).
9. D. Ziessow, *J. Chem. Phys.* 984 (1971).
10. R. L. Lichter, P. R. Srinivasan, A. B. Smith, R. K. Dieter, C. T. Denny, and J. M. Schulman, *J. Chem. Soc. Commun.* 1024 (1977).
11. M. Witianowski, L. Stefaniak, and G. A. Webb, *Ann. Rep. NMR Spectrosc.* **7,** 117 (1977).
12. G. E. Hawkes, E. W. Randall, J. Elguero, and C. Martin, *J. Chem. Soc. Perkin II* 1024 (1977).
13. H. Schultheiss and E. Fluck, *Z. Naturforsch.* **32b,** 257 (1977).
14. P. Buechner, W. Maurer, and H. Ruterjans, *J. Magn. Res.* **29,** 45 (1978).
15. W. McFarlane and B. Wrackmeyer, *J. Chem. Dalton Trans.* 2351 (1976).
16. J. Briggs, F. A. Hart, G. P. Moss, and E. W. Randall, *Chem. Commun.* 364 (1971).
17. H. J. Schneider and W. Bremser, *Tetrahedron Lett.* 5197 (1970).
18. G. C. Levy and G. L. Nelson, *Carbon-13 Nuclear Magnetic Resonance for Organic Chemists*, Wiley-Interscience, New York, 1972.
19. Ditchfield and P. D. Ellis, in G. C. Levy, ed., *Topics in Carbon-13 Spectroscopy*, Vol. 1, Wiley, New York, 1974.
20. G. C. Levy and R. L. Lichter, *Nitrogen-13 NMR*, Wiley-Interscience, New York, 1979.

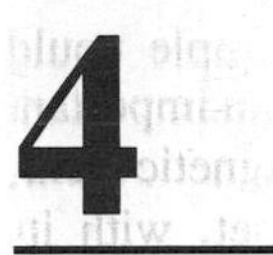

ENVIRONMENTAL EFFECTS

4.1 THE SAMPLE

4.1.1 General Considerations

The two most important factors that control the shape and size of an NMR sample are the *homogeneity* of the applied and that of the molecularly generated magnetic field that the nuclides in the sample experience and generate, respectively. The latter is accomplished by permitting the various magnetic interactions among the molecules in the sample to average. This averaging provides resonance lines as narrow as possible for the particular nuclide being studied.

The strength of a magnetic field varies across the pole pieces of the magnet, consequently molecules of a sample when placed into this field will experience slightly different magnetic fields depending on their relative positions within the pole gap.

The *effective* magnetic field homogeneity acting on the NMR sample could clearly be improved by decreasing the sample volume. Although this is an important consideration if an iron core magnet, with its relatively nonstable magnetic field, is employed, it is of less importance when a superconducting magnet, with its superbly stable and linear magnetic field, is used.

A further consideration that controls sample size is the NMR sensitivity and the natural abundance of the different magnetically active nuclei. In a given experimental arrangement, at constant temperature, where the T_1 and T_2 values are the same, the signal strength is proportional to the expression given in equation 4.1:

$$N(I + 1)\mu\omega_0^2 \tag{4.1}$$

For a particular field strength this expression becomes equation 4.2:

$$N\frac{I + 1}{I^2}\mu^3 H_0^2. \tag{4.2}$$

On the assumption that an equal number of nuclei, N, are present in the two samples, these equations permit the calculation of the relative NMR sensitivity of different nuclides. These calculations, supported by experimental results, have demonstrated that the most NMR-sensitive *common* nuclei are H^1 and F^{19}. All others show relative signal strengths of less than 40% that of an equal number of protium and fluorine nuclei (see Table 2.3).

4.1.2 Liquids and Solutions

In samples of low viscosity, whether pure liquids or solutions, the molecules are relatively free to move in their matrix so that an averaging of molecular alignments, with respect to the applied magnetic field can be accomplished under appropriate circumstances. One of these is the averaging caused by application of Bloch's technique [1, 2] of sample spinning. The spinning of the sample tubes provides motions to the molecules so that a given one experiences an *average* magnetic field. This average is optimized when the spinning rate, t, fulfills the relationship given in equation 4.3:

$$t \approx \frac{1}{\gamma \Delta H}. \tag{4.3}$$

on the reasonable assumption that the magnetic field intensity varies within 0.001 gauss in a particular magnet gap (this value is considerably smaller in the supercon magnets). Based on this, t is of the order of 0.25 s for proton samples. This speed is readily obtained by means of an air-driven spinner.

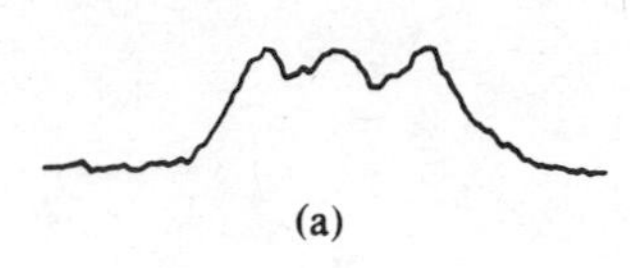

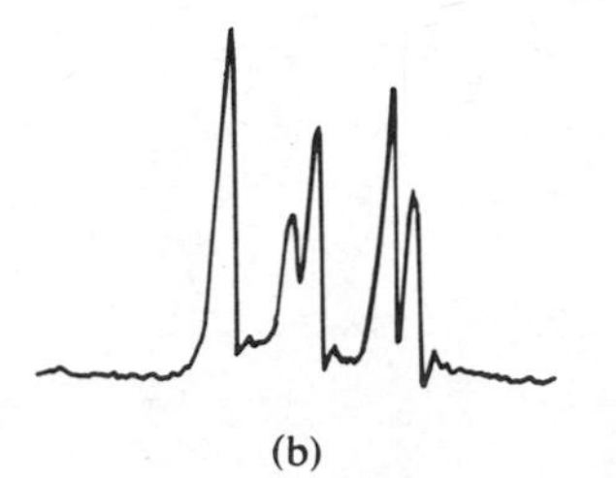

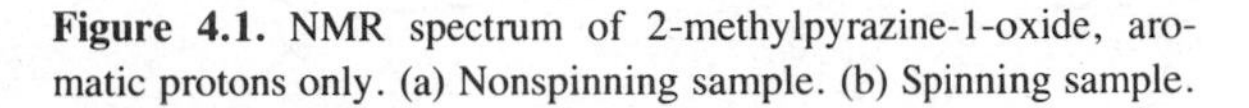

Figure 4.1. NMR spectrum of 2-methylpyrazine-1-oxide, aromatic protons only. (a) Nonspinning sample. (b) Spinning sample.

The drastic effect that sample spinning has on the NMR spectrum of a compound, in this instance, 2-methylpyrazine 1-oxide, is shown in Figure 4.1. The ideal shape of a sample to minimize any homogeneity problems in NMR spectroscopy is spherical, since this shape is *isotropic*. Spherical sample cells, although available, have found little popularity, since they are not only expensive but are difficult to fill and to place into the optimum position in the magnetic field.

Although sample volumes of as little as 0.01–0.03 mL have been used for protium and fluorine spectra, satisfactory sample volumes for studies of these two nuclides are about 0.25–0.50 mL. The solution is placed into a *cylindrical* tube with 5–8-mm o.d., depending on the spectrometer being used. Generally, concentrations down to 0.01 mol% and less can be examined.

When nuclei other than protium or fluorine are studied, sample tubes with diameters of 15 mm and larger are frequently required. The use of sample tubes of this and larger sizes is readily accomplished in supercon-driven spectrometers.

4.1.3 Solids (Magic-Angle Spectra)

Although NMR spectra of liquids can be readily obtained by the sample spinning procedure as long as the sample is not too viscous, spectra of viscous liquids provide, at best, only poorly resolved spectra. This is largely a result of nonaveraged dipolar interactions and inappropriate relaxation times. These effects become *extreme* in solid samples, and a different technique, *magic-angle* spinning, is employed [3]. To develop the concept of *magic-angle* high-resolution NMR spectroscopy, it is appropriate to discuss some fundamental principles.

The *chemical shift* of a nucleus depends on, among other things, the orientation of the molecule in the applied magnetic field. If nucleus X in a general structure A–X–B is being examined, the magnetic field at X in this molecule will be different when it is aligned with the magnetic field (parallel) then when the molecule is situated at right angles to the applied field. In the former case X will experience a greater magnetic field than in the latter instance. In a solid sample, or in a highly

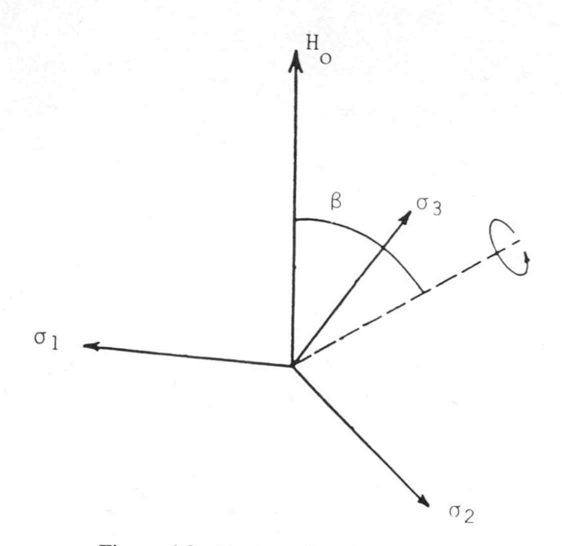

Figure 4.2. Magic-angle spinning diagram.

viscous sample, the molecules will align themselves at these two extremes as well as at all of the possible intermediate angles. The result of this is that the chemical shift of X is spread over a wide range (500 ppm for many carbon nuclides).

The *chemical shift anisotropy* (*CSA*), already described in Chapter 1, is defined as the width of this *spread* (chemical shift of nucleus when the system is parallel to the applied magnetic field minus the chemical shift when it is at right angles to it). It is this value that is added to the isotropic chemical shift to describe the chemical shift of a nuclide in the solid state. The anisotropic part of the chemical shift is $1/3$ [CSA] $[3\cos^1(\theta)-1]$. The angle θ is the angle that, in this instance, the A–X–B axis of the molecule makes with the applied magnetic field. When this angle is $54.7°$ and the sample spinning rate is greater than the CSA, the *anisotropic* contribution to the chemical shift becomes *zero* and the spectrum that is obtained approximates that of a solution spectrum in terms of resolution.

In *real life* this means that, to obtain high-resolution NMR spectra of solids or highly *viscous* materials, the sample spinning axis has to be $54.7°$ from the axis of the applied magnetic field. The spinning speed of the samples, for carbon spectral analysis, for example, is 15K.

Figure 4.2($\theta = \beta$) is a graphic representation of the magic-angle relationship to the applied field, along with the various reference axes.

4.2 NMR SOLVENTS

The selection of the appropriate solvent for NMR spectral studies depends not only on the solubility of the *solute* but also on the magnetic properties of the *solvent*. Ideally, either the solvent should contain none of the nuclides of the species being

TABLE 4.1 Selected Solvents for NMR Spectroscopy

Solvent	Residual Absorptions H	Residual Absorptions C	Comments
Acetone-D_6	2.05	30.4, 204.1	Hydrogen bonding
Benzene-D_6	7.20	128.5	Solvent effects
Chloroform-D	7.25	77.2	Complex formation
Carbon tetrachloride		96.0	
Cyclohexane-D_6	1.40	27.5	
Deuterium oxide	4.75		Concentration-dependent
Dimethylformamide-D_6	2.76, 2.94, 8.05	162.7	Hydrogen bonding
Dimethylsulfoxide-D_6	2.50	40.5	Hydrogen bonding effects
Methyl alcohol-D_4	3.35, 4.84	43.3	Concentration-dependent
Pyridine-D_5	8.50, 6.98, 7.35		Hydrogen bonding
Tetramethylene-D_8-sulfone	2.92, 2.16		Hydrogen bonding
Hexafluorobenzene			Solvent effects
CF_3Cl and isomers			
Trifluoroacetic acid		166.0	
Carbon disulfide		192.8	
Perdeuterotoluene	2.34, 7–9 M		Solvent effects
Sulfur dioxide			Useful for low temperatures

examined, or it should not have resonances in the region of interest in the examination of the solute. It is for these reasons that CCl_4 and $CDCl_3$ have long been the solvents of choice in H NMR spectroscopy. Among the *deuterated* solvents that are commercially available are D_2O, C_6D_6, CD_3COCD_3, $CDCl_3$, CD_3SOCD_3, CH_3OD, CD_3OD, CH_3CN, and CD_3CN. Since solvents generally have an effect on the chemical shifts of many nuclides, the selection of the proper one for a particular solute becomes a matter of some importance.

The data in Table 4.1 provide some of the necessary information to permit selection of an appropriate solvent for a particular sample. If spectra are to be taken at elevated temperatures and the sample properties permit, no solvent is used. However, if the boiling point of the sample is too high for the particular instrument in use, compounds such as biphenyl, naphthalene, dimethylformamide, dimethylsulfoxide, or sulfolane often serve as good high-temperature solvents.

Low-temperature spectra are frequently obtained in carbon disulfide, sulfur dioxide, difluorodichloromethane, thionyl chlorofluoride, or the *perdeutero*-derivatives of methyl alcohol, toluene, acetone, and dimethyl ether.

4.3 SOLVENT EFFECTS

The chemical shifts of many magnetically active nuclei often depend on the nature of the surrounding molecules. These molecules might be either solvent or solute.

TABLE 4.2 Concentration Dependence of the Chemical Shift of H_3 of 2,6-Dimethylquinoline [3]

Conc. (%w/v)[a]	16	8	4	2	1
δ(ppm)	7.04	7.07	7.10	7.11	7.12

[a] CCl_4 solutions.

The concentration of the solute is the most frequently observed solvent effect. The data in Table 4.2 demonstrate this concentration effect on the chemical shift of H_3 in 2,6-dimethylquinoline (structure **1**):

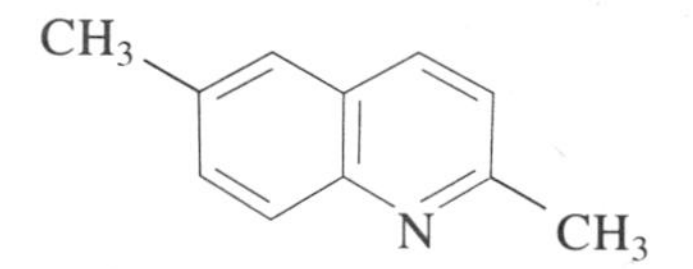

There exist two important and different *solute–solvent* effects that can have an effect on the NMR behavior of a particular solute: (1) chemical shift changes caused by the diamagnetic susceptibilities of the solute and/or solvent, and (2) chemical shift changes arising from solute-solvent and/or *solute–solute* interactions. The extent of the chemical shift changes caused by the former effect can be estimated by expression 4.4:

$$\tfrac{2}{3}\pi\Delta X_{v} \tag{4.4}$$

The volume magnetic susceptibility is X_v, and any chemical shift difference observed beyond this amount is attributed to effect 2.

An example of a solvent effect on some proton chemical shifts is given in Figure 4.3. Solvent effects on many metal nuclides far exceed those observed for protons! For example, the cadmium chemical shift in dimethyl cadmium differs by 100 ppm between solutions in tetrahydrofuran and cyclohexane.

The most commonly used deuterated solvent, deuteriochloroform, usually contains a trace amount of protio chloroform with a proton chemical shift at δ 7.27 and a carbon chemical shift of δ_c 77.2 (the carbon chemical shift in the deuterio compound is 76.9 ppm with respect to the carbon nuclei in TMS). These values can be used as convenient *internal* proton and carbon standards. Since chloroform is relatively unstable in the presence of moisture, oxygen, and light, it is a good idea to pass it through a small silica gel or alumina column to purify it immediately before use. This process also removes the small amount of ethanol often added to the chloroform as a stabilizer.

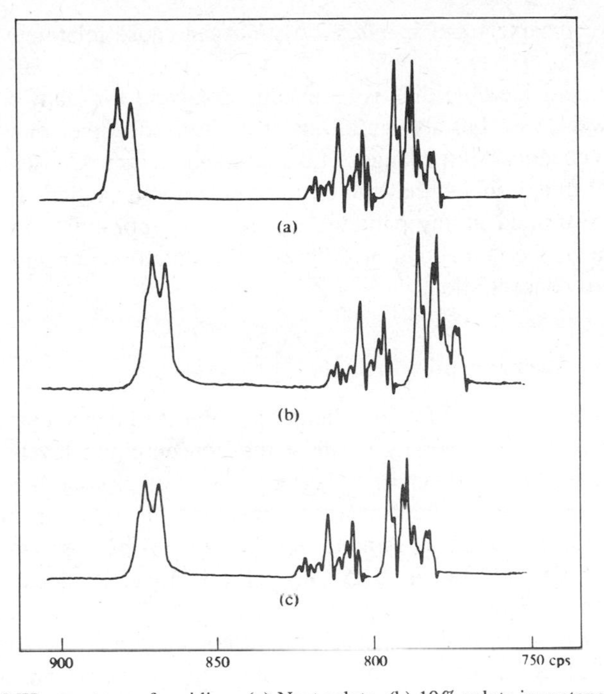

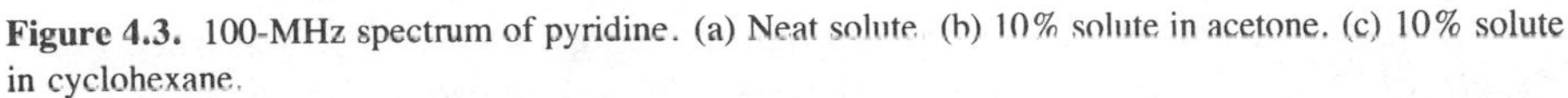
Figure 4.3. 100-MHz spectrum of pyridine. (a) Neat solute. (b) 10% solute in acetone. (c) 10% solute in cyclohexane.

Perdeuteracetone, perdeuteroacetonitrile, and perdeuterodimethylsulfoxide are excellent solvents for polar solutes. The latter solvent is, however, rather *hygroscopic* and somewhat viscous. The residual protons in all three of these solvents are coupled to two deuterium atoms and, consequently, will appear as closely spaced *quintets* (deuterium has a spin quantum number of 1). The carbon atoms in all of these solvents can, of course, serve as convenient internal references in C NMR studies. Among these, the methyl carbon in acetone resonates at 30.4 in the protio compound and at 29.2 in the perdeutero derivative. The dimethylsulfoxide carbon resonates at 40.5 and 39.6 ppm in the protio and deutero compounds, respectively.

Deuterotrifluoroacetic acid, readily prepared by the careful addition of deuterium oxide to trifluoroacetic anhydride, is an excellent solvent for basic compounds. However, the strong acidity of this solvent may cause chemical shift changes (deshielding) of nuclei that are situated close to any protonation sites in the solute.

Deuterium oxide and perdeuteromethanol are both excellent solvents for very polar solutes. Since deuterium oxide often contains trace amounts of ferric oxide, it is advisable to pass the solvent through a short silica gel column prior to use. The spectral results obtained from solutions in aromatic solvents have to be interpreted with caution, since both *diamagnetic anisotropy* (to be described in a later section of this chapter) and *specific solvation* can cause relatively large chemical shift changes.

Although perdeuteropyridine is an excellent solvent for many compounds, it is not only a weak base but also an aromatic compound, either of which effect can cause proton chemical shift changes of 0.3 ppm and larger. Consequently, H NMR data obtained in this solvent have to be interpreted with caution. If accurate chemical shifts are desired in any particular study, it is appropriate to obtain three or more spectra of the compound at different concentrations and to extrapolate the values to zero concentration.

4.3.1 Readily Exchangeable Protons

While the chemical shifts of many nuclei are subject to concentration and temperature effects, they are generally small at the concentration levels (about 0.1 *M*) normally employed. However, the chemical shifts of protons bonded to oxygen, sulfur, or nitrogen atoms exhibit marked concentration and temperature dependencies. For example, studies by Arnold and Packard [4] have shown that a temperature change from −110 to +75°C causes the hydroxyl protons of ethyl alcohol to become more shielded by 1.5 ppm.

The H NMR spectrum of methanol should contain a three-proton doublet and a one-proton quartet, if the principles described in Chapter 2 are applied. Yet, a methanol spectrum obtained under normal conditions shows one- and three-proton *singlets*. When the temperature of the sample is reduced to about −50°C, the predicted doublet and quartet appear. The only reasonable way to explain this behavior is to stipulate that the proton bonded to the hydroxyl group becomes *detached* from the oxygen atom, and, consequently, it cannot spin–spin interact with the methyl group proton. Even if the proton is continuously *removed* and *reattached*, the spin information becomes completely randomized, and the methyl group protons will appear as a singlet.

However, if the *exchange* process of these protons becomes slow enough, as in the −50°C spectrum, a critical rate is reached at which the proton resides on the oxygen sufficiently long to permit spin–spin coupling between the hydroxyl and the methyl group protons.

The concentration dependence of the chemical shift of a proton bonded to a nitrogen is shown in Figure 4.4, where the proton singlet becomes more shielded in going from a dilute solution of piperidine (structure **1**) in deuteriochloroform to a "neat" sample. The N^{15} chemical shift in this compound is also concentration- and solvent-dependent, as shown by the change in the N^{15} chemical shift from 36.7 to 38.66 in going from a dilute cyclohexane solution to a neat sample.

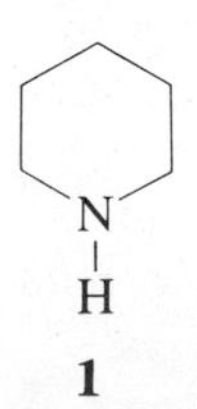

1

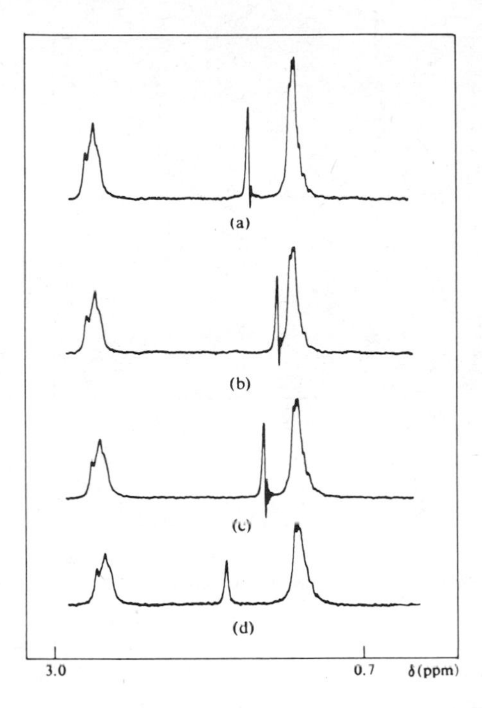

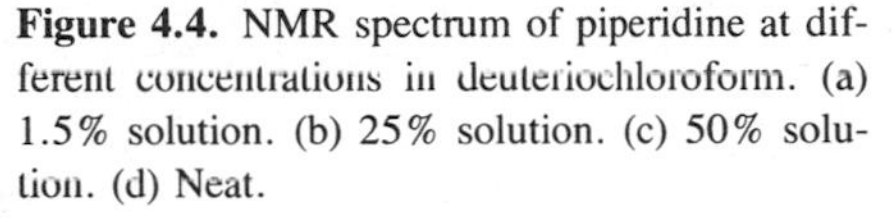

Figure 4.4. NMR spectrum of piperidine at different concentrations in deuteriochloroform. (a) 1.5% solution. (b) 25% solution. (c) 50% solution. (d) Neat.

It is of interest that, in this example, the chemical shift of the N-bonded proton in the 1.5% solution is at $\delta_H 1.6$, in the 25% solution at $\delta_H 1.2$; while in the more concentrated sample (50%) the chemical shift is at $\delta_H 1.4$ it and becomes more deshielded in the neat sample.

In addition to these temperature and concentration effects, which so clearly identify protons bonded to strongly electronegative atoms, another means exists of causing rapid proton–proton exchanges. If, for example, some deuterium oxide is added to a spectrum containing a peak suspected of belonging to a proton bonded to a sulfur, oxygen, or nitrogen atom, the peak will disappear. This disappearance is caused by the replacement of the proton by a deuteron. It is, of course, necessary to add sufficient deuterium oxide to accomplish this by forcing the following equilibrium to the right:

$$R{-}XH + D_2O \rightleftharpoons R{-}XD + HDO$$

If this requirement is not met, the peak will decrease only to the extent permitted by the ratio of proton to deuteron. The exchange process is demonstrated for 2-aminopyridine, in Figure 4.5, where the sample in spectrum *b* has had deuterium oxide added to it in sufficient quantity to obliterate the amino protons.

The exchange rate of hydroxyl, amino, mercapto, and similar, readily exchangeable protons can be slowed by using various specific solvents. Among these are perdeuterodimethyl sulfoxide and deuterated fluorosulfonic acid. These solvents form hydrogen bonds with the exchangeable protons but do not cause proton ex-

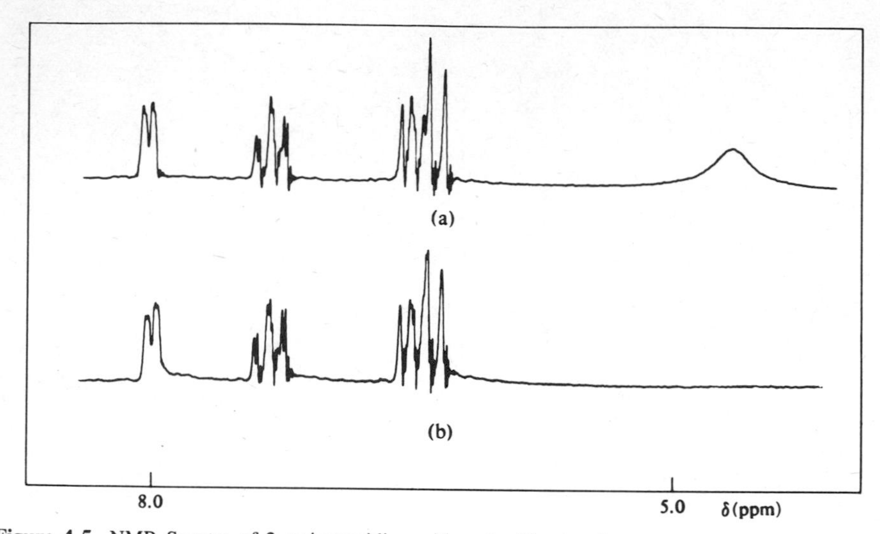

Figure 4.5. NMR Spectra of 2-aminopyridine with and without amino group protons exchanged by deuterium. (a) Solute in deuteriochloroform. (b) Solution (a) with deuterium oxide added.

change. Consequently, coupling with these protons by spin–spin interaction with other magnetically active nuclei in the molecule can take place.

4.3.2 Calculation of Exchange Rates

The discussion in the previous paragraph has pointed to yet another potentially important application of the NMR technique—namely, the study of exchange rates of different nuclei and consequently the determination of kinetics of fast reactions.

In a *qualitative* sense, for a particular coupling to be observed in an exchange process, the rate (s^{-1}) must be slower than the size of the coupling constant once the exchange process has been slowed. A proton could still exchange a few times per second and maintain spin coupling.

If the exchange rate (τ) between two protons is long in comparison to the time required for a transition from one magnetic energy state to another, two sharp resonance lines are visible, separated by υ_{AB}. On the other hand, when τ is short in comparison to the magnetic transition times, a single sharp line centrally located between ν_A and ν_B is observed. In those instances where τ is of the same order of magnitude as the transition time, a broad line, centrally located between ν_A and ν_B, will appear.

Gutowsky [5, 6] has demonstrated that there exists a correlation between the smallest time (τ) for which two separated states can be distinguished and the separation of the corresponding resonance lines ($\Delta\nu$). Equation 4.5 expresses this relation.

$$\tau = \sqrt{2}/\pi\Delta\nu \quad (4.5)$$

The study of NMR spectral changes to examine various rate processes, which result in obtaining kinetic information, is referred to as the "dynamic nuclear magnetic resonance method (DNMR)." One example of DNMR H (i.e., the technique applied to proton NMR) demonstrates this method (see Chapter 8 for a detailed discussion).

The methyl group protons of N,N-dimethyl nitrosamine (structure **2**):

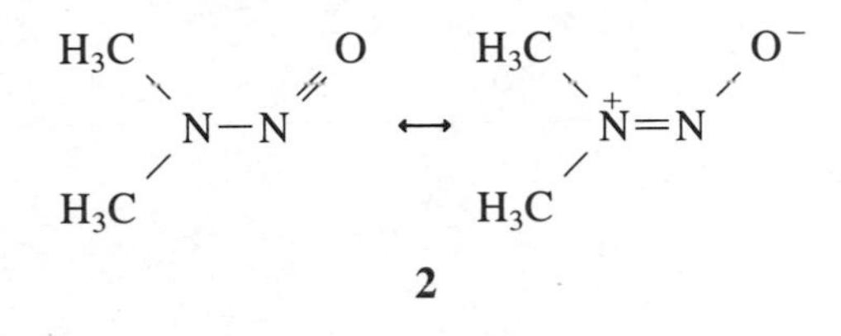

2

appear as two singlets, separated by 26 Hz, when the spectrum is obtained at 40 MHz and at room temperature. These signals collapse slowly into a singlet as the temperature of the sample is raised. Finally, at 193°C the two methyl groups appear as *one* singlet. Thus, at this temperature, the rotation about the C–N bond occurs at a rate of 0.01 s^{-1} ($\sqrt{2}/(\pi \times 26)$).

The application of DNMR to C^{13} spectroscopy has a number of advantages over DNMR H analyses. Among these are the fact that frequency differences between *exchanging* carbon sites are greater in C NMR than in H NMR, carbon resonances are singlets (often as a result of the proton decoupling process applied) and the spectra are considerably less complex than the corresponding proton spectra. For example, while it is not readily possible to examine the cyclononane conformational mobility by H NMR, the C NMR spectrum at −162°C shows two singlets in a ratio of 1:2, separated by 9 ppm. At −65°C these two singlets *collapse* into one, indicating rapid conformational interchange among the different conformers.

These exchange processes can also be used for the *quantitative* analyses of mixtures of compounds. For example, in the spectrum of a dilute chloroform solution of phenol, the phenolic hydroxyl proton resonates at δ_H5.82, whereas the hydroxyl group proton in a dilute chloroform solution of methanol absorbs at δ_H2.84(ppm) (see Fig. 4.6). When these two solutions are mixed and the H NMR spectrum of the mixture is recorded, only one exchangeable proton peak is left. The chemical shift of this peak is δ_H4.54. This value lies almost in the middle between the hydroxyl group absorptions of the pure compounds. In fact, the chemical shift is a function of the mole percent of the two compounds present in the mixture (in this instance, a 50:50 mixture). Thus, if the line position of the absorptions of exchangeable protons is known for the pure components, this technique is, indeed, readily applicable to quantitative analyses!

The single peak of the hydroxyl proton observed in the mixture is caused by the rapid exchange between the two types of hydroxyl protons. The observed signal is actually a *time-averaged* one.

The addition of small amounts of strong acids or strong bases to a sample that shows two different absorption peaks caused by two different exchangeable protons

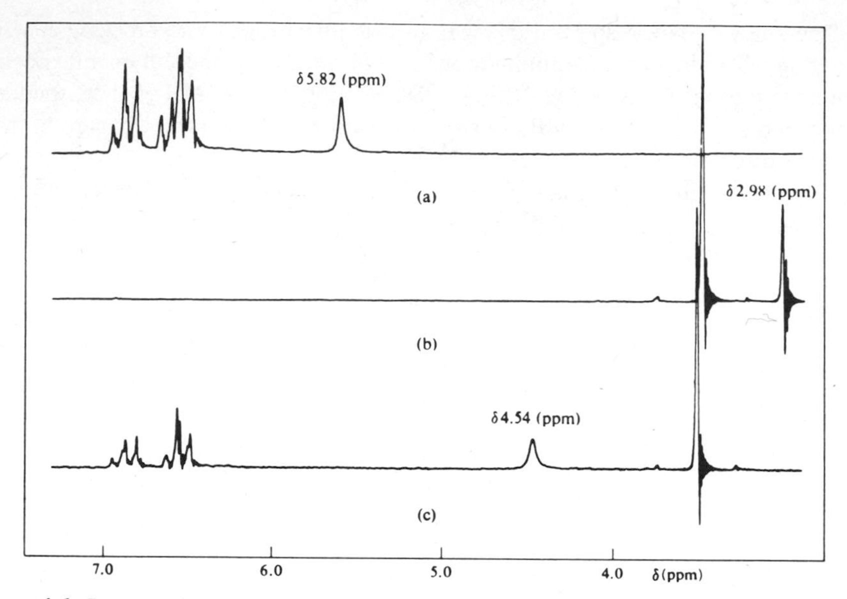

Figure 4.6. Proton exchange phenomena. (a) Pure phenol in $CDCl_3$. (b) Pure methanol in $CDCl_3$. (c) Mixture of solutions (a) and (b).

greatly accelerates the exchange process between these species and generally causes the two peaks to collapse to a time-averaged singlet. An example will serve to demonstrate this process and the application of equation 4.5 to it. If the separation between two different hydroxyl group protons in a particular compound is 30 Hz, equation 4.5 demonstrates that these protons will be exchanged only when the *mean* lifetime of a particular proton on either site is less than about 0.015 s^{-1}.

4.3.3 Hydrogen-Bonding Effects

4.3.3 General Discussion

The observation that the chemical shifts of hydroxyl and related protons are temperature- and concentration-dependent was explained by Liddel and Ramsey [7] on the basis of hydrogen bonding in these substances. A proton involved in a hydrogen bond experiences a rather strong deshielding. Thus, the free hydroxyl proton in ethanol resonates at δ_H0.7, whereas in neat ethanol it does so at 5.3. If the proton exchange rate between the hydrogen-bonded and non-hydrogen-bonded species is rapid (about 0.0002 s), then only one absorption peak corresponding to the average chemical shift of the two proton species will be observed. The difference between the chemical shift of the protons in the associated and nonassociated states is referred to as ''hydrogen bond shift''. If hydrogen bonding is the cause of the chemical shift differences, then an increase in temperature and a decrease in concentration of the compound should cause a corresponding decrease of the hydrogen-bonded species and an increase in the nonassociated molecules. These effects will, of course, be reflected by changes in the chemical shifts of the proton being examined.

The chemical shift of a nonassociated proton is readily determined by extrapolating the results obtained from solution spectra at different concentrations, to zero concentration, or, more ideally, by measuring in the gas phase, the chemical shift of the proton in question. To determine the chemical shift in the associated state, the spectrum should be obtained in the crystalline state of the compound. It has been common practice to obtain the associated chemical shifts at just above the melting point of the compound, although solid-state NMR spectra can be obtained by application of the magic-angle technique (see Chapter 10).

Typically, the measured hydrogen bond shift for saturated hydrocarbons is zero; for olefins, it is about 0.4 ppm, and for acetylene, the value is 1.30 ppm.

The acidic nature of the acetylene proton, as demonstrated by the size of the hydrogen bond shift, is consistent with the fact that these protons are readily replaced by metals.

The hydrogen bond shifts of a number of hydride molecules are as follows: H_2O (4.58); HF (6.65); HCl (2.05); HBr (1.78); HI (2.55); HCN (1.65); NH_3 (1.05). As these data demonstrate, proton magnetic resonance signals are generally displaced to lower fields by the formation of a hydrogen bond. This effect can be explained by consideration of two major effects.

The hydrogen bond shift arises as a result of modifications of the *intramolecular* electronic effects experienced by a proton when it becomes part of a hydrogen-bonded system, *X*–H . . . *Y*.

Thus, the proton in *X*H will experience a change in magnetic field owing to the currents induced in the donor molecule *Y*. If this contribution is nonzero in all directions, a net contribution will be made to the proton chemical shift. The presence of the donor molecule, *Y*, will, concurrently, disturb the electronic structure of the *X*H molecule and will cause a change in the chemical shift as a result of this change of its magnetic susceptibility.

4.3.3.2 Alcohols

The *true* OH signal for the *isolated* ethanol molecule is obtained from measurements in the gas phase. Relative to the methane reference standard, the hydroxyl proton resonates at −0.48 ppm. The chemical shift of the OH group in liquid ethanol at −114°C, close to its melting point, was also obtained. When this information is combined with the gas-phase data, the hydrogen bond shift for ethanol is determined to be 6.15 ppm. By way of comparison, the value for water is 4.58 ppm. This is consistent with the fact that the hydrogen bond energy in ethanol is greater than that of water. Generally, the hydroxyl protons of dilute solutions of alcohols resonate between δ_H4.0 and 5.5(ppm).

4.3.3.3 Phenols

The chemical shifts of phenolic protons show marked concentration dependencies. This behavior and the observation that only one absorption peak assignable to a

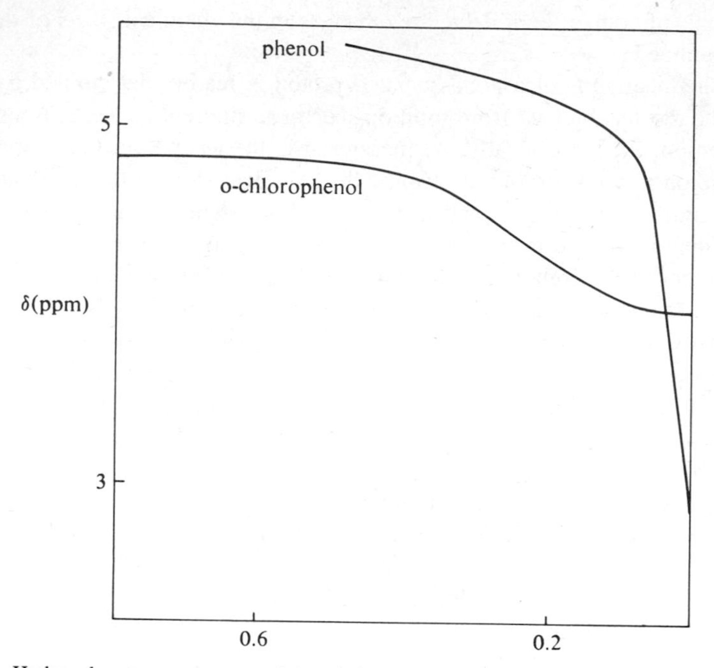

Figure 4.7. Hydroxyl proton resonances of phenols in carbon tetrachloride. [Reprinted with permission from C. M. Huggins, G. C. Pimentel, and J. N. Shoolery, *J. Phys. Chem.*, **60** 1311 (1956). Copyright (1956) American Chemical Society.

phenolic proton can be observed are clear evidence of the existence of various molecular species that coexist in a rapid equilibrium with each other:

$$n\mathrm{ArOH} \rightleftharpoons (\mathrm{ArOH})_n \qquad (4.6)$$

The hydrogen bond shift of phenol can be estimated from the data given in Figure 4.7 to be 2.8 ppm [8]. This value, which is lower than the corresponding values for ethanol and water, is in agreement with the lower hydrogen bond strength of phenol.

The dilution curve for *o*-chlorophenol (see Fig. 4.7) is quite different from that of phenol itself in that there is no change in the hydroxyl proton chemical shift up to a concentration at which it is present to an extent greater than a mole fraction of 0.06. Another interesting feature of this dilution curve is that, at infinite dilution, the chemical shift of the phenolic proton is 1.33 ppm higher than that of phenol. This difference is in the same direction as that associated with intermolecular hydrogen bonding and must, almost certainly, be due to the presence of intramolecular hydrogen bonding in *o*-chlorophenol:

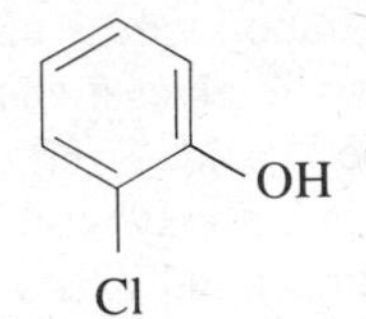

Generally, the phenolic group protons resonate between 6.5 and 7.5(ppm). This range is clearly in a more deshielded region than the absorption region of the hydroxyl group protons in alcohols.

4.3.3.4 Carboxylic Acids

These compounds are unique insofar as their H NMR spectral behavior is concerned. When carboxylic acids are dissolved in moderate concentrations in nonpolar solvents such as carbon disulfide or carbon tetrachloride, the carboxylic acid proton resonance is concentration-independent. On the other hand, when the solvent being used has a functional group that can hydrogen-bond, such as acetone, then very pronounced dilution shifts occur (see Figure 4.8).

The hydrogen bonding that exists in nonpolar solvents involves largely a monomer-dimer equilibrium:

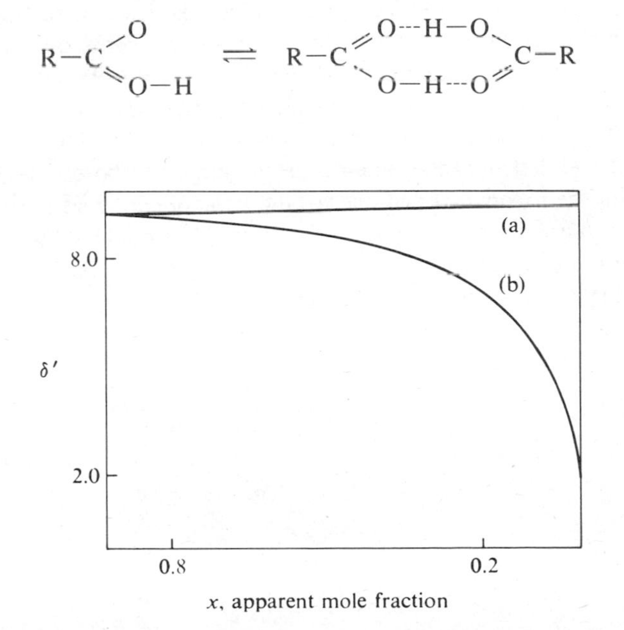

Figure 4.8. Hydroxyl proton chemical shifts of acetic acid in (a) carbon tetrachloride and (b) acetone. [Reprinted with permission from C. M. Huggins, G. C. Pimental, and J. N. Shoolery, *J. Phys. Chem.*, **60** 1311 (1956). Copyright (1956) American Chemical Society.

Extrapolation of the concentration versus carboxylic acid proton chemical shift curve of acetic acid in cyclohexane gives a chemical shift at zero concentration of −0.65 ppm relative to pure acetic acid. From these data it has been estimated that the strongest hydrogen bonds in carboxylic acids occur in the dimeric species. It appears very likely that the pure liquid exists as associated species composed of two or more molecules.

Generally, the carboxylic acid protons of dilute solutions in nonpolar solvents resonate in the region δ_H7.5–9.1 ppm.

4.3.3.5 Amines

Since the N^{14} nuclei have a quadrupole moment and a spin quantum number of 1, those protons that are bonded to this nucleus are unique among the various readily exchangeable nuclei. The quadrupole moment gives rise to a very strong relaxation mechanism, which, if the electron charge distribution around the N^{14} nucleus approaches symmetry, causes the proton resonance line to become considerably broadened.

This line broadening can, at times, become severe enough to prevent any detection of the absorption signal. Line broadening of this type is an example of "quadrupole relaxation" [9].

Since N^{14} has a spin quantum number of 1, a proton bonded to it will appear as a triplet. This spin–spin interaction is observable only under conditions where the normally rapid proton–proton exchange process is slow on the NMR time scale.

Akin to hydroxyl proton behavior, proton chemical shifts on secondary amine nitrogens are also concentration-dependent [9] (see Fig. 4.9).

In all of the secondary amines studied, only one NH proton signal has been observed. Thus, proton–proton exchanges in secondary amines occur at a sufficiently rapid rate that only the time-averaged signal can be observed, rather than one signal for the *associated* and one for the *nonassociated* species.

Generally, *basic* amine protons, when observed in dilute solutions, resonate at 1.4–2.0(ppm), and *acidic* amine protons, such as those found in pyrroles, absorb at more deshielded positions (δ_H8–14(ppm)).

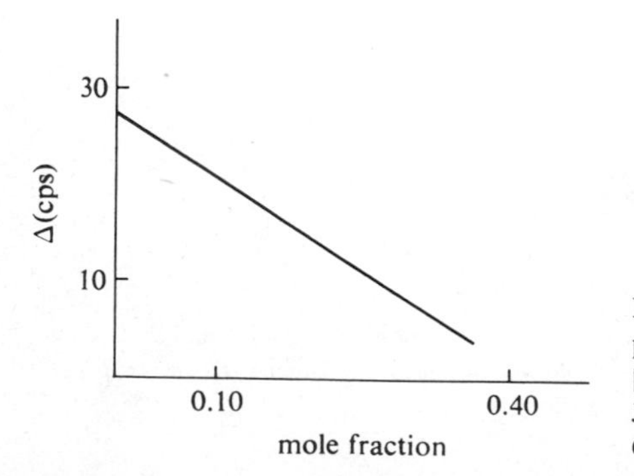

Figure 4.9. Variation of N-H chemical shift of piperidine with concentration changes in cyclohexane [Reprinted with permission from R. A. Murphy and J. C. Davis Jr., *J. Phys. Chem.*, **72** 3111 (1968). Copyright (1968) American Chemical Society.

4.3.4 Aromatic Solutes

The aromatic proton chemical shifts of many aromatic compounds are strongly concentration-dependent when these compounds are dissolved in different nonaromatic solvents [10–15]. For example, the benzene proton chemical shift extrapolated to *infinite* dilution when compared to a 90% solution in cyclohexane is −0.30 ppm.

On the other hand, the carbon chemical shifts of the aromatic ring carbons are rather insensitive to the nature of the solvent. For example, the C^{13} chemical shift (relative to TMS at zero) of neat benzene is δ_c128.35, in deuteriochloroform 128.48, and in trifluoroacetic acid 128.59 [15].

A *similar but opposite* phenomenon is observed in proton chemical shift behavior when a nonaromatic compound is dissolved in varying concentrations of an aromatic solvent. In these instances, the solute protons in a dilute solution absorb at more shielded positions than those in a concentrated solution.

These effects are a result of the large diamagnetic anisotropy of aromatic compounds (see Chapter 5), which is caused by the circulation of the π electrons induced by the applied magnetic field in an NMR experiment. The *ring current* causes protons, or any other magnetically active nuclide, situated above or below and close to the center of the ring to become significantly more shielded. Any protons or other magnetically active nuclide located within the plane and close to the periphery of the benzene ring will resonate at significantly more shielded positions than is observed in nonaromatic solvents.

Since aromatic molecules are somewhat disk-shaped, there exists a greater probability that another aromatic molecule will approach above or below the plane of the ring than there does for an approach in the plane of the ring. Thus, if all of these various possibilities are averaged, there will be a net shielding effect. The extent of this diamagnetic shielding will decrease as the number of solute molecules is increased (a dilution of the aromatic compound), since this will tend to separate the associated benzene molecules from each other. This explanation is also applicable to the *reverse* shielding effect that occurs when a nonaromatic solute is present in ever increasing concentrations in an aromatic solvent.

In a dilute solution there will be fewer available solute molecules, and consequently more aromatic solvent molecules can crowd around them. As a result of this, protons of nonaromatic solutes when dissolved in an aromatic solvent will be more shielded the more dilute the solution.

The proton chemical shift differences of a solute dissolved in an inert solvent such as carbon tetrachloride or cyclohexane and those dissolved in benzene (or perdeuterobenzene, if the solute protons resonate in the benzenoid region) have been utilized in the elucidation of the structure, stereochemistry, and conformation of many compounds. The utility of the benzene-induced solvent shifts is based on the existence of stereoselective solvation of the solute molecules by benzene. Bhacca [10], Williams [11], Connelly [12], and others have analyzed a large number of compounds. These analyses have allowed Williams to formulate a set of generalizations that are of considerable value in explaining the solvent shifts of proton resonances induced by benzene. It is common practice to express these solvent shifts by the following equation:

$$\Delta = \delta_{\text{inert solvents}} - \delta_{\text{benzene}}$$

The following generalizations apply [13]:

1. A benzene molecule will interact with electron-deficient sites of solute molecules.
2. Benzene-solute interaction may occur in a transient 1 : 1 collision complex, but phase-independent 1 : 1 associations may occur at *each* electron-deficient molecule.
3. The orientation of benzene molecules is caused by local dipole-induced interactions.
4. A benzene solvent molecule will probably be oriented in a nonplanar collision complex with the partial positive charge of a local dipole in such a fashion that the benzene ring lies as far as possible from the negative end of the local dipole.

Some examples will serve to demonstrate the application and utility of these rules. A solution of dimethylaniline in benzene will contain collision complexes arranged whereby the positive end of the amine dipole will be situated with respect to a benzene molecule as follows:

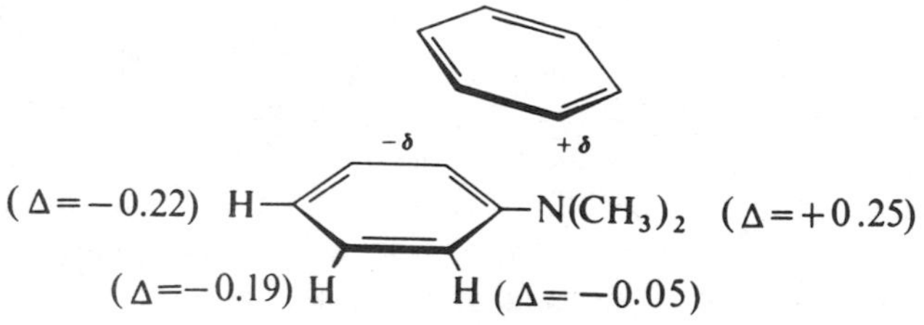

$-\delta$ $+\delta$

($\Delta = -0.22$) H—[ring]—$N(CH_3)_2$ ($\Delta = +0.25$)

($\Delta = -0.19$) H H ($\Delta = -0.05$)

This sketch shows that the ring protons of the dimethylaniline will be located in the deshielding region of the benzene molecule such that the largest deshielding will be observed for the *para*- and the *meta*-protons. The *ortho*-protons are, clearly, still close to the deshielding region of the benzene molecule and should thus be least affected. The methyl protons, however, are quite close to the shielding region of the benzene molecule and should resonate at a more shielded position when benzene is the solute. The numbers in parentheses indicate the observed solvent shifts for dimethylaniline and are in agreement with the predictions.

The application of Williams' rules to the prediction of solvent shifts in camphor serves to demonstrate the effect that a carbonyl group has on the shape of the collision complex with benzene. The benzene molecule involved in this complex would tend to arrange itself so that the negative end of the carbonyl dipole, the oxygen atom, is as far from it as possible, whereas the positive end would be as close as possible. This would place the nonbridgehead methyl groups into the benzene shielding region. The bridgehead methyl group, on the other hand, would be in an essentially neutral region of the benzene molecule. The data in diagram 4.2 are clearly in agreement with these predictions.

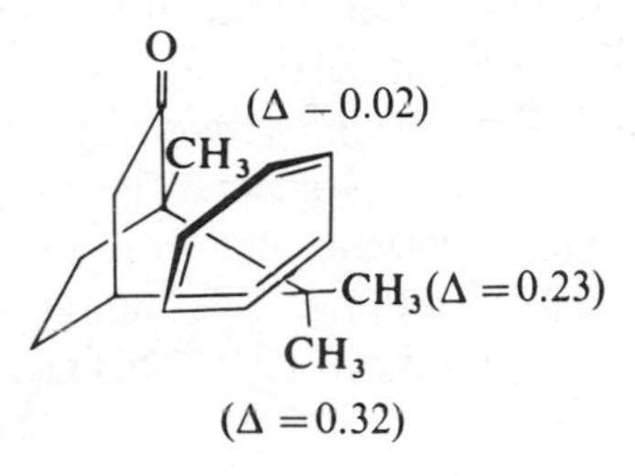

An extension of these analyses to nuclei other than protons is, of course, possible. For example, the solvent shift (benzene vs. cyclohexane) for the carbon in chloroform, is 0.47 ppm, whereas for the proton it is −0.75 ppm. Thus the solvent effect on the carbon is deshielding, whereas it is shielding on the proton in chloroform. The reader is encouraged to provide a rationale for this observation.

Although Williams' rules do not allow quantitative analyses, they are, nevertheless qualitatively very useful.

4.3.5 Liquid Crystal Effects

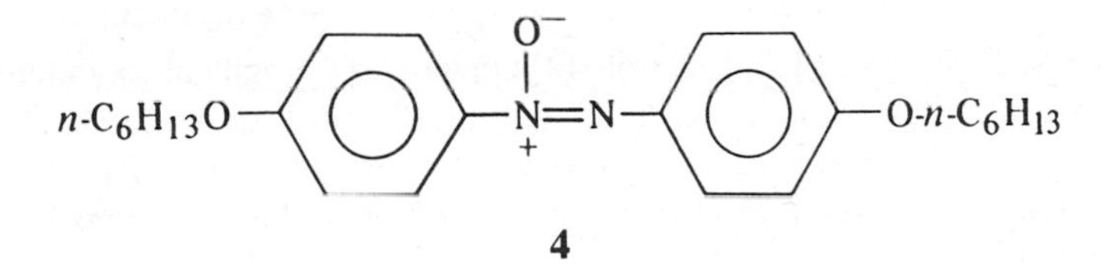

4

Molecules such as the azoxy compound **4** exist as nematic liquid crystals within certain temperature regions [16]. The properties of molecules in this phase lie intermediate between those expected for a liquid and those expected for a solid. These rodlike molecules tend to align themselves along their long axis to form a "domain." This domain can be oriented, when placed into a magnetic field, so that its long axis is perpendicular to the applied field. Any molecules dissolved in this phase will thus also be aligned. Since the molecular ordering in these nematic phases is not nearly as rigid as in the solid phase, dissolved compounds will maintain some degree of freedom of motion. For example, a benzene molecule will be aligned with its plane parallel to the applied field but will be able to freely rotate around the *sixfold* axis perpendicular to this plane.

The NMR spectral result of this arrangement is that one observes neither the narrow (1 Hz) singlet expected for the benzene protons nor the broad, featureless singlet observed when the spectrum of solid benzene is examined. Instead, a very complex spectrum of about 40 relatively narrow lines covering a width of about 2500 Hz is observed.

Since the molecules ensconced in a liquid crystal matrix cannot tumble freely, direct dipole–dipole ($D_{1,3}$) as well as indirect ($J_{1,3}$) couplings can be determined. The liquid crystal technique also permits the determination of the absolute *signs* of coupling constants in general and the value of the spin–spin coupling constants of chemically identical nuclei in particular.

The H NMR spectrum of CH_3CN in a *nematic* phase shows a triplet as a result of each proton being coupled to two others. The fluorine dipole–dipole ($D_{1,3}$) coupling constants in hexafluorobenzene, obtained in a nematic phase, are *ortho*, −1452.7; *meta*, −271.6; *para*, −194.2. The indirect coupling values are J1,3, −22 (*ortho*); −4 (*meta*), and +6 (*para*) [17]. This technique has also established that bicyclobutane is nonplanar to the extent of 35° [18].

4.4 PROBLEMS

1. Calculate the necessary spinning speed for C^{13} nuclei when they are placed into a magnet gap whose field intensity variation is about 0.001 gauss.
2. Calculate the chemical shift of H_3 in 2,6-dimethylquinoline in carbon tetrachloride for an infinitely dilute solution (see text).
3. Explain the concept "hydrogen bond shift."
4. Predict the direction and relative extent of the benzene-induced solvent shifts for the different protons in each of the following compounds: pyridine: pyrrole; *p*-benzoquinone; 4-methylcyclohexanone; 3-methylindole.
5. The N–H proton of 1,2,4-triazole [**4**] appears as a singlet at room temperature both in the pure liquid and in anhydrous dimethylsulfoxide-D_6. While this proton remains a singlet, even at lower temperatures, the singlet C–H proton present at room temperature becomes two peaks at temperatures below 0°C. Explain [19]!

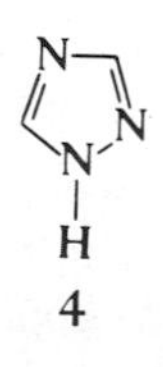

6. Predict the relative changes of the benzene-induced chemical shifts of the *beta* protons of tetrahydrofuran, tetrahydrothiophene, tetrahydroselenophene, cyclopentanone, and cyclopentane. Is there a relationship between these changes and the dipole moments of the compounds [20]?
7. Under conditions of slow exchange, the following equation is valid:

$$\frac{\text{Coupling constant at small } \tau^*}{\text{Coupling constant at large } \tau^*} = \left(1 - \frac{1}{2\pi^2 \tau^{*2} (J_{\text{at large } \tau})^2}\right)^{1/2}$$

At 46.8°C, J_{HNMe} in protonated dimethylaniline is 5.11 Hz, whereas at 26.8°C and below it is 5.22 Hz. Calculate the NH exchange rate at 46.8°C. Given the further information that the coupling constant is 5.02 and 4.67 Hz at 50.6 and 55.5°C, respectively, estimate the activation energy for this exchange reaction [21].

REFERENCES

1. J. A. Pople, W. G. Schneider, and H. J. Bernstein, *High-Resolution Nuclear Magnetic Resonance*. McGraw-Hill, New York, 1959, p. 39ff.
2. F. Bloch, *Physiol. Rev.* **102,** 104 (1956).
3. J. R. Lyerla and C. S. Yannoni, *IBM J. Res. Dev.* **27,** 302 (1983), and C. S. Yannoni, *High Resolution NMR in Solids*, **15,** 201 (1982).
4. J. Ronayne and D. H. Williams, *J. Chem. Soc. (B)* 805 (1967), and J. T. Arnold and M. E. Packard, *J. Chem. Phys.* **19,** 1608 (1951).
5. H. S. Gutowsky and C. H. Holm, *J. Chem. Phys.* **25,** 1228 (1956).
6. R. J. Gillespie, *Accounts Chem. Res.* **1,** 202 (1968).
7. U. Liddel and N. F. Ramsey, *J. Chem. Phys.* **19,** 1608 (1951).
8. C. M. Huggins, G. C. Pimentel, and J. N. Shoolery, *J. Phys. Chem.* 1311 (1965).
9. R. A. Murphy and J. C. Davis Jr., *J. Phys. Chem.* **72,** 3111 (1968).
10. N. S. Bhacca and D. H. Williams, *Tetrahedron Lett.* **42,** 3127 (1964).
11. D. H. Williams and J. Ronayne, *J. Chem. Soc. (C)* **5,** 2642 (1967), and earlier papers.
12. J. D. Connolly and R. McCrindle, *Chem. Industry* 379 (1965).
13. J. Ronayne and D. H. Williams, *Chemical Commun.* 712 (1966).
14. G. C. Levy and G. L. Nelson, *Carbon-13 Nuclear Magnetic Resonance for Organic Chemists*, Wiley-Interscience, New York, 1972, p. 200.
15. J. B. Lambert, H. F. Shurvell, L. Verbit, R. G. Cooks, and G. H. Stout, *Organic Structural Analysis*, Macmillan, New York, 1976, p. 19 ff.
16. R. L. Lichter and J. D. Roberts, *J. Phys. Chem.* **74,** 912 (1970).
17. P. Diehl and W. Niederberger, *Nucl. Magn. Resonance* (special report) **3,** 368 (1974).
18. G. R. Luckhurst, *Q. Rev. (Lond.)* **22,** 179 (1968).
19. L. T. Creagh and P. Truitt, *J. Org. Chem.* **33,** 2956 (1968).
20. E. T. Strom, B. S. Snowden, Jr., H. C. Custard, D. E. Woessner, and J. R. Norton, *J. Org. Chem.* **33,** 2555 (1968).
21. W. F. Reynolds and T. Schaeffer, *Can. J. Chem.*, **41,** 540 (1963).

5

CHEMICAL SHIFT—QUANTITATIVE CONSIDERATIONS

5.1 FACTORS RESPONSIBLE FOR CHEMICAL SHIFTS

The relationship interrelating the frequency (ν) at which a particular magnetically active nucleus will resonate when placed into a magnetic field, H_0, is given by equation 5.1:

$$\nu = \frac{\gamma H_0}{2\pi} \tag{5.1}$$

If a nucleus placed into the magnetic field is assumed to be *bare*—that is, stripped of its electrons—the magnetic field experienced by it is identical to the applied field H_0. However, an isolated atom experiences a local magnetic field different from the *applied* field because of the shielding caused by the surrounding electrons. The extent of this shielding is expressed by equation 5.2:

$$H = H_0(1 - \sigma) \tag{5.2}$$

In this expression σ is referred to as the "shielding parameter" or, more commonly, the "screening constant."

When an atom is placed into a magnetic field, both its electron cloud and its nucleus are subject to the *Larmor precession*. Lenz's law states that the magnetic field created by the motion of the diamagnetic electrons opposes the applied magnetic field. It is this opposing field that gives rise to the shielding phenomenon, and is maximum when the electron cloud is *symmetrical*. This is actually only correct for free atoms (e.g., Li) and some tetrahedrally symmetrical molecules (e.g., NH_4^+). The diamagnetic contribution to the shielding can be computed by application of the Lamb [1] equation (5.3),

$$\sigma = \frac{4e^2}{3mc^2}\int_0^\infty r\rho(r)\,dr \tag{5.3}$$

which establishes that the atomic screening constant is proportional to the electrostatic potential energy between the electrons and the nucleus being studied. To counteract this shielding effect, the applied magnetic field, or frequency, must be increased to cause the resonance phenomenon to occur!

Lamb's equation is totally valid only for atoms in *s* states—that is, atoms with zero angular momentum—and may not be at all adequate for atoms in a molecular environment. The numerical sizes of the *screening* constants of different atoms are related to their atomic number and vary from 1.8×10^{-5} for hydrogen and 1.42×10^{-3} for a fluorine atoms to 8.82×10^{-3} for thallium atoms. Within the recognized limitations, these constants are of qualitative use when applied to computations involving simple molecular systems.

Since the calculation of screening constants for atoms in molecular systems is mathematically quite complex, it is customary to divide the screening effects, artificially, into four different categories:

1. Atom-centered *diamagnetic* effects
2. Atom-centered *paramagnetic* effects
3. Screening contributions caused by neighboring atoms
4. Screening contribution resulting from interatomic currents

The diamagnetic screening effect is exemplified by the difference in chemical shifts between a *proton* and a *hydrogen atom*. To achieve resonance of the latter, a stronger field will be required than for a proton, since the electrons surrounding a hydrogen atom will shield the proton nucleus.

The precessional motion of the electrons that cause the diamagnetic shielding is impeded in unsymmetrical molecules and alters the shielding in a negative (opposite to the diamagnetic effect) manner. This opposite effect is referred to as "paramagnetic" shielding. The diamagnetic as well as paramagnetic effects operating within the magnetic sphere of a particular atom will also affect neighboring atoms, giving rise to *contribution 3*, as described above.

In those molecules where bonding involves π electrons, interatomic currents become of significance, since the various magnetic effects can then be transmitted by the electrons that flow from one atom to another. These *ring-current effects* are important contributors to the chemical shifts of aromatic protons and will be discussed in some detail [2].

5.1.1. Local Diamagnetic Effects (The Lamb Term)

Since protons have only *s* electrons, the surrounding electronic cloud is reasonably symmetrical, and the diamagnetic contribution, which rarely exceeds 100 ppm, dominates the "screening constant." A pictorial representation of this effect is presented in Figure 5.1.

Bailey and Shoolery [3] have studied the changes in the proton chemical shifts of a series of substituted methanes and ethanes and analyzed the results on the reasonable assumption that, in a series of closely related molecules, the paramagnetic contribution in a C–H grouping will be zero and that the neighboring atom contribution in this arrangement will be constant. Additionally, in the absence of any π electrons, any interatomic current contributions will be zero. Since the diamagnetic effect arises largely from the bonding electron during the time that it occupies an atomic *s* orbital on a proton, a correlation should exist between the electron-withdrawing power, the well-known *inductive effect*, of a given functional group (X) and the proton bonded to the carbon bearing this substituent (X–C–H). To minimize possible solvent and/or other effects, Bailey and Shoolery employed

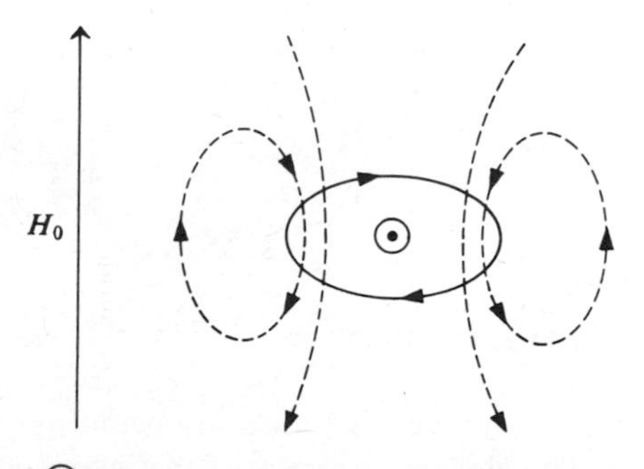

Figure 5.1 Diamagnetic shielding of a nucleus.

the proton chemical shifts of the methyl group in a number of substituted ethane derivatives as an internal reference standard.

The data listed in Table 5.1 demonstrate the very satisfactory relationship that exists between the proton chemical shifts of a number of substituted ethanes and the Huggins [4] electronegativity values. The correlation line, which permits the calculation of electronegativities of substituents, is

$$\text{Electronegativity} = 0.02315(\Delta CH_3 - \Delta CH_2) + 1.71$$

This correlation shows that as the inductive effect of a neighboring electronegative atom increases, the secondary field around an adjacent proton decreases. Consequently, the proton will resonate at a lower applied field (become more deshielded). This relationship also explains the observation that the protons of a methoxyl group resonate at a lower applied field than do those of a methyl group bonded to nitrogen and that the protons of a C–CH_3 group are the most shielded of these three types of compounds.

Musher [5] has described an approach that correlates, semiquantitatively, the linear variation of the nuclear magnetic shielding with the electric field generated by a given substituent. However, discussion of the Bailey and Shoolery correlation is more than adequate to demonstrate the effects of the diamagnetic shielding on proton chemical shifts.

Isolated nuclei—that is, nonbonded species such as Na^+, for example—have associated with them nearly symmetrical electron clouds, and, as a result, the diamagnetic contribution predominates in these instances.

In covalently bonded atoms, involving nuclei such as C^{13}, N^{14}, O^{17}, F^{19}, P^{31}, and all other nonhydrogen nuclei of interest, the presence of *p* and *d* electrons causes large deviations from spherical symmetry, and, as a result, the diamagnetic shielding effect is overpowered by the paramagnetic one [6].

TABLE 5.1 Electronegativities of Some Substituent Groups[a]

Group (X)	$(\Delta CH_3 - \Delta CH_2)$[b] for $CH_3CH_2{-}X$	Calculated electro-negativity	Electro-negativity of X
$-CN$	35	2.52	2.6
$-COOH$	37	2.57	2.6
$-I$	42	2.68	2.65
$-Br$	53	2.94	2.95
$-NH_2$	55	2.99	3.05
$-Cl$	64	3.19	3.15
$-OH$	77	3.51	3.5
$-NO_2$	95	3.91	3.5
$-F$	96	3.93	3.90

[a]Another such correlation table developed by J. C. Muller is reported in *Bull. Soc. Chim. France*, (1964) 1815.
[b]Units are cps.

It has been computed, for example, that only 1%, or 29 ppm, of the chemical shift range (F^- to F_2) in fluorine compounds is due to the diamagnetic effect; the remainder, some 2000 ppm, is due to other causes.

5.1.2 Paramagnetic Effects

The dominance of the paramagnetic effect on the shielding constant of all nonhydrogen and nonisolated ionic nuclei is largely responsible for the width of many hundreds of ppm observed for the chemical shift ranges of these nuclei. Because of this dominance, simple electronegativity considerations are generally obliterated.

Adequate correlations exist between this effect and the ionic character of a fluorine atom bonded to another molecule [7]. In the spherically symmetrical fluoride ion the paramagnetic contribution approaches zero, whereas in a totally covalently bonded fluorine the shielding contribution is at a maximum for this nucleus. This effect, calculated to be -20×10^{-4}, compares favorably with the observed difference between the fluorine chemical shifts in F_2 and the partial covalent bond in HF. In carbon, nitrogen, phosphorus, and most other magnetically active nuclei it is this, the "paramagnetic contribution," that dominates the chemical shift. It arises from contributions of higher electronic states (especially low-level excited states) to the description of the *ground*-state. Proportionality 5.4 expresses the three structurally related contributions:

$$\sigma^{p}_{\text{loc}} \propto \frac{1}{\Delta E} \times \frac{1}{r^3} \times \sum Q \tag{5.4}$$

In this expression ΔE is related to the electronic excitation energy between the *ground state* and appropriately weighted excited states. It is a measure of the availability of low-lying excited states. The $1/r^3$ term is the average inverse cube of the orbital radius (not *s*) describing electrons influencing the chemical shift. The Q summation term is derived from the charge densities and the bond orders of the bonding electrons; it represents a measure of multiple bonding to the nucleus being studied.

These considerations make it apparent that a particular nucleus is deshielded when all or some of the following conditions exist:

1. Low-lying states are available involving electrons at the nucleus being studied.
2. Electrons are in orbitals with larger *s* character (smaller radius!).
3. Multiple bonding to the nucleus exists (larger Q).

For example, since the paramagnetic contribution is approximately inversely proportional to the electronic excitation energy, magnetically active nuclei in organic compounds with low-lying excited states, as found in olefinic and aromatic systems, should be less screened than the same type of nuclei in saturated molecules. This is indeed the case. For example, both the C^{13} and hydrogen atoms in olefinic compounds resonate at more deshielded positions (at lower field) than do these atoms in saturated molecules.

Acetone has a low-lying electronic state arising from an n to π^* transition and, as anticipated, the O^{17} in acetone resonates at a lower field than does the oxygen in an alcohol, for example.

The lower-lying resonance position of the nitrogen in aniline versus a dialkyl amine is attributable to the presence of n to π^* transitions in the aromatic compound and an increased π bond character to nitrogen (increased Q).

The broad applicability of the concept of paramagnetic shielding is demonstrated by the linear relationship that exists between the absorption frequencies of the lowest-lying optical absorption maxima and the response frequencies of a series of cobalt complexes (the Co^{59} isotope, with a $\frac{7}{2}$ spin, is the only naturally occurring cobalt isotope) [8].

5.13 Contributions from Neighboring Atoms

The diamagnetic as well as paramagnetic effects of a nucleus can also affect the shielding constant of an atom bonded to it. This generated *secondary* magnetic field will, of course, affect the shielding constant of the nucleus to which the atom is bonded. The effect that the secondary field of a particular nucleus (X) has on an atom bonded to it will depend on the geometry of the situation. Figure 5.2 is a schematic representation of the two geometric possibilities.

The secondary magnetic field experienced by atom H when it is parallel with the applied field is opposite to it. Consequently this atom will experience additional shielding. If atom H is situated *perpendicular* to the applied field, it experiences an enhanced magnetic field and consequently will resonate at a more deshielded position. In a general sense, these considerations lead to the conclusion that, if atom X has a greater diamagnetic susceptibility along the X–H axis than it does

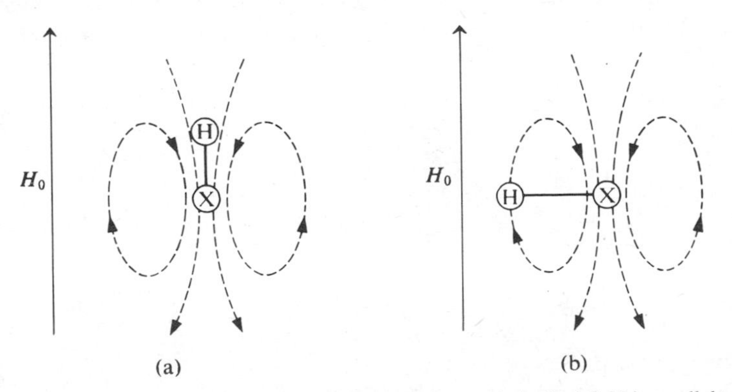

Figure 5.2 Diamagnetic contributions from neighboring atoms. (a) Applied field is parallel to the bond axis (shielding effect). (b) Applied field is perpendicular to the bond axis (deshielding effect).

perpendicular to it, atom H will resonate at a more shielded position than it would if the perpendicular axes were preferred.

These effects, referred to as neighbor anisotropic contributions, do not operate through chemical bonds, as is the case with the local diamagnetic and paramagnetic effects, but rather are transmitted through space. This anisotropic contribution is dependent only on the spatial relationship between the atom being studied and atom X. The effect is essentially independent of the nature of the atom being examined!

The acetylenic protons resonate at rather shielded positions (see Chapter 3) when compared to the much more deshielded olefinic protons. This unexpected behavior is a result of the anisotropy phenomenon. The axially symmetrical acetylene can be envisioned to line up parallel with the applied field, as pictorially represented in Figure 5.3.

Since the electron distribution is axially symmetrical about the molecular axis, there will be no paramagnetic contribution to the shielding if the applied field is parallel to the axis. If this is not the case, the paramagnetic contribution will be large. The significant shielding of the acetylenic protons is in agreement with the suggestion that the magnetic susceptibility of the acetylenic linkage is greater along the carbon–carbon axis than it is perpendicular to it. Thus, the secondary field caused by the symmetric acetylenic bond electrons will cause shielding of the acetylenic protons [9, 10]. An extension of this argument suggests that any nucleus situated above or below the acetylenic bond would experience a deshielding effect. This has been experimentally confirmed [3].

Since C^{13}, N^{14}, and most other nuclei chemical shifts occur over a wide range, the numerically small anisotropic effect in NMR spectroscopy is largely limited to proton work.

It is customary to indicate the shielding (+) and the deshielding (−) zones associated with various structural features by drawing cones around them that indicate the variation of these shielding effects with distance. The shielding and deshielding zones for a triple bond can then be represented as shown in Fig. 5.4.

The susceptibility axis in the nonaxially symmetrical olefin, carbonyl, imine, and alkane bonds is such that the functional groups align perpendicular to any applied field in the manner shown in Figure 5.5. As a result, olefinic protons, and

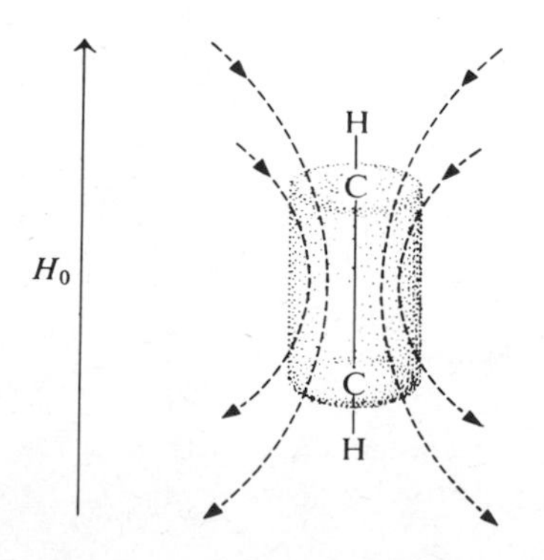

Figure 5.3 Diamagntic shielding of acetylenic protons.

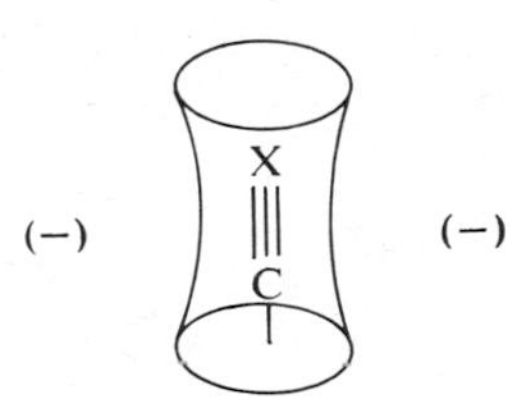

Figure 5.4 Shielding "cone" associated with triply-bonded functional groups.

other substituents on these functional groups, will be subject to ansiotropic deshielding. Figure 5.6 is a graphic representation of the shielding and deshielding zones in the various functional groups under discussion.

5.1.4 Interatomic Currents

Pauling [11] suggested some years ago that the six π electrons in benzene, when subjected to a magnetic field, will precess in a plane perpendicular to this applied field so that their angular frequency (ω) will be as expressed in equation 5.4:

$$\omega = eH_0/2mc \tag{5.4}$$

The current that is generated may be considered as flowing in a ring having a radius equal to that of the benzene ring. Equation 5.5 expresses this relationship:

$$I = -ne^2H_0/4mc\pi \tag{5.5}$$

The number of π electrons in the system is n, and e is the charge on each atom. The direction of the generated current leads to a diamagnetic moment which is opposed to the applied field. Figure 5.7 indicates this ring current effect as it occurs in benzene.

It is evident that any nucleus situated within the plane of the benzene ring, like

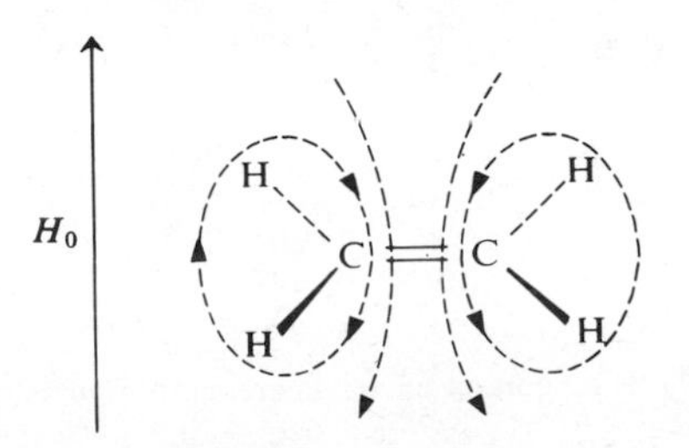

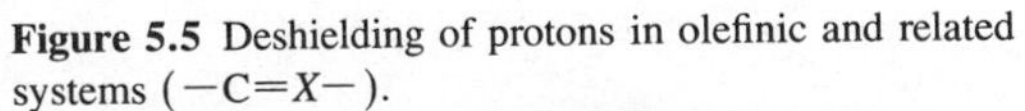

Figure 5.5 Deshielding of protons in olefinic and related systems (—C=X—).

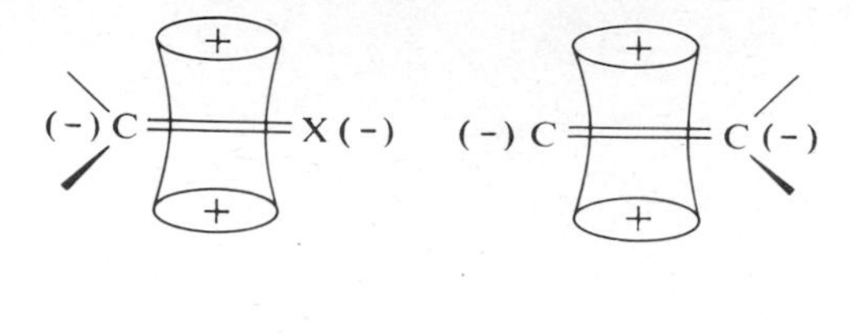

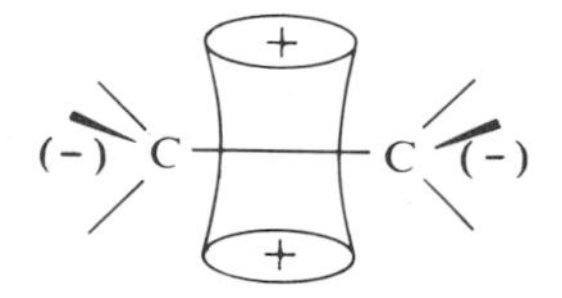

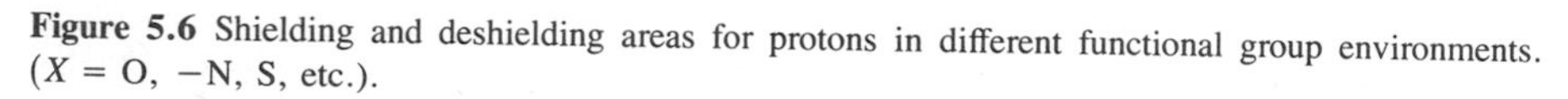

Figure 5.6 Shielding and deshielding areas for protons in different functional group environments. (X = O, −N, S, etc.).

the protons in benzene itself, will reinforce the applied magnetic field and will deduct from the contribution to the screening constant, thus causing a deshielding effect. On the other hand, any nucleus situated either above or below the plane of the benzene ring will experience a shielding effect (detract from the applied magnetic field). Clearly, a region in space will also exist where these two opposing effects cancel each other, and any nucleus located in this region will be neither shielded nor deshielded as a result of the ring current effect.

An estimate of these effects can be obtained if the current I is replaced by a magnetic dipole $3e^2 H_0 a^2/2mc^2$ at the center of the ring (a = radius of the ring) and perpendicular to it [2]. The distance between the nucleus subjected to the ring current and the center of the ring is expressed by R. Inclusions of these considerations lead to the following expression for the ring current contribution to the screening constant of the effected nucleus:

$$\Delta\sigma = -\frac{e^2 a^2}{2mc^2 R^3}$$

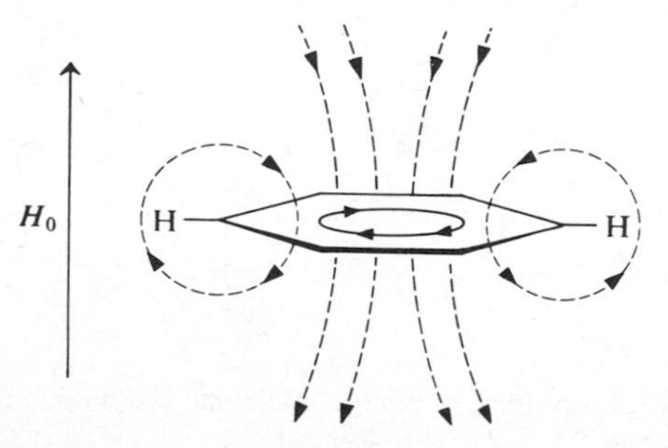

Figure 5.7 Ring current effects in aromatic systems.

This expression, with $R = 2.5 \times 10^{-8}$ and $a = 1.4 \times 10^{-8}$, gives $\Delta\sigma = -1.75 \times 10^{-6}$ for the ring current contribution to the deshielding of the protons in benzene.

If one of the protons were replaced with another magnetically active nucleus (e.g., N^{15} or C^{13}), the same deshielding effect would operate on these nuclei. However, proportionally, the ring current effect would contribute much less to the total screening constant of these nuclei (the paramagnetic effect is numerically much greater in these instances), and as a result it would affect their chemical shifts in a minor way only.

The existence of this type of diamagnetic circulation of electrons in benzenoid systems is observed for all molecules that obey the Hückel rule—that is, molecules with $4n + 2\pi$ electrons in a planar, circular, and conjugated spatial arrangement. It has also been shown that a paramagnetic circulation of electrons is generated in molecules where $4n\pi$ electrons exist in a circular and conjugated arrangement. In these instances, nuclei above and below the plane of the ring are deshielded, and those situated along the plane of the ring are shielded, the reverse of the effects in aromatic systems.

An example of the paramagnetic ring current effect is the observation that the inner protons in [16]annulene resonate at δ 10.3, whereas the outer ones do at 5.2.

To determine the ring current contribution experimentally, an appropriate reference system has to be chosen. The ideal one would be a benzenoid ring without a ring current. This is, of course, not feasible. The best non-ring-current-influenced olefinic systems available are cyclooctatetraene and cycloctatriene, with olefinic hydrogen chemical shifts of δ 5.67 and 5.74 ppm, respectively. Although, these molecules are not planar, they can serve as reasonably approximate references. The difference between the benzene proton chemical shifts (δ_H 7.27) and the average of these olefinic proton chemical shifts (δ_H 5.72) is 1.55 ppm and represents the experimental estimate of the ring current effect on an aromatic proton or any other magnetically active nucleus. This value compares reasonably well with the 1.75 ppm contribution, computed by means of the Pople approximation of the ring current model.

The dipole approximation, although adequate for the estimation of ring current effects, suffers from the fact that the electron density in benzenoid systems is not located in the center of the ring but rather is in the well-known, *doughnut*-shaped π clouds, above and below the carbon skeleton of the benzene ring.

Any attempt to calculate ring current effects that include this consideration requires that the question of the distance between the π electron clouds be known. Johnson and Bovey [13] have estimated this separation to be 1.28 Å and have, with the aid of equation 5.6, developed by them and by Waugh and Fessenden [14], calculated the ring current effects experienced by nuclei located at different positions in space with respect to the benzene ring

$$(5.6) \quad \delta' \times 10^{-6} = \frac{ne^2}{6mc^2a} \times \frac{1}{[(1+\rho)^2 + z^2]^{1/2}} \times \left[K + \frac{1 - \rho^2 - z^2}{(1-\rho)^2 + z^2} \right] E$$

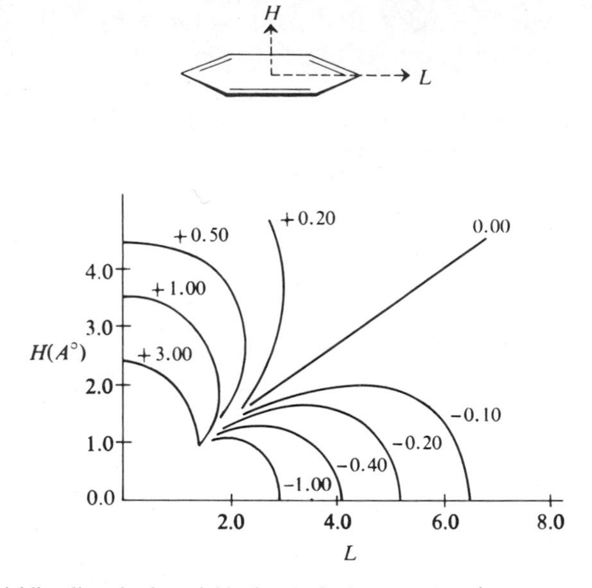

Figure 5.8 Isoshielding lines in the neighborhood of a benzene ring. (H = location of proton above benzene ring, and L = location in the plane of the benzene ring (Å units); to convert the shifts into cps they must be multiplied by MHz of spectrometer ($-0.10 = -6$ cps at 60 MHz).) [Reprinted with permission from C. E. Johnson and F. A. Bovey, *J. Chem. Phys.*, **29**, 1012 (1958). Copyright (1958) American Chemical Society.

Figures 5.8 and 5.9 are the graphically presented results of these computations. The curve for $\delta' = 0$ (Fig. 5.8) is the cross section of the nodal surface that separates the diamagnetic region (δ' = positive) from the paramagnetic one (δ' = negative). Clearly, the diamagnetic region covers a greater area than the paramagnetic one. This is in agreement with the results obtained from benzene-induced solvent shifts, where the diamagnetic solvent effects greatly outweigh the paramagnetic ones. Application of the Johnson and Bovey model to a number of aromatic systems (see Table 5.2) demonstrates the validity of this approach.

5.1.5 Van der Waals and Electric Dipole Shielding Effects

The deshielding effect caused by inductive electron withdrawal of a substituent decreases rapidly as the number of bonds between the substituent and the nucleus being examined increases. The contribution becomes effectively negligible if more than three bonds are involved.

If the distance between a rigidly held substituent that is not directly bonded to a resonating nucleus is less than the sum of the Van der Waals radii of two entities, the substituent atom will repel electrons from the vicinity of the resonating nucleus. The net effect of this interaction is, generally, a decrease in the diamagnetic shield-

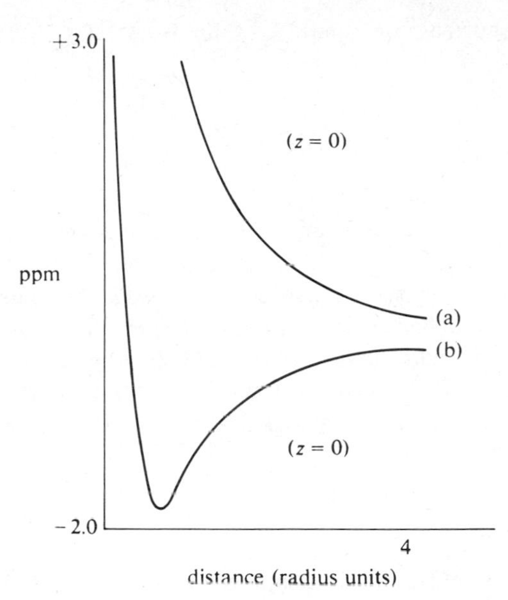

Figure 5.9 Chemical shifts of protons in the neighborhood of a benzene ring. (a) Ppm as a function of the distance along the hexagonal benzene axis. (b) Ppm as a function of the distance in the plane of the ring, measured from its center. [Reprinted with permission from C. E. Johnson and F. A. Bovey, *J. Chem. Phys.*, **29,** 1012 (1958). Copyright (1958) American Chemical Society.

ing contribution. Consequently, the resonating nucleus will become more deshielded. This phenomenon, a result of the mutual repulsion of the induced dipoles, decreases rapidly with increasing distance (r) between the two interacting centers. It depends to a great extent on the size and polarizability of the nuclei.

The degree of Van der Waals shielding between a proton and a group X ($H \dots X$) can be computed by application of equation 5.7, where Q is a measure of the

TABLE 5.2 Observed and Calculated H Shifts of Some Aromatic Compounds

		Chemical Shift	
Compound	Group	Observed	Calculated
Toluene	$-CH_3$	2.34	2.48
Cumene	$-CH_3$	1.23	1.21
Tetraline			
Alpha H	$-CH_2-$	2.70	2.66
Beta H	$-CH_2-$	1.78	1.93
Dibenzyl	$-CH_2-$	2.95	2.89
Naphthalene			
Alpha H	$-CH=$	7.73	7.97
Beta H	$-CH=$	7.37	7.54

effectiveness of the nucleus, or group, X (1 for H; 3.1 for F; 14 for Br) in deshielding the proton

$$\Delta\sigma_W = -\frac{60Q}{r^6} \tag{5.7}$$

An example of the occurrence of this effect is found in the observation that the *tert*.-butyl protons in *ortho*-di-*t*-butylbenzene experience a downfield shift of 0.2 ppm with respect to the *t*-butyl resonances of the *meta*- and *para*-isomers [15]. The larger chemical shift changes in F NMR tend to magnify these effects and are responsible for the large number of electric field shielding effects observed in F NMR. For example, the F chemical shift in *ortho*-chlorofluorobenzene is 20 ppm more deshielded than expected from a consideration of the chlorine inductive effect alone.

5.1.6 Shielding Effect Summary

The shielding and deshielding effect contributions to the shielding constant of a particular nucleus depend on the induced magnetic field strengths. The contributing components are the diamagnetic and paramagnetic shielding effects and a term composed of the contributions due to the magnetic susceptibilities of neighboring groups and atoms.

While the diamagnetic term is generated by the electrons of the nucleus being studied, the paramagnetic effect arises largely from contributions of low-level excited states to the ground state of the nucleus. The electron circulations on neighboring atoms and the spatial field effects are the major contributors to the anisotropic effects.

Where differences in hybridization of the nucleus are possible, this effect must also be kept in mind. For example, sp^3-hybridized carbons resonate at higher fields than sp^2-hybridized ones, with sp carbons at intermediate fields. All of these effects are summarized in expression 5.8 proposed by Saika and Slichter [11] and using the notations developed by Pople [12]

$$\sigma_A = \sigma_{\text{diam.}}{}^{AA} + \sigma_{\text{param.}}{}^{AA} + \sum_{B \neq A} \sigma^{AB} + \sigma^{A,\text{ring}} \tag{5.8}$$

The shielding constant operating on an atom, A, is thus composed of the diamagnetic and paramagnetic contributions caused by the electrons of the atom in question, the sum of the electronic circulation effects of atom or atoms, B on atom A, and any contributions caused by interatomic ring currents.

5.2 H^1, C^{13}, AND N^{15} CHEMICAL SHIFTS IN AROMATIC SYSTEMS

The various contributions to the shielding of a particular nucleus, as shown in expression 5.8, are of differing significance in carbocyclic aromatic compounds, depending on whether H or C^{13} NMR spectra are being examined. In heteroaromatic systems the effects on the chemical shifts of the heteroatoms can be expected to be different, as well.

5.2.1 Proton Chemical Shifts

The theoretical ring current contributions to the proton chemical shifts of a number of different aromatic and heteroaromatic ring systems have been calculated [16] and are listed in Table 5.3. These data show that for geometrically closely related aromatic ring systems, the ring current contribution to the shielding constant of the various protons in rings of equal size is essentially the same and is independent of the extent of heteroatom substitution (this does *not* mean that the heteroatoms are not effecting the chemical shifts of the ring protons!). The differences between the chemical shifts of the protons of any aromatic compound being studied and those of benzene itself are, consequently, due to differences in the diamagnetic and/or paramagnetic contributions to the total shielding constant.

TABLE 5.3 Ring Current Contributions to the Proton Chemical Shifts in Some Aromatic Compounds

Compound	δ (ppm)[a]	δ (ppm) (corrected)
Monocyclics		
benzene	−2.58	−1.55
pyridine	−2.48	−1.49
polyazabenzenes (2 to 4 N atoms)	−2.46 to −2.48	−1.48 to −1.49
pyrrole and aza derivatives	−1.36 to −1.40	−0.82 to −0.84
furan and aza derivatives	−1.20 to −1.26	−0.72 to −0.76
Bicyclics		
naphthalene and aza derivatives	−3.18 to −3.23 (α protons)	−1.91 to −1.94
	−2.84 to −2.88 (β protons)	−1.70 to −1.73

SOURCE: P. J. Black, R. D. Brown, and M. L. Heffernan, *Australian J. Chem.*, 20 (1967), 1305. Reprinted with permission of CSIRO, publishers.

[a]These data are self-consistent with respect to each other. To bring them into numerical agreement with the proton chemical shifts of benzene, we must multiply them by the empirical factor −1.55/−2.58. D. W. Davies, *Chemical Communications*, (1965) 84, and J. A. Elvidge, *Chemical Communications* (1965), 160, have derived somewhat different values for the ring-current contribution for 5-membered ring heterocyclic compounds.

Any or all of the following can contribute to the shielding constant of a particular aromatic proton:

1. The polarity of the C–H bond, representing the amount of *s* character that the p_X carbon orbital will contribute to the σ^{AB} contribution terms. One of the principal factors that contributes to this effect is the ring-size of the aromatic compound.

2. Changes in the magnetic polarizability of C–C bonds in the aromatic ring. In aromatic systems, this contribution is negligible since π electronic changes outweigh these minor effects.
3. Changes in the π electron densities of the carbon atom bearing the proton under study are expected to alter the local diamagnetic shielding contribution in a proportional manner.
4. The replacement of an aromatic carbon by a heteroatom (e.g., benzene changed to pyridine) will cause polarization changes of the sigma-bond dipoles. While no totally reliable figure is available for this effect, experimentally determined estimates account for this contribution in a consistent manner.

Among all of these effects, only the π electron changes, effect 3, remain as a prime variable as long as the relationships being developed deal with related aromatic systems only [16, 17].

5.2.1 Monocyclic Compounds

The relationship between the proton chemical shifts and the electron densities of related aromatic systems can be expressed by equation 5.9.

$$\delta(\text{ppm}) = k\Delta\rho \tag{5.9}$$

For benzene itself $\Delta\rho = 0$, since all of its carbon atoms have a π electron density of 1. Schaeffer and Schneider [17] have adopted the convention of designating $\Delta\rho$ as positive for protons appearing at higher fields (more shielded) than benzene protons. Conversely, protons resonating at lower fields (less shielded) than benzene protons will be designated by negative values. These conventions cause k to be a positive constant, and an electron-deficient carbon atom will be represented by a negative $\Delta\rho$ value (with respect to the protons in benzene itself).

The evaluation of the constant, k, can be accomplished by two different means. If the $\Delta\rho$'s are known for a number of derivatives of a particular system, the proton chemical shifts can be determined and the value of k calculated.

An alternate method is based on the chemical shifts of the protons in a positively and a negatively charged aromatic ring system. The total π electron densities are one electron less and one electron more with respect to the corresponding neutral benzene ring. Consequently, the sum of the chemical shifts of all of these protons, when measured relative to benzene, corresponds to $\Delta\rho = 1$, and the chemical shift of the proton will be equivalent to the value of k.

The measured chemical shifts, corrected for changes in ring size (-0.10 and $+0.07$ ppm, respectively), are $+1.73$ and -1.98 ppm, respectively, for the ions $C_5H_5^-$ and $C_7H_7^+$. These values are equivalent to $+0.20$ (6 electrons distributed

TABLE 5.4 Proton Chemical Shifts and Electron Densities of Some Aromatic Compounds

Compound	Position	δ (ppm) (relative to benzene)	ρ^a experimental	ρ^a (HMO) calculated
NH_2 (aniline; positions 2, 3, 4)	2	+0.75	1.07	1.08
	3	+0.20	1.02	1.00
	4	+0.62	1.06	1.06
OCH_3 (anisole; positions 2, 3, 4)	2	+0.42	1.04	1.05[b]
	3	+0.04	1.00	0.97
	4	+0.37	1.04	1.04
Pyridinium (N^+–H; positions 2, 3, 4)	2	−1.56	0.85	0.90[c]
	3	−0.91	0.92	0.93
	4	−1.48	0.86	0.83

[a] π-electron densities.
[b] HMO calculations from the author's laboratory.
[c] VESCF calculations.

over 5 carbons) and −0.143 ppm (6 electrons distributed over 7 carbons), respectively. The generally accepted value for k, obtained from these and related computations, is 11.0 ± 1.0 (ppm/electron).

Table 5.4 lists the electron densities computed by molecular orbital computations and compares them with those obtained by applying equation 5.9 to the experimentally determined chemical shifts of the protons in a number of aromatic compounds. It is encouraging that this technique can be used to determine π electron densities of these and other aromatic ring systems.

5.2.1.2 Bicyclic Compounds

When dealing with bicyclic aromatic compounds, the same approach used for monocyclic aromatics can be employed. The reference compound in this instance is naphthalene, with two different aromatic protons, which must be considered in any computations.

Consideration of the ring current effect of ring 2 on the protons of ring 1 in naphthalene, a relationship based on the dipole model, discussed earlier, generates equation 5.10,

$$\delta(\text{ppm}) = 12.0(a^2/R^3) \tag{5.10}$$

In this expression, a is the radius of the neighboring ring, and R is the distance of

TABLE 5.5

	Position	δ (ppm) relative to benzene	δ^a	$\delta_{cor.}$	$\rho_{exp.}$	$\rho_{(HMO)}$
	2	−0.58	+0.29	−0.29	0.97	0.98
	1,3	−0.06	+0.55	+0.49	1.05	1.09
	4,8	−0.94	+0.52	−0.43	0.96	0.88
	5,7	+0.28	+0.17	+0.45	1.04	1.05
	6	−0.17	+0.12	−0.05	0.99	0.95

[a] This is the ring-current correction term.

the proton being studied from the center of the second ring. While this approximation is reasonably reliable for bicyclic systems, it is much less so for tricyclic and larger ones.

When this technique is applied to the computation of the proton chemical shifts of azulene, the data presented in Table 5.5 are obtained. Recognizing the number of approximations that have been made in developing equation 5.10, the agreement between experimental and theoretical chemical shifts is quite remarkable.

5.2.1.3 *Heteroaromatic Compounds*

The third term in equation 5.8, the contribution from the electronic circulations on atom B to atom A, is significant only when atom B or group B is magnetically anisotropic—that is, when, in an organic molecule, a carbon that is part of the backbone molecular skeleton is replaced by an atom other than carbon. It has been shown that this becomes important only where the replacement atom has a lone pair of electrons, as is the case for a nitrogen atom.

The contribution of an sp^2 nitrogen atom to a proton bonded to an adjacent carbon (HCN—) has been estimated to be in the range −0.20 to −0.33 ppm [18–23].

When a value of −0.33 ppm for the nitrogen anisotropic effect is included in the calculations for pyridine, the results given in Table 5.6 are obtained. In this instance and in similar situations as delineated in articles by the author referenced

TABLE 5.6 Computed and Experimental Chemical Shifts in Pyridine

	Position	δ (ppm) relative to benzene	$\rho_{exp.}$	SCF ρ	VESCF ρ
	2	−1.31	0.91	0.95	0.95
	3	+0.16	1.01	1.01	1.00
	4	−0.26	0.98	0.98	0.98

at the end of this chapter, a nitrogen anisotropic value of -0.33 ppm appears generally appropriate.

It is accepted that the effect an sp^2 nitrogen atom exerts on hydrogen atoms further removed than on an adjacent carbon is small enough to be neglected in these approximations.

There remains one additional significant nitrogen anisotropic contribution to be considered, the *peri* effect, as shown in the following generalized structure.

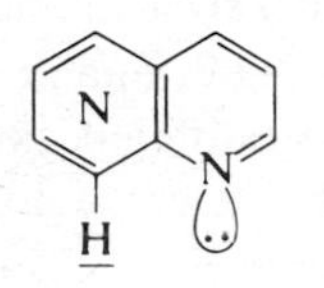

This anisotropic effect is estimated to be about -0.15 to -0.20 ppm [21].

In compounds with either a secondary or bridgehead-type nitrogen such as found in pyrrole (**1**) and indolizine (**2**) or an sp^2 atom that is protonated, as in pyridinium ions (**3**), the lone pair of electrons is either involved in the π system or is covalently bonded.

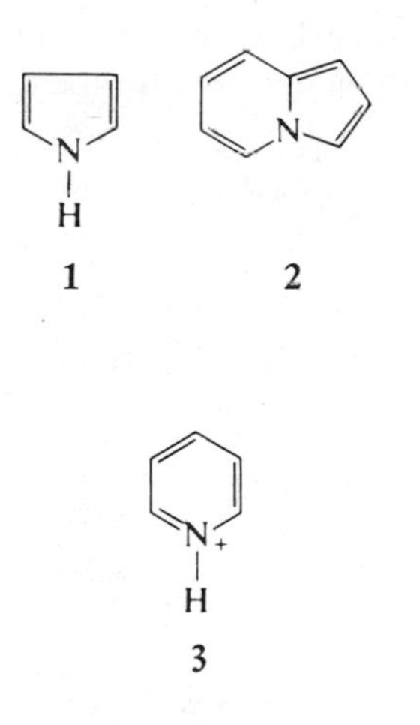

The anisotropic contribution of these nitrogens is essentially equivalent to that of a =C—H grouping.

5.2.2 Carbon Chemical Shifts

The C^{13} chemical shifts of aromatic compounds are subject to the same shielding effects as protons. However, in C NMR it is the paramagnetic term that dominates the chemical shift [24]. The size of this contribution is determined by the charge polarization, the variations in bond order, and the average excitation energy. The anisotropic contributions arising from the magnetic susceptibility of neighboring atoms and groups is minimal when compared to the size of the paramagnetic effect in C NMR.

Since the paramagnetic shielding effect is the dominant contributor to chemical shift changes in C NMR, it is not surprising that a plot of total charge density versus observed C^{13} chemical shifts for the *para* position of a number of benzene derivatives results in a linear correlation with a correlation coefficient of 0.95 [6] (see Figure 5.10).

A very significant correlation of C^{13} chemical shifts with aromatic π electron densities, obtained from simple Hückel calculations, has been described for positive as well as negative species containing 2, 6, and 10 π electrons in conjugated cyclic molecules. Since the chemical shift differences between these various species are rather large, it is not surprising that as crude a calculation as a simple Hückel one is sufficient to give reasonable correlations. This interrelationship does, however, point to the importance of π electron densities in C^{13} NMR spectroscopy [25] (see Figure 5.11).

Heteroaromatic compounds can be divided into *five-membered* and *six-membered* ring groupings. Among the five-membered ring heteroaromatic compounds, the carbon chemical shifts of the β carbons in furan and pyrrole resonate at 109–108 ppm, whereas the α carbons, situated adjacent to the heteroatom, resonate at 142 and 118 ppm, respectively. This chemical shift difference is essentially nil in thiophene, where both the α and β carbons resonate at 126 $\pm$ 1 ppm. These chemical shifts are close to those observed for the carbon atoms in benzene (128.5

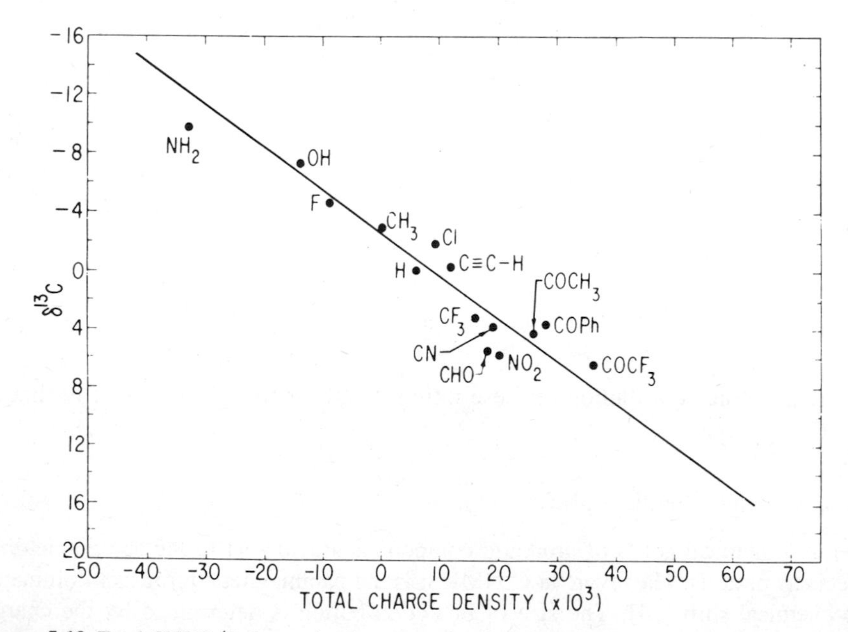

Figure 5.10 Total CNDO/2 charge density versus chemical shift for the *para* position of substituted benzenes. [ref. 6]

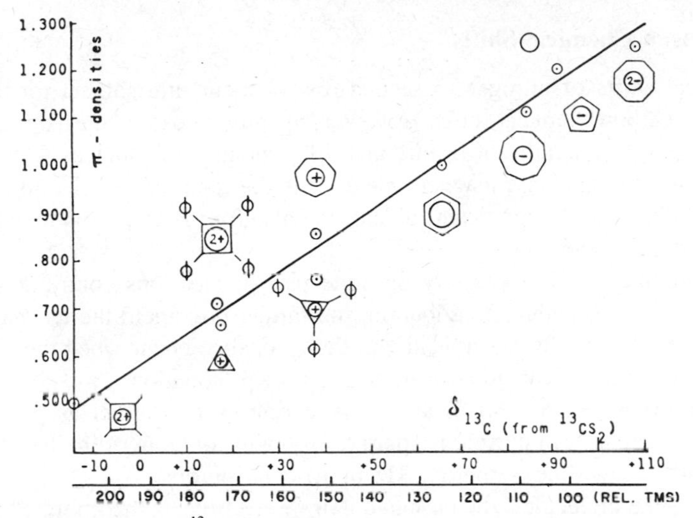

Figure 5.11 Correlation of ^{13}C chemical shift and π-electron density in some aromatic systems.

ppm). By comparison, the carbons of the aromatic five-membered ring carbocyclic analog, the cyclopentadienyl anion, resonate at 103 ppm.

On a qualitative basis, these data show, not surprisingly, that none of these heteroaromatic five-membered rings have a π-electron distribution as extensive as exists in the cyclopentadienyl carbanion. On the other hand, the similarity of the carbon chemical shifts in thiophene (125.4 and 127.2 ppm, respectively) with those in benzene (128.5 ppm) is supportive of the well-established fact that these two compounds have similar degrees of aromaticity.

Among the six-membered ring heteroaromatic systems, pyridine is the best-known and most-studied heterocyclic compound. The three different carbons in this molecule resonate at lower frequencies than those in the five-membered ring heteroaromatic systems. If one considers the various resonance-contributing structures in pyridine, it is expected that the β carbon will resonate at a position similar to that of the carbons in benzene; their chemical shift (124.8) is in agreement with this. The α and β carbons resonate at 150.6 and 136.4, respectively. Thus, the fact that these two positions are more electropositive than the β position is demonstrated by C NMR spectroscopy, as well. Substitution of any of the carbons by a nitrogen atom in pyridine causes an additional downfield (deshielding) shift of about 10 ppm at the alpha positions and approximately 7 ppm at the gamma position. A small shielding effect (approx. 3 ppm) is observed at the β positions. All of these effects are consistent with expectation and have been related to the usual electron density data. [26, 27].

In addition to the various electronic effects, C^{13} chemical shifts are also subject to steric effects, a process that is referred to as the *steric compression shift*. This steric effect is, in fact, a powerful mechanism available for structure elucidation studies and is demonstrated by the difference in chemical shifts between the methyl groups in *ortho*-xylene which resonate 1.9 ppm upfield from those in the *meta*-isomer.

5.2.3 Nitrogen Chemical Shifts

The chemical shifts of nitrogen, a second-row element, are subject to the same effects as the carbon chemical shifts. However, the presence of an unshared electron pair on nitrogen is a significant feature that differentiates this nucleus from carbon in terms of its behavior in a magnetic field. It is the lone pair of electrons that is responsible, not only for significant solvent, but also for pH effects on the chemical shifts of nitrogen nuclei.

In general, the more extensively the lone pair of electrons contributes to the local paramagnetic shielding contribution, the farther downfield the resonance position. Conversely, the further upfield will the resonance occur when the lone pair of electrons is removed by either quaternization or protonation.

Nitrogen atoms can be part of an aromatic ring system either as pyrrole-like (the N^{15} chemical shift in pyrrole is approx. 145 ppm) or as pyridine-like (the N^{15} chemical shift in pyridine is approx. 318 ppm) contributors.

In pyrrole-like molecules, the unshared pair of electrons is incorporated into the aromatic π system, and, consequently, the nitrogen NMR behavior (not the chemical shift values, *per se*) in these instances approaches that of the nitrogen in aniline. On the other hand, the lone pair of electrons in pyridine is orthogonal to the aromatic sextet of electrons, and the additional electronic transition leads to further deshielding of this type of nitrogen.

The nitrogen chemical shifts of pyrrole and its derivatives are considerably influenced by solvent interactions. Hydrogen-bonding solvents deshield the nitrogen considerably (up to 10 ppm and more in going from a chloroform solution to a solution in diemethylsulfoxide!). Consideration of resonance-contributing structures allows qualitative predictions regarding changes in chemical shifts of pyrrole-like nitrogens. For example, the ring nitrogen chemical shift in 3-nitropyrrole is 150 ppm more deshielded by 4 ppm than the nitrogen in pyrrole itself.

Not only was pyridine the first compound whose nitrogen chemical shift was studied in great detail, it remains the compound that provides the *backbone* for all theoretical studies dealing with of N^{15} NMR spectroscopy. Not only are pyridine chemical shifts very strongly solvent dependent, but the addition of small amounts of internal reference standards are known to cause changes in the nitrogen chemical shift of this compound. Pure pyridine in the gas phase is deshielded by 6.3 ppm relative to its liquid phase. In fact, any type of intermolecular interaction involving the pyridine nitrogen causes shielding. This behavior is consistent with the recognition of the importance of the $n \rightarrow \pi^*$ transition in influencing nitrogen chemical shifts.

Although the nitrogen chemical shifts in pyrroles are effected by methyl substituents (approx. 12 ppm shielding due to the presence of a methyl group in position 2 in indoles, for example), methyl group substitution has only a minimal influence in pyridine systems. The exception is a methyl group in the γ position (the position *para* to the nitrogen), where *hyperconjugation* has been invoked as a rationale [28].

Protonation of pyridine shields the nitrogen by approximately 100 ppm depending on the solvent, concentration, and nature of the counterion. The resonance position of the nitrogen in N-methylpyridinium salts (119.5 ppm) is relatively independent of the solvent, concentration, and nature of the counterion. The chemical shift of the nitrogen in pyridine N oxides (292 ppm) is at a higher field than it is in pyridine itself, a situation consistent with removal of the lone pair of electrons. However, the chemical shift change is attenuated because of the large electronegativity of the oxygen atom.

The shifts of pyridines substituted with conjugating or electron-polarizing substituents are consistent with expectations. Thus, conjugative electron donation by substituents in the *ortho* and *para* positions shield the nitrogens, whereas electron withdrawal deshields them.

5.3 AROMATICITY AND RING CURRENT

The description of a compound as either aromatic, anti-aromatic, or non-aromatic has been addressed in great depth ever since the aromaticity concept was initiated during the time of Kekule, Dewar, and others of that era. Quantum mechanically, Hückel's $(4n + 2)$ and Craig's (symmetry and π electron-based) rules have attempted to place these concepts on a quantitative basis. While it is relatively easy to define an aromatic compound when it is clearly a benzenoid structure, the definitions, when based on chemical reactivities only, become much less clearly defined.

In the earlier discussion of the concept of ring current anisotropy, it was already noted that this contribution is of considerable importance only when six π electrons are present in a planar conjugated system (i.e., an aromatic compound). This behavior of benzenoid compounds has led to the reasonable suggestion [29] that *a compound can be classified as aromatic if it can sustain an induced ring current.*

This qualitative definition has been placed on a somewhat more quantitative footing by relating the extent of the ring current of a particular compound to that found in the reference compound benzene. This approach permits a semiquantitative assignment of the percent of aromaticity of a particular compound with reference to benzene. The approach is exemplified by the determination of the aromaticity of N-methyl-2-pyridone (**4**).

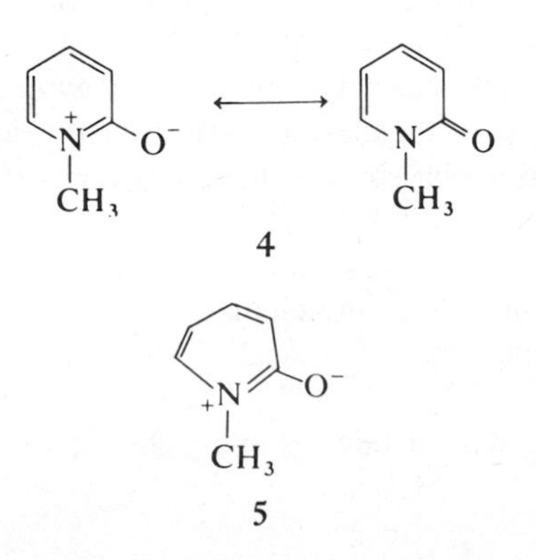

In principle, the proton chemical shifts of compound **4** should be compared with a structurellay identical molecule that does not sustain a ring current (in benzene calculations, the selected reference compound was cyclooctatetraene). This molecule would be represented by structure **5**, where nonconjugated single and double bonds are indicated as being present. Additionally, such a reference compound should also have the same electron densities as the real molecule (**4**). Again, recourse must be taken to a related molecule.

Since π electron density changes have little effect on the chemical shifts of methyl group protons attached to the carbon (e.g) a nitro group in either the *meta* or *para* position of a toluene causes a deshielding effect of only 0.09 ppm!), it is more appropriate to refer to the proton chemical shift of the methyl groups in appropriate derivatives of N-methyl-2-pyridone and suitably substituted reference compounds. Any differences between the proton chemical shifts of methyl derivatives of N-methyl-2-pyridone and the reference compounds will be due to the ring current effect in the pyridone compounds.

The ring current effect of the benzene ring on the chemical shift of the methyl protons in methylated benzenes has been shown to be a deshielding one of 0.40 ppm (the difference between the chemical shift of the methyl group protons in methylated polyolefins (1.98 ppm) and those of the toluene methyl protons (2.34 ppm)). Thus, the benzene molecule with 100% aromaticity has a 0.40 ppm deshielding effect on the protons of substituent methyl groups.

The observed chemical shifts for the methyl protons in 3-, 4-, and 5-methyl-1-methyl-2-pyridone are, respectively, 2.15, 2.11, 2.08 ppm. The difference between these values and those of the methyl groups of substituted polyolefins (1.94 ppm) are 0.21, 0.17, and 01.4 ppm, respectively. The average difference of 0.17 ppm corresponds to 44% (0.17/0.40) of that caused by the benzene ring current. Thus, it appears that 1-methyl-2-pyridone is 44% (probably $\pm$ 10%) as aromatic as benzene (this value is somewhat higher than that given in ref. [22] since slightly different reference methyl shifts were used in this approximation).

5.4 INORGANIC IONS

One of the fundamental considerations that go into any NMR experiment of any one nuclide concerns its sensitivity to the NMR technique. As already mentioned in Chapter 3, in a general sense, the greater the *gyromagnetic* ratio, the more sensitive is the nucleus to NMR analysis. Additionally, nuclei with spin quantum numbers $>\frac{1}{2}$ (e.g., quadrupolar species) have greater intrinsic nuclear sensitivities than the nuclei with smaller spin values.

All of these sensitivity-affecting factors are, of course, also subject to the abundance of the particular nuclide in the elemental isotope mixture. These considerations are of special importance when dealing with inorganic species, such as the spin $\frac{1}{2}$ nuclides of Tl^{205}, Sn^{119}, and Ag^{109}, which represent examples of very high, medium- and low-sensitivity nuclides. Among the nuclides with quadrupolar spin quantum numbers, Li^7, K^{39}, and Ca^{43} are good representative examples of nuclides of decreasing NMR sensitivity.

Quantitatively, the term receptivity expressed by

$$R_c = 9.45 \times 10^{-6} \times \nu_0 \times I \times (I + 1) \times P$$

combines the major contributors to the sensitivity of a magnetically active nuclide. In this expression, the value is computed relative to C^{13} for a field of 4.7 T. The unit P is the percent natural abundance, and ν_0 is the Larmor frequency of the nuclide being examined. Based on this expression, the sensitivity ratio of $C^{13}:Sn^{119}:Pb^{207}:Tl^{205}$ is 1:25:12:769!

NMR spectroscopy of metal nuclides differs significantly from that of the non-metal nuclides for a number of reasons, one of which is the large range of chemical shifts observed in many of the metal ones. While C^{13} spectra usually fall within a range of 200 ppm, the solvent-dependent differences of Tl^{205} species cover a range of close to 3000 ppm. The apparently largest range so far observed deals with C^{59}, which has a ligand-depending shift of 12,000 ppm [9, 10].

In going *across* the periodic table within one row, the chemical shift ranges found for different nuclei increase from left to right. For example, the Na^{23} shift range is about 50 ppm, and in Al^{27} it is 300 ppm and becomes 1000 ppm for Cl^{35} [11, 12].

There is also a reasonable relationship within any one group going down the periodic table. The solvent-dependent alkali metal ion chemical shifts for Li^7, Na^{23}, K^{39}, and Cs^{133} are, respectively, 5, 17, 30, and 130 ppm. The data presented in Table 2.3 serve to illustrate these relationships [13–14].

Among the main group elements, with spin quantum numbers other than zero, there appears to exist a fairly linear relationship between the local symmetry of the nuclide being studied and the chemical shift. For example, the aluminum nuclides in tetrahedral aluminum compounds resonate about 100 ppm downfield from the corresponding octahedral aluminum derivatives. Similarly, tetrahedral lead compounds resonate at higher field when solvent-coordinated [16–18].

Although these behavioral patterns cannot be accurately computed, it is possible to describe many of the trends in terms of three factors: the mean excitation energy, the radius of the nuclear sphere, and the local symmetry of the nuclide. Gutowsky and Jameson have analyzed these trends in chemical shifts in terms of the radial factor.

$$\sigma_p = \frac{2e^2\hbar^2}{3\Delta m^2c^2}\left(\left\langle\frac{1}{r^3}\right\rangle_p P_u + \left\langle\frac{1}{r^3}\right\rangle_d D_u\right)$$

The effect of the bonded neighbor atom on the chemical shift of a particular nuclide has also been successfully investigated. An electronegative neighboring atom or group increases the effective nuclear charge on the nuclide being studied and consequently increases both the radial term and the σ_p value. Consequently, as the neighbor atoms come further and further from the right of the periodic table, the nuclide chemical shift is at a lower field. In going down within a particular family, the *d* orbitals of the larger neighbor atoms can cause greater electron delocalization, causing a decrease in the radial factor, which, in turn, may be responsible for the high field shifts observed when a nuclide is bonded to a heavy neighbor atom such as iodide.

Although many of these considerations are only semiquanitatively correct, they nevertheless offer a better understanding of the NMR behavior of metal nuclides.

5.5 PROBLEMS

1. The difference in chemical shift between the methyl and methylene protons of a substituted ethane derivative is 24 cps. Estimate the electronegativity of the substituent.
2. Offer an explanation to account for the observation that cyclopropyl protons resonate at much more shielded positions than do the protons of cyclopentane.
3. Predict the chemical shift difference between the protons on C_3 and those on C_6 in nonyne.
4. How would you distinguish between C_6H_6 and $C_7H_7^+$ and **6** by means of proton and C^{13} NMR spectroscopy?
5. Estimate the ring current contribution to the deshielding of the protons in:

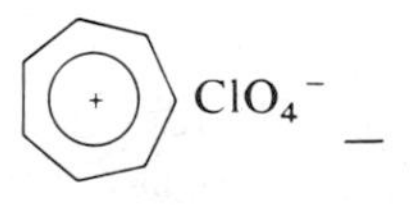

6. Predict whether the protons of the cyclopentadienyl carbanion are more or less shielded than those of the protons in benzene. Suggest reasons for this difference, and estimate its numerical value.
7. Predict the behavior of the carbon chemical shifts of the two compounds mentioned in problem 6.
8. The chemical shift of the inner protons in [18] annulene is -1.8 ppm, whereas that of the outer protons is 8.9 ppm. Explain!
9. The methylene protons bonded to the benzene ring in [5] paracyclophane resonate at 2.0 ppm, whereas those situated directly above the benzene ring resonate at -1.0 ppm. Using the Johnson–Bovey model, suggest the stereochemical arrangement between the benzene ring and the methylene groups in this molecule.
10. Would the carbon of the methylene proton that is situated above the benzene

ring in [5] paracyclophane be more or less shielded than the gamma carbon in butylbenzene? Explain.

11. Explain why the carbon chemical shifts are more readily related to the total electron densities than to the π electron densities.
12. Predict the change in chemical shift of the nitrogen in pyrimidine (1,3-diazabenzene) when the spectrum is obtained in aqueous trifluoroacetic acid compared to a solution in $CDCl_3$.
13. The N^{15} chemical shift in pyrrole appears at a considerably more shielded position than in pyridine. Explain!
14. A particular proton in a benzene derivative is located 2.5 Å above the plane of the benzene ring and 1.8 Å removed from the center of the ring. Calculate the ring current contribution to the shielding of this proton.
15. What would be the ring current effect on a carbon located at the coordinates given for the proton in problem 14?
16. Precict the changes in the proton, carbon, and nitrogen chemical shifts, if any, that occur when the NMR spectra of aniline in D_2O and in hydrochloric acid solution are compared.
17. Estimate the differences in the chemical shifts of the methyl carbons and protons in 2-, 3-, and 4-methylpyridine. How would these chemical shifts change in N-methylated derivatives of these compounds?
18. Outline the experimental approach you would take to estimate the aromaticity of compound **6**.

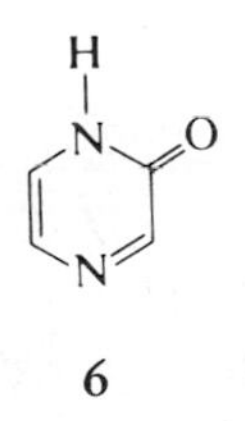

6

19. Consider the following data in light of the possible aromatic character of compound **7**. The compound shows an 8-proton A_2B_2 system (δ_H6.8–7.5 ppm) and a sharp signal at −0.5 ppm [30].

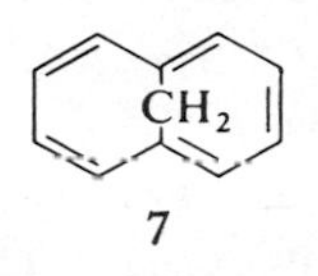

7

20. The aldehyde proton of benzaldehyde resonates at 0.23 ppm to low-field relative to benzaldehyde. The contribution to its chemical shift caused by

the benzene ring current, and calculated by the method of Johnson and Bovey, is + 0.58 ppm. Explain this difference [31].

REFERENCES

1. W. E. Lamb, *Phys. Rev.*, 1941, **60,** 817.
2. W. C. Dickinson, *Phys. Rev.*, 1950, **80,** 563.
3. B. P. Bailey and J. N. Shoolery, *J. Amer. Chem. Soc.*, 1955, **77,** 4328.
4. M. L. Huggins, *J. Amer. Chem. Soc.*, 1953, **75,** 4123.
5. J. I. Musher, *J. Chem. Phys.*, 1962, **37,** 34.
6. G. C. Levy and G. L. Nelson, *Carbon-13 Nuclear Magnetic Resonance for Organic Chemists*, WILEY-INTERSCIENCE, John Wiley and Sons, New York, 1972.
7. J. A. Pople, W. G. Schneider, and H. J. Bernstein, *High-Resolution Nuclear Magnetic Resonance*, McGraw-Hill Book Co., New York, 1959, p. 180.
8. R. Freeman, G. R. Murray, and R. E. Richards, *Proc. Roy. Soc. (London)*, 1957, **A-242,** 455.
9. P. T. Narasimham and M. T. Rogers, *J. Phys. Chem.*, 1959, **63,** 1388.
10. J. A. Pople, *J. Chem. Phys.*, 1956, **24,** 1111.
11. A. Saika and C. P. Slichter, *J. Chem. Phys.*, 1954, **26,** 22 (quotation therein).
12. J. A. Pople, *Discussions Faraday Soc.*, 1962, **34,** 7.
13. C. E. Johnson, and F. A. Bovey, *J. Chem. Phys.*, 1958, **29,** 1012.
14. R. W. Fessenden and J. S. Waugh, *J. Amer. Chem. Soc.*, 1957, **79,** 846, and 1958, **80,** 6697.
15. G. C. Levy, R. L. Lichter, *N-15 Nuclear Magnetic Resonance*, John Wiley and Sons, New York, 1979, p. 33.
16. P. J. Black, R. D. Brown, and M. L. Heffernan, *Australian J. Chem.*, 1967, **20,** 1305.
17. T. Schaeffer and W. G. Schneider, *Canad. J. Chem.*, 1963, **41,** 966.
18. J. D. Baldeschweiler and E. W. Randall, *Proc. Chem. Soc.,* 1961, 303.
19. V. M. S. Gil and J. N. Murrell, *Trans. Faraday Soc.*, 1967, 248.
20. P. J. Black, R. D. Brown and M. L. Heffernan, *Australian J. Chem.*, 1967, 1325.
21. W. W. Paudler and T. J. Kress, *First Intern. Congress of Heterocyclic Chemistry*, 1967, Albuquerque, N.M.
22. W. W. Paudler and J. E. Kuder, *J. Heterocyclic Chem.*, 1966, 3, 33.
23. K. Tory and M. Ogata, *Chem. Pharm. Bull. (Tokyo)*, 1964, **12,** 272.
24. M. Karplus and J. A. Pople, *J. Chem. Phys.*, 1963, **38,** 2803.
25. G. A. Olah and G. D. Mateescu, *J. Amer. Chem. Soc.*, 1970, **92,** 1430.
26. J. P. Larkindale and D. J. Simkin, *J. Chem. Phys.*, 1971, **55,** 5048.
27. J. Baldeschweiler and E. W. Randall, *Proc. Chem. Soc.*, 1961, 303.
28. A. J. DiGioia, G. T. Furst, L. Psata and R. L. Lichter, *J. Phys. Chem.*, 1978, **82,** 1644.
29. J. A. Elvidge and L. M. Jackman, *J. Chem. Soc.*, 1961, 859.
30. E. Vogel and H. D. Roth, *Augen. Chem. Intern. Edit.* **3,** 228 (1964).
31. R. E. Klinck and J. B. Strothers, *Canad. J. Chem.* **40,** 1071 (1962).

6

COUPLING CONSTANTS—QUANTITATIVE CONSIDERATIONS

6.1 GENERAL CONSIDERATIONS

Nuclei are chemically equivalent if they can be interchanged by a symmetry operation of the molecule. For example, the protons in methylene chloride can be interchanged by a C_2 operation. This relationship is referred to as ''homotopic.'' Nuclei of this type are chemically equivalent.

Those nuclei that are interrelated by other symmetry features are referred to as ''enantiotopic.'' For example, the methylene protons in α-chloroacetic acid (structure **1**) are enantiotopic and chemically equivalent:

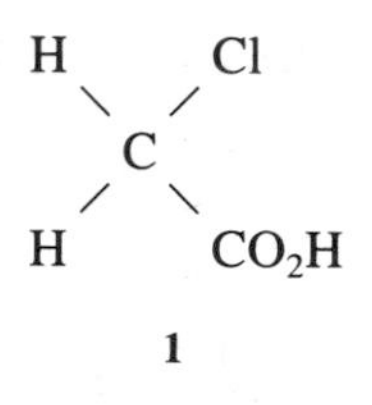

This equivalence is a result of the relationship of the two protons with the plane generated by Cl—C—CO_2H. It is interesting to note that nuclei that are enantiotopic, with respect to each other, are potentially chemically nonequivalent when dissolved in a chiral solvent!

Nuclei that are chemically equivalent are referred to as ''isochronous''—they are frequency equivalent. Thus, for example, the protons in methylene chloride (CH_2Cl_2) are said to be isochronous. On the other hand, the *ortho*-protons in toluene are nonisochronous with the *meta*- and *para*-protons.

Symmetry equivalent nuclei are magnetically equivalent if they are equally coupled to each member of any other symmetry-equivalent set of nuclei. An example of *magnetically nonequivalent* nuclei, as determined by the coupling constant criterion, are the fluorine atoms in 1,1-difluoroethylene ($CF_2{=}CH_2$). In this molecule each fluorine is coupled by different degrees to the protons, depending on whether it is interacting with the *cis*- or *trans*-situated proton. The same argument holds, of course, for the proton magnetic nonequivalence in the molecule.

Another requirement for magnetic equivalence is the need for identical chemical shifts among the equivalent nuclei. Thus, *anisochronous nuclei are magnetically nonequivalent by the chemical shift criterion* [1].

The splitting patterns for different interacting nuclei have been experimentally as well as theoretically analyzed by a number of workers, among them Gutowsky [2], Karplus [3], McConnell [4], Pople [5], Musher [6], and many others (Ref. [15], Chapter 5).

The general mechanism giving rise to the phenomenon of spin–spin coupling has been described in Chapter 3. The fact that a particular spin–spin coupling constant can be either positive or negative, although implicit, has not been discussed.

By definition, it has been established that application of the Fermi contact principle will generate a positive coupling constant when the preferred arrangement of two coupled spins is antiparallel. Consequently, a numerically negative coupling constant is obtained when the preferred arrangement of coupled spins is parallel.

It has been demonstrated that most two-bond proton–proton (H—C—H) coupling constants are in fact negative, whereas three-bond coupling (H—A—B—H) constants are positive. Generally, it can be expected that coupling through an odd number of bonds will be positive, whereas coupling through an even number will be negative.

6.2 COUPLING BETWEEN DIRECTLY BONDED NUCLEI

The contact coupling between two nuclei involves only *s* orbitals, as predicted by the Fermi contact model and as experimentally demonstrated by the observation that, for example, one-bond coupling between C^{13} and H nuclei increases as the percent *s* character of the C^{13} nucleus increases ($sp^3 < sp^2 < sp$). The sizes of these coupling constants (J_{AB}) are directly proportional to the product of the gyromagnetic ratios of the two interacting nuclei ($\gamma A \times \gamma B$) [7–10].

If the gyromagnetic ratio for nucleus *X*, in an *X*–H interacting system, happens to be negative, the resulting *X*–H coupling constant will be negative as well. For example, all N^{15}–H coupling constants are negative!

In those instances where the interaction between two nuclei does not involve a hydrogen, the size of the coupling constant depends on the product of the *s* character of the orbitals of the two nuclei that form the *X*–*Y* bond (e.g., $s(\text{C–N}) \times s(\text{N–C})$ is proportional to $80 \times J_{CN}$). While no simple relationships have been developed for these instances, it is generally true that coupling between two nuclei, when neither is a proton, is very large. An example, albeit extreme, is the 5700-Hz coupling observed between Pt^{195} and P^{31}.

On a quantitative level, equations 6.1 and 6.2 give the interrelationships between the degree of *s* character of the nitrogen and carbon, respectively, in C^{13}–H and N^{15}–H spin–spin systems:

$$s(\text{C}^{13}\text{–H}) = 0.2J_{\text{C-H}} \tag{6.1}$$

$$s(\text{N}^{15}\text{–H}) = 0.43J_{\text{N-H}} - 6 \tag{6.2}$$

The data given in Table 6.1 show these interrelationships for nitrogen, carbon, and phosphorus examples. The hybridization values computed for cyclopropene and cubane show experimentally that fractional hybridization, as predicted by HMO computations, is supportable by NMR results. Although as already mentioned, no simple quantitative relationships for directly bonded nuclei, when neither one is a hydrogen, exist, relationships with some reasonable applicability have been developed for C–C and C–N coupling constants (cf. equations 6.3 and 6.4) [11–13]:

$$s(\mathbf{C^{13}}\text{–C})s(\text{C–}\mathbf{C^{13}}) = 17.4J_{\text{C-C}} + 60 \tag{6.3}$$

$$s(\mathbf{C^{13}}\text{–N})s(\text{C}^{13}\text{–}^{15}\mathbf{N}) \sim 80J_{\text{C-N}} \text{ (known exceptions)} \tag{6.4}$$

TABLE 6.1 Vicinal Coupling and Degree of Hybridization

Structural Feature	Hybridization	Coupling Constant (Hz)
$-C-H$	sp^3	125
$=C-H$	sp^2	157 ± 5
$\equiv C-H$	sp	249
Cyclopropane	sp^2 (computed)	160
Cyclopropene	$sp^{1.2}$ (computed)	226
Cubane	$sp^{1.8}$ (computed)	179
CH_3NH_2	sp^3	73
$R_2C=\overset{+}{N}H_2$	sp^2	93
$CR\equiv\overset{+}{N}H$	sp	130
PH_2PH_2	sp^3	186
H_3PO_3H	?	707

6.2.1 Carbon–Hydrogen Coupling Constants

While C^{13}–H coupling constants are not normally obtained in C NMR, they can be extracted from *high-resolution* studies. In some molecules, it is possible to obtain these coupling constants from C^{13} satellites in proton spectra (see Chapter 3). As already mentioned, there exists a good relationship between the percent *s* character of a particular carbon and the size of the C–H coupling constant (see Table 6.2). Additionally, the size of these coupling constants also depends on the electronegativity of any substituent. For example, the C—H coupling constant in CH_3NH_2 is 133 Hz, in CH_3OCH_3 it is 140 Hz, and in CH_3F it is 149 Hz (see Table 6.3).

There is also considerable variation in the C—H coupling constant with angular strain imposed on the particular carbon. For example, while the C—H coupling constant in methane is 125 Hz, it increases to 134 Hz in cyclobutane.

The C—H coupling constant in olefinic systems is approximately 30 Hz larger than the equivalent constant in the corresponding saturated analogs. As is true in the alkane series, bond angle distortion and substitution by electronegative groups increases the size of these coupling constants. For example, the =C—H coupling constant in 1,1-dimethylcyclohexene is 221 Hz as compared to a value of 156.4 Hz in ethylene. Similarly, the =C—H coupling constant in *cis*-di-iodoethylene is 187.9 Hz, again a much larger value than that in ethylene.

The C—H coupling constants of acetylenes are rather large (170–180 Hz) and change, although to a smaller extent than olefins, with the electronegativity of any substituent. For example, the C—H constant of acetylene is 249 Hz and increases to 251 Hz in phenylacetylene.

6.2.2 Carbon–Fluorine Coupling Constants

One of the unique features of C—F coupling constants lies in the observation that the C—F and =C—F constants are negative (approximately −167 and −245 Hz,

TABLE 6.2 C^{13}-H Coupling Constants Dependence on Hybridization of C^{13}

Compound	Hybridization of C^{13}	$J_{C^{13}H}$	% s character
Cyclopentane (C–H)	sp^3	128	25
$C(CH_3)_4$	sp^3	120	25
Cyclohexene (vinyl C–H)	sp^2	159	33.33
Benzene (C–H)	sp^2	159	33.33
$H_2C{=}CH_2$	sp^2	157	33.33
$H_3C{-}C({=}O){-}H$	sp^2	174	33.33
Cyclopropane (C–H)	?	161	25
$H_3C{-}C{\equiv}C{-}H$	sp	248	50

respectively). A correspondence between the size of the C—F coupling constants and the nature of the substituent in *para*-substituted fluorobenzenes has also been noted (see Table 6.4). Clearly, any other physical property that affects bond lengths, such as *ionic character* of a bond, will also be correlatable with the size of these coupling constants [14].

TABLE 6.3 Correlation between Electronegativities of Substituents and $J_{C^{13}H}$ Values

Compound	Pauling electronegativities	$J_{C^{13}H}$
$CH_3{-}F$	4.0	149
$CH_3{-}OCH_3$	3.5	140
$CH_3{-}N(CH_3)_2$	3.0	131
$CH_3{-}C(CH_3)_3$	2.5	124
$CH_3{-}Li$	1.0	98

TABLE 6.4 Changes in $J_{C^{13}H}$ with Changes in Substituents in Fluorobenzenes

F
X

X	$J_{C^{13}H}$ (average)	$J_{C^{13}F}$
$-NH_2$	162.3	233.3
$-OH$	160.2	240.7
$-OCH_3$	165.6	236.8
$-CHO$	166.4	251.6

6.2.3 Nitrogen–Hydrogen Coupling Constants

The one-bond coupling constants between nitrogen and protons are the largest and are generally dominated by the Fermi contact mechanism. As equation 6.2 demonstrates, the size of these coupling constants increases with increasing *s* character: *pyramidal* N—H constants are about 75 Hz, *trigonal* ones are about 90 Hz, and linearly bonded N—H bonds have constants of about 135 Hz [14].

Within each type of coupling are variations that are not related to the degree of hybridization. While the reason for the increase (in terms of absolute values) of these coupling constants with increase in charge is not well understood, it is a general phenomenon. The geometrical dependence of these coupling constants (the *syn* N—H coupling constant in formamide is -88 Hz, while the *anti* one is -92 Hz) can be used to establish configurational aspects of protein structures.

6.2.4 Phosphorus–Hydrogen Coupling Constants

The size of P–H coupling constants vary from 186 Hz in PH_2–R to 707 Hz in H_3PO_3. Generally, the same factors that control the N–H constants are also operative in this nucleus.

6.3 COUPLING BETWEEN GEMINALLY BONDED NUCLEI

Theoretical calculations of the spin–spin coupling constants in two-bond (geminal; from the Latin *geminus*, ''twin'') H–C–H coupling systems are based on molecular-orbital and valence-bond theoretical principles. The former approach, fairly rudimentarily applied, indicates that there should be a linear relationship between geminal coupling constants and the proton–proton bond order in these systems (Eq. 6.5) [4]:

(6.5) $$J_{AA'} = 200 * (\text{bond order between } A \text{ and } A')$$

In the hydrogen molecule instance, the bond order is 1, resulting in a *predicted* coupling constant of approximately 200 Hz, a value that compares reasonably with the experimental one of 280 Hz. This theory predicts, incorrectly, that the coupling constants will always be positive. In fact, more complete calculations by Pople and Santry [5] and others [11] correctly predict that the H–H and H–F coupling constants in comparable systems will always be of opposite signs [7–11].

6.3.1 Proton–Proton Coupling in H–C–H Systems

The difference in the geminal coupling constants between two protons bonded to a sp^3 carbon (-12.4 Hz in methane) and those bonded to a sp^2 carbon ($+2.3$ Hz in ethylene) is 15 Hz. This numerical difference is ascribed to the hybridization, and consequently the structural nonequivalence, between these two bonding states. Any substituent that withdraws electrons from the methylene group by inductive electron withdrawal will make the coupling constant more positive, whereas the reverse, electron donation, will make the coupling constant more negative.

The changes in the proton–proton coupling constants in going from methane ($J = -12.4$ Hz) to methylene bromide ($J = -5.5$ Hz) to malonic acid ($J = -15.9$ Hz) confirms the theoretically predicted behavior [15–17].

The systematic variation of geminal coupling constants with changes in the electronegatitivity of nearby substituents is exemplified by the changes in the geminal coupling constants (J_{AB}) in the bicycloheptenes shown in Table 6.5.

In a manner similar to the electronegativity (E) relationships, it has been shown [18, 19] that the π electron contribution to *geminal* coupling constants is additive, the value of 1.9 Hz being added to the base value for each adjacent π electron bond (see Table 6.6)

The coupling constant calculations based on the valence–bond approach of Karplus, Anderson, Farrar, and Gutowsky [2, 3] have demonstrated the existence of a smooth relationship between the *geminal* angle and the size of the coupling constant in H–C–H systems. This correlation, reproduced in Figure 6.1, has been very useful in a large number of structure determination studies.

Most of the geminal coupling constants through sp^2 carbon atoms fall between the large H–H coupling constant in formaldehyde (H–CO–H) ($+42$ Hz) and the small negative value (-1 Hz) for allenes ($CH_2{=}C{=}C$). For vinyl compounds of general structure $CH_2{=}CXY$, the H–H coupling constants are proportional to the sum of the electronegativities of the X and Y substituents and are expressed by equation 6.5:

(6.5) $$J_{H\text{-}C\text{-}H} = a(E_X + E_Y) + b$$

TABLE 6.5 Variation of Geminal Coupling Constants with Changes in the Electronegativity of Substituents

Cl Cl
Cl
Cl
H_X
R
H_B
H_A

I

R	J_{AX}	J_{BX}	J_{AB}	Electro-negativity of R
−CN	4.6	9.3	−12.6	2.49
−COOH	4.4	8.5	−12.6	2.60
$-C_6H_5$	4.2	8.9	−12.7	2.75
−Cl	3.2	8.0	−13.2	3.25
−OH	2.4	7.4	−12.6	3.43
−OAc	2.5	7.6	−13.3	3.80

SOURCE: From K. L. Williams, *J. Am. Chem. Soc.*, **85** (1963), 516, copyright 1963 by the American Chemical Society. Reprinted by permission of the copyright owner.

Although the methylene proton coupling constants are generally close to 2.5 Hz, the presence of a strongly electron withdrawing group, as found in formaldehyde (H–CO–H), along with the oxygen atom's ability to be involved in hyperconjugative interactions increases the coupling constant to the amazing size of +42 Hz!

A lone pair of electrons on oxygen can serve as a π donor to the CH_2 group and thereby make this H–H coupling constant more positive as well. For example, the H–H coupling constant of the methylene protons in the methylenedioxybenzene

TABLE 6.6 Effect of π Electrons on Geminal Coupling Constants

Compound	$J_{HH}(gem)$
$NC-CH_2-CN$	20.3 ±0.3 cps
$CH_2=CH-CH_2-CN$	19.4 ±0.3
CH_3-CN	16.9 ±0.3
$CH_3CO-CH_2CH_2-COCH_3$	14.3 ±0.2
CH_3NO_2	13.2 ±0.2
CH_4	12.4 ±0.6

SOURCE: From M. Barfield and D. M. Grant, *J. Am. Chem. Soc.*, **85** (1963), 1899, copyright 1963 by the American Chemical Society. Reprinted by permission of the copyright owner.

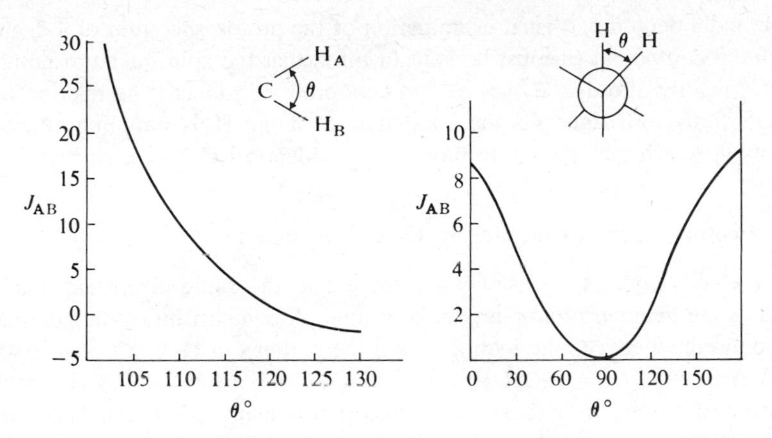

Figure 6.1. The angular dependence on the *geminal* and *vicinal* proton–proton coupling constants.

(**1**) is ±1.5 Hz, a much more positive value than that observed for the protons in methane (−12.4 Hz).

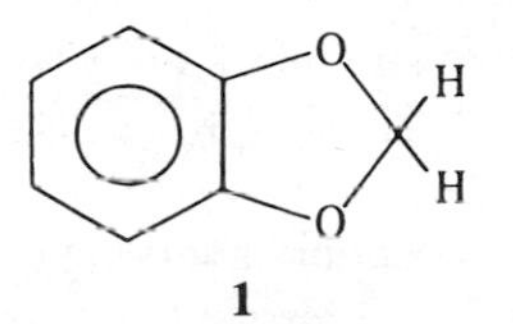

The extent of the influence of the filled lone-pair orbital of oxygen on the coupling constant depends on the angle between this orbital and the hydrogen atom. The optimal overlap occurs when the dihedral angle between a C–H bond and the lone pair of electrons is close to either 0(a) or 180°(b).

The least-favored overlap occurs when this angle is 45°, 90°, or 135°—that is, the staggered configuration [11].

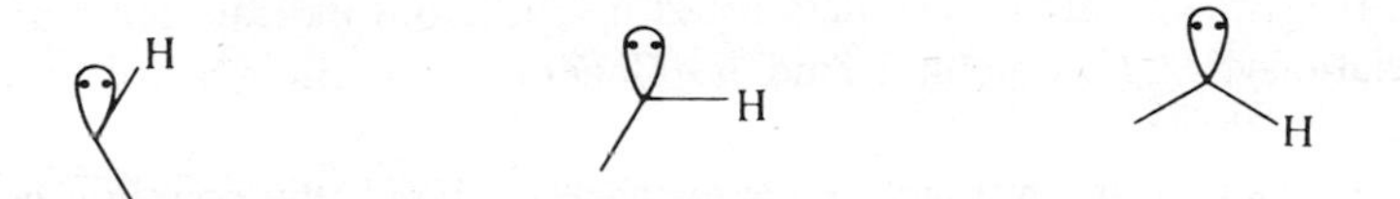

6.3.2 Proton–Nonproton (*X*) Coupling in H–C–*X* Systems

Geminal coupling constants between protons and other nuclei ($J_{H\text{-}C\text{-}X}$) are, of course, also well known. The simplest of these systems is the coupling between a proton and a deuteron. Visual examination of the proton spectrum of a deuterium-containing compound (it must be kept in mind that the spin quantum number for D is 1!) readily affords the various H–D coupling constants. The relationship $J_{HH} = 6.55 \times J_{HD}$ also permits the computation of the H–H coupling constant for systems in which the interacting protons have identical chemical shifts.

6.3.3 Proton–Carbon Coupling in H–*X*–C Systems

The H–C–C^{13} coupling constants are subject to the same electronic and spatial effects as the *geminal* proton–proton constants. A comparison of the geminal coupling constants between the hydrogen and the carbons in **H–C–C** ($J = -4.5$ Hz) versus **H—C=C** ($J = -2.4$ Hz) versus **H—C≡C** ($J = +49.3$ Hz) structural systems testifies to the existing relationship between the size of these coupling constants and the degree of hybridization of the interacting nuclei in these systems. The amazing size of the geminal coupling constants in acetylenes is noteworthy.

Within the **H–C–C** system, electron-withdrawing groups increase the size of the coupling constants (see Table 6.7). As in the analogous proton–proton coupling constants, a carbonyl carbon, as in acetates ($\mathbf{CH_3CO_2}$–R), for example, affects the geminal coupling constant more strongly (+6.5 Hz) than does a saturated carbon substituted by two halogen atoms (+1.2 Hz), as in 1,1-dichloroethane ($\mathbf{CHCl_2CH_3}$).

The geminal proton–carbon coupling involving olefinic carbons of general structure **H—C=C** (see Table 6.7) are also subject to electronegativity effects. The dramatic increase in the geminal coupling constants between a *cis*- ($J = -7.9$ Hz) and a *trans*- ($J = +7.5$ Hz) configuration in a structure such as **H—C=C**—Cl is expected and very useful in structure determination studies. When this structure is part of an aromatic system (see *o*-dihalobenzenes listed in Table 6.7), the coupling constants are *less* positive than in the corresponding olefinic structure. Furthermore, the effect is less for the least electronegative halogen (I) and greater for the more electronegative halogen (Cl).

6.3.4 Proton–Nitrogen Coupling in H–C–N Systems

The geminal coupling constants involving N^{15} in the H–C–N– systems, although numerically somewhat different from the correpsonding carbon analogs, follow the same patterns, with added effects caused by the presence of the lone pair of electrons on the nitrogen atom. The data listed in Table 6.8 indicate the usefulness of these relationships in structural studies. Generally, coupling across a saturated carbon atoms tends to be the smallest in this series (approx. −2 Hz). If the lone pair of electrons on the nitrogen is *cis* to the H–C bond, the coupling constant is more negative ($J = -16$ Hz in *syn*-oxims) than in the *trans* configuration ($J = -3$ Hz in *anti*-oximes).

Electronegative substituents tend to make these coupling constants larger. For example, the geminal coupling across trigonal carbons is numerically larger, more negative ($J = -10$ to -14 for *cis* and -2.5 to $+2.5$ for *trans* structures) than

TABLE 6.7 Geminal Proton–Carbon Coupling Constants

Structural Feature:	$J_{H,C}$
H \\ C—C	−4.5
H \\ C—C—$(X)_2$	+1.2
H \\ C—C=O | OR	+6.5
H \\ C=C	−2.4
H, X, X on benzene ring (H and X on adjacent ring carbons; second X on the next carbon)	Cl = −3.5 Br = −3.2 I = −1.8
H \\ C=C \\ Cl	+7.5
H \\ C=C / Cl	+6.9
H \\ C=C / \\ Cl Cl	+16.0
H \\ C=C / \\ I I	+11.0
H Cl \\ / C=C / Cl	+0.8

TABLE 6.7 (*Continued*)

Structural Feature:	$J_{H,C}$
H(I)C=C(I) (H and I on one carbon, I on the other)	−1.4
H–C≡C	+49.3

TABLE 6.8 Geminal H–N[15] Coupling Constants

Structural Feature	$J_{H,N^{15}}$
H–C=N–	−10 to −14
H–C=N–	−2.5 to +2.5
H–C=N⁺–	−2
H–C–N	0 to −2 (more negative if *cis*)
H–C=N⁺=N	0.14 to 2.8
H–C–N⁺≡C	+3.2 to 3.8
H–C≡N	+8.7

similar interactions across saturated carbons (J = 0 to −2 Hz). As in the corresponding carbon geminal constants, coupling across an sp carbon is among the largest (J = +8.7 Hz) in this system as well. The size of this interaction is decreased in the absence of the lone pair of electrons, as exemplified by the H-C-N^+≡C system (J = 3.2 to 3.8 Hz).

6.3.5 Proton–Phosphorus Coupling in H–C–P Systems

Geminal coupling interactions involving H-C-P^{31} systems are controlled by the same factors as in the analogous nitrogen analogs. The magnitude of the proton–

phosphorus coupling constant in H–C–P systems is also strongly dependent on the dihedral angle α between the lone pair of electrons on phosphorus and the C–H bond. The maximum coupling (approx. +25 Hz) is observed when this angle is zero. It decreases to −5 Hz when the angle is approximately 115° and becomes 0 Hz at 180° [1, 14].

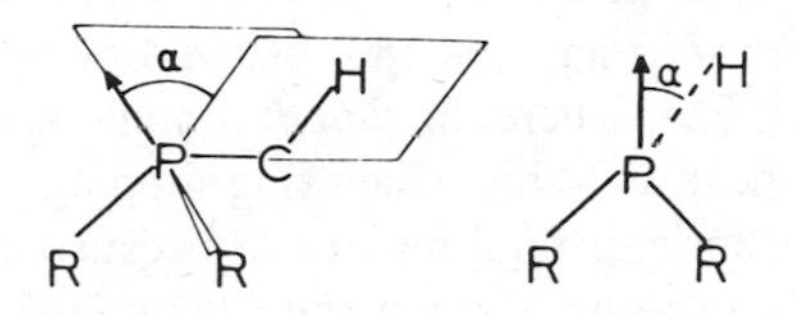

6.3.6 Proton–Fluorine and Fluorine–Fluorine Coupling in H–C–F and F–C–F Systems

The geminal coupling constants between fluorines and protons are about +50 Hz when these atoms are bonded to a saturated carbon (HFC) and increase to approximately +80 Hz when bonding occurs across an olefinic carbon (HFC=). The same pattern seems to prevail in F–C–F systems, where the values are quite large (150 to 250 Hz for saturated carbon systems; 50 to 150 Hz for olefinic ones).

6.4 COUPLING BETWEEN VICINALLY BONDED NUCLEI

The coupling between nuclei (X,Y) on adjacent atoms (C,C), as in X–C–C–Y, is referred to as vicinal (from the Latin *vicinus* = neighbor). This type of coupling is, perhaps more than any other NMR phenomenon, responsible for the great success of this technique in molecular structural studies.

The discussions dealing with the quantitative aspects of geminal coupling constants showed that these interactions are effected by the electronegativity of nearby atoms, groups of atoms, the hybridization of the carbon atoms, and the stereochemical arrangement of these various interacting features. It is perhaps not surprising that similar forces act on the size of the vicinal coupling constants. In the H—C—N system, where the lone pair of electrons on the nitrogen can be considered as emulating nucleus X or Y in the vicinal configuration (X–C–C–Y), the maximum coupling constant is obtained when the proton and the lone pair of electrons are eclipsed. Similarly, in the X–C–C–Y systems, the XY coupling is large when the two groups are eclipsed. When either XY or the H and lone pair of electrons are *anti* to each other, the corresponding coupling constants are large as well. Intermediate angles cause correspondingly smaller coupling constants.

The mathematical relationships interrelating the dihedral angle between the X–C and the C–Y bonds in vicinally interacting systems were developed [2, 3] and are given, for the general cases, by equations 6.6 and 6.7:

$$J = A \cos^2(\theta) + C \quad (\theta = 0\text{–}90) \tag{6.6}$$

$$J = A' \cos^2(\theta) + C \quad (\theta = 90\text{–}180) \tag{6.7}$$

The additive constant C in these equations is normally neglected, since its value is estimated to be less than 0.3 hz. The constants A and A' depend on the system being examined and are in the range of 8–14 Hz when X and Y are protons in an X–C–C–Y system. Figure 6.1 is a general graphic representation of the angular relationship in these vicinal systems.

These steric relationships predict that the *axial–axial* coupling constant in cyclohexane, where the *dihedral* angle is 180°, should be approximately 10 Hz and that the *axial–equatorial* one, where the *dihedral* angle is close to 60°, should be about 1.5 Hz. In agreement with this, when ring flipping of the cyclohexane molecule is impaired, either for structural reasons or because the spectrum is obtained at a low temperature, the experimentally obtained coupling constants for the axial–axial protons are between 8 and 12 Hz and the axial–equatorial protons are coupled to the extent of 1–3 Hz.

The HF coupling in H–C–C–F systems also follows this Karplus relationship. As is true for all *heteronuclear* coupling constants, these values are significantly larger than those in the corresponding proton homonuclear situations. With a dihedral angle of 0°, the HF coupling constant is approximately 30 Hz and increases to 45 Hz when the angle is 180°.

The general applicability of the Karplus relationship to *vicinal* systems is demonstrated by the observations that systems of general structure H–C–Z–Y also obey the relationship (where Z = O, N, S, Se, Te, or Si). For example, since the proton on sulfur in protonated thiane (**5**) is coupled to the extent of 14.1 and 2.3 Hz, respectively, to the vicinally located axial and equatorial protons, it must be situated axially. Interestingly, the fluorine–fluorine coupling constants in the F–C–C–F systems do not obey a consistent angular dependence with the size of the *vicinal* coupling constants.

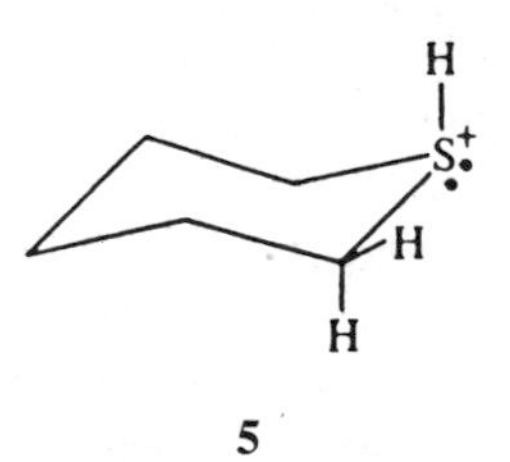

5

This and other limitations of the relationship, recognized and described by Karplus and others [7, 18], point to the fact that the length of the carbon–carbon bond, the H–C–C angle, the electronegativity of any substituents in addition to the dihedral angle will effect the *vicinal* coupling constants [14].

As the C–C bond lengths in these *vicinally* interacting systems decrease, so does the size of the coupling constants. It is this change that most drastically effects the size of the A constant in the relationship. This bond length and, logically, also bond order dependence are exemplified by the observation that the *ortho*-proton–proton coupling constants in naphthalene (J = 8–9 Hz) are larger than those in benzenoid systems ($J = 6$–7 Hz). The bond order in the former system is, indeed, greater than in the latter one.

The decreasing *vicinal* coupling constants with increasing X–C–C angle are demonstrated by the systematic decreases in the $J_{1,2}$ values observed in a series of cycloalkenes: cycloheptene (10.8 Hz), cyclohexene (8.8 Hz), cyclopentene (5.1 Hz), cyclobutene (3.0 Hz), cyclopropene (1.3 Hz). It is important to note that the dihedral angles between the interacting protons in all of these ring systems are the same! Fortunately, this angular dependence can be disregarded when comparisons of structurally similar systems are made.

The effect that the electronegativity of a substituent has on the size of vicinal coupling constant is shown by determining these constants in some substituted, sterically rigid, bicycloheptenes (see Table 6.5). The only variance that occurs in these particular compounds is that caused by the substituents; all other contributors to the coupling constant are kept constant.

The following equations, developed by Karplus [18], allow the estimation of coupling constants not only in this bicyclic ring system but in any nonfreely rotating system of the general type —CHCHX—. The expression is applicable to most freely rotating systems containing this structural fragment.

$$J^{\text{substituted}}_{H_AH_B}(cis) = J^{\text{unsubstituted}}_{H_AH_B}(cis) \times (1 - 0.60\ \Delta X)$$

$$J^{\text{substituted}}_{H_AH_B}(trans) = J^{\text{unsubstituted}}_{H_AH_B}(trans) \times (1 - 0.25\ \Delta X)$$

$$J^{\text{substituted}}_{H_AH_B}(\text{average}) = J^{\text{unsubstituted}}_{H_AH_B}(\text{average}) \times (1 - 0.07\ \Delta X)$$

6.5 LONG-RANGE COUPLING: GENERAL CONSIDERATIONS

Coupling constants describing interactions between nuclei (*X*, *Y*) that are more than three bonds removed from each other (*X*–C–C–C–*Y*, for example) are referred to as *long-range* coupling constants. They are, normally, much smaller than the corresponding *geminal* and *vicinal* constants.

Spin–spin coupling that occurs between nuclei that are not directly bonded to each other appears to be dominated by a coupling mechanism that involves the nucleus (the proton, if hydrogen coupling is being examined) with the electron spins of the various intervening bonds [14, 19]. These essentially electrostatic interactions among the electron spins are transmitted through the magnetic interactions between the nuclear spins of the interacting nuclei and adjacent electrons. This coupling can take place through sigma and pi electron bonds. The proton–proton sigma bond transmitted coupling interactions are rarely greater than 1 Hz. Interestingly, the carbon–carbon long-range coupling between the methyl group carbon in toluene and the carbon *para* to it is 0.86 Hz, a rather large value considering that it involves four intervening conjugated bonds (C—C=C—C=C). The proton–carbon coupling constant in benzenoid systems between a particular proton and a carbon *para* to it (H—C=C—C=C) is similarly numerically large, but negative (−1.1 Hz). When the *para*-carbon is replaced by a nitrogen, as in pyridine (H—C=C—C=N), the coupling constant is 0.2 Hz, and increases to 0.7 Hz if the nitrogen is protonated (H—C=C—C=N^+—). Long-range coupling between N^{15} and C^{13} is also well known. For example, the carbon *para* to the nitrogen in aniline is coupled to it by 0.27 Hz. This value increases to 0.60 Hz in nitrobenzene, where the nitrogen is more positive as a result of the electron-withdrawing influence of the two oxygen atoms.

The various long-range coupling constants can be divided into contributions caused by π and σ electronic effects:

$$J_{AB} = J_{AB}^{\pi\text{-electronic contribution}} + J_{AB}^{\sigma\text{-electronic contribution}}$$

The four paths that have been suggested to account for the long-range coupling phenomenon are the following:

1. A zigzag pathway involving sigma bonds
2. Through-space pathways supported by non-bonded electron pairs
3. Double-bond (π bond) pathways involving conjugated systems
4. Triple and cumulative-bonds involving pathways

Paths 1 and 2 are caused by saturated carbon systems; paths 3 and 4 involve π electronic backbones.

6.5.1 Sigma Bond Coupling

6.5.1.1 Zigzag Paths

When the interacting nuclei *X* and *Y* are separated by three bonds, as in *vicinal* systems (*X*–C–C–*Y*), the coupling interactions are reasonably large in those instances where the gyromagnetic ratios of *X* and *Y* are large (8–12 Hz in H systems) but are essentially zero when the ratio is small (essentially zero in N^{15} systems). The sigma bond interaction mechanism falls off quite rapidly as the separation between the *X* and *Y* nuclei is increased [20]. This suggests that coupling across four bonds (the so-called long-range coupling) in saturated systems should be very small or nonexistent. This is indeed the case when *X* and *Y* are neither fluorine nor hydrogen.

However, spin–spin coupling between protons separated from each other by four bonds is numerically larger when the bonding path between the two protons occurs through a *zigzag* (or *H*) connection. The indicated protons in structures **2** and **3** are coupled to the extent of 1.3 and 1.8 Hz, respectively [5]. The long-range proton–proton coupling observed in 1,6-anhydro-D-hexapyranose (**5**) involves only H_A and H_B (1.3–1.8 Hz) and none of the other potential protons [21–25].

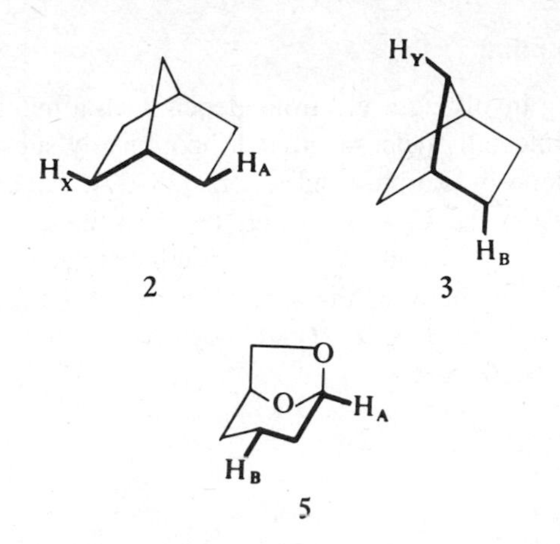

In all of these instances it is the existing zigzag path between the interacting nuclei that is used to explain these unusually large coupling constants. It has been suggested that these long-range interactions may not occur through the sigma bonds at all but rather through overlap between the small orbitals of the carbon atoms present in these structures (indicated by the dotted line in structure **6**) [21, 23]. This proposal accounts for the existence of the zigzag rule rather ingenuously! A similar proposal [24] suggests that there may well be enough π electron density on the carbon atoms to allow some thru π electron spin–spin coupling in these systems.

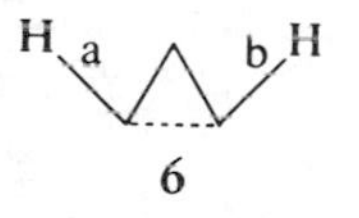

6

6.5.1.2 Through-Space Pathways

Although coupling information is generally expected to be transmitted by electron-involving pathways, molecule **7** represent an example where the coupling is augmented by the presence of a suitably situated lone pair of electrons. The six-bond coupling between the indicated protons in this compound is 1.1 Hz. This is one of a number of examples of through-space coupling. The term "lone-pair-mediated coupling" has also been employed to describe this particular phenomenon. Similar long-range coupling effects have been described between hydrogen and fluorine as well as between fluorine nuclei [24, 26, 27].

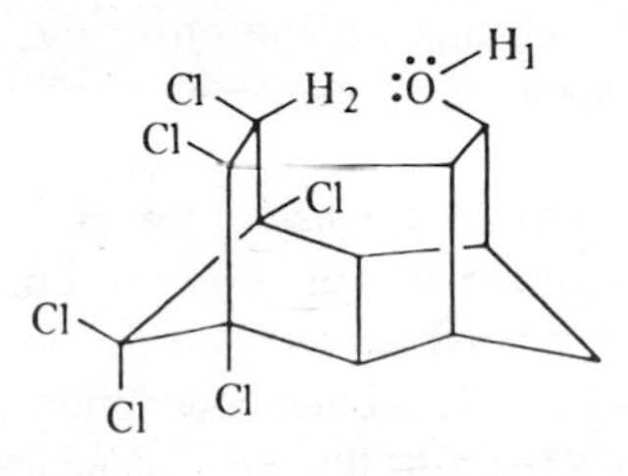

7

6.5.2 Pi Bond Coupling

Long-range coupling involving π electrons depends on a number of different contributions, as experimentally determined and theorctically substantiated by valence bond calculations done by Karplus and others [28–37]. Among these are (1) the dihedral angle between the C–X bond and the π orbitals and (2) the coupling constant, when involving an odd number of bonds between the interacting nuclei is positive (e.g., coupling between the italized nuclei in the following structures is always positive (C^{13}–H and H–X–Y–H) and negative when the intervening number of bonds is even (e.g., H–X–C).

6.5.2.1 Aromatic Compounds

In contrast to the behavior of sigma bonds, the pi electronic contribution to the coupling constant in aromatic systems, although generally small (approx. 2 Hz), is not strongly *attenuated* with increasing numbers of intervening bonds [28]. It has been established that the π bond contribution to coupling constants is related to the hyperfine splitting constant for the C–X bond and the bond order between the interacting nuclei, as well as a term describing an effective electronic excitation energy (about 4 eV in C–H systems). The relationship, given by equation 6.8, expresses these relationships.

$$J^{\pi}_{\mathrm{AB}} = \frac{\beta^2 Q^2 p_{\mathrm{AB}}{}^2}{h\,\Delta E}, \tag{6.8}$$

Where β = Bohr magneton, Q = hyperfine splitting constant (30 $\pm$ 3 gauss for C–H), p_{AB} is the bond order between the interacting nuclei and E is an effective electronic excitation energy (approx. 4 eV for H,H coupling). With these values substituted in equation 6.18 it is reduced to

$$J^{\pi}_{AB} = 1.8p^2_{AB}$$

The results of the application of this expression to the calculation of the various proton–proton coupling constants in benzene and naphthalene, along with the experimentally determined values, are given in Table 6.9.

The important information to be gleaned from these calculations is that the σ bond-caused coupling contribution to the coupling constant of nearby nuclei is considerably larger than the π bond-caused interaction. To confirm this relationship, coupling interactions where the σ bond contributions are essentially zero must be used. Acenapthylene lends itself to this type of analysis [29].

TABLE 6.9 Calculated and Observed Benzenoid Coupling Constants

Compound	p_{AB}		J_{AB} (calculated)	J_{AB} (observed)
Benzene	p_{12}	0.67	0.80	8 to 9
	p_{13}	0.00	0	2 to 5
	p_{14}	−0.17	0.05	0.5
Naphthalene	p_{12}	0.72	0.95	8.6
	p_{23}	0.60	0.66	6.0
	p_{13}	0	0	1.4
	p_{14}	−0.36	0.23	0.6[a]

SOURCE: Taken, in part, from H. M. McConnell, *J. Mol. Spectr.*, **1** (1957), 11 and reproduced with permission of the publisher.

[a] From some substituted naphthalenes (unpublished results from the author's laboratory).

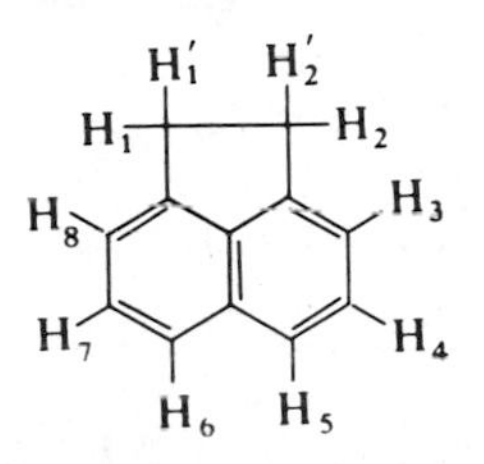

To apply the McConnell equation (Eq. 6.7) to the long-range coupling constant computations in this molecule, the hyperfine constant for the fragment H–C–C is used to compute the coupling constants involving H_1 and H_2. Hyperfine constants vary approximately as $\cos^2 \varphi$, where the angle is defined as that between the plane of H–C–C and the axis of the π orbital. The value Q is then given by $-50 \cos^2 \varphi$. Since φ is estimated to be 25° in acenaththene, Q becomes 41 gauss. When this constant is used in the McConnell expression, the data given in Table 6.10 are obtained.

The bond orders for J_{16}, J_{17}, and J_{18} in this compond are all less than 0.1 Hz and as such cause the computation of J^{π}_{AB} values of less than 0.1 Hz. In agreement with this, no coupling is observed between H_1 and H_6 and between H_1 and H_7. These results are remarkable, especially considering that the bond orders were computed by the simple HMO method!

TABLE 6.10 Long-Range Coupling Constants in Acenaphthene

	J_{AB} (theoretical)	J_{AB} (experimental)
J_{16}	0.3	0.5
J_{17}	0	0
J_{18}	1.1	1.5

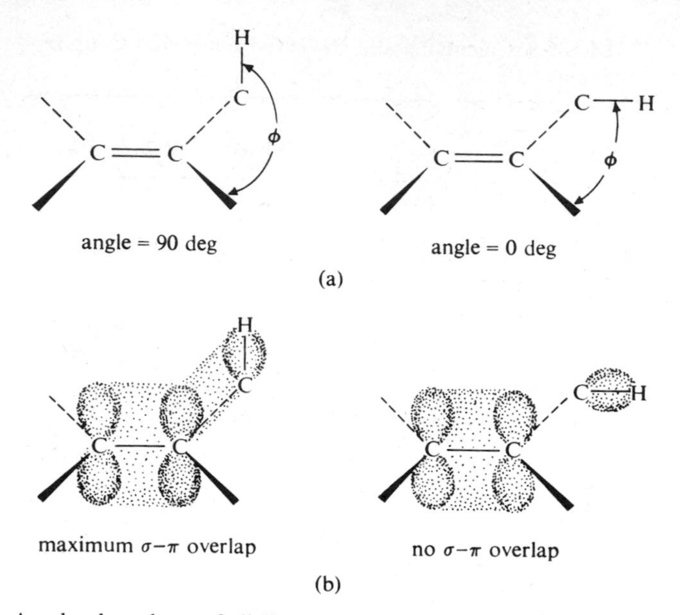

Figure 6.2. Angular dependence of allylic coupling constants. (a) Conventional representation. (b) Molecular orbital estimation.

6.6 LONG-RANGE COUPLING: SPECIFIC CONSIDERATIONS

6.6.1 Allylic Coupling

The most frequently observed long-range coupling involves the interaction between two nuclei (X and Y; often protons) separated by one double and one single bond ($X{-}Z{-}C{=}C{-}Y$). In this general structure, Z is most frequently a saturated carbon. When this is the case and X and Y are protons (i.e., the *traditional* allylic system in organic chemistry), the coupling constants that are negative vary between 1 and 4 Hz. The larger values are observed when the C–H bond in these structures is parallel to the π orbitals (see Fig. 6.2; $\theta = 90°$), where interactions between the C–H σ bond and the π electrons of the olefinic bond are maximum. When the C–H bond is orthogonal to the π orbital (θ in Fig. 6.2 is 0°), there is no σ–π interaction, and the coupling constant is typically small [30, 31]. The lactone **7** is an excellent example incorporating an allylic *cisoid* and *transoid* structure [33].

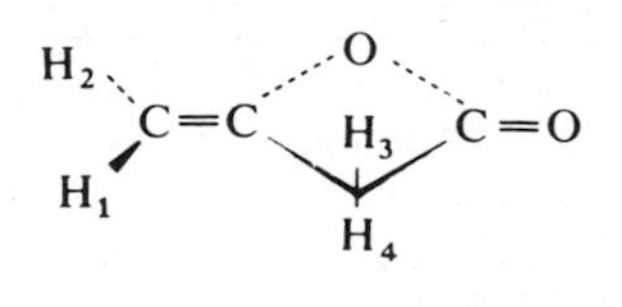

7

$$J_{13} = J_{14} = J_{\text{cisoid}} = 1.82 \text{ cps}$$

$$J_{23} = J_{24} = J_{\text{transoid}} = 1.34 \text{ cps}$$

In the structurally flexible noncyclic allylic systems, the dihedral angle is averaged so that the coupling constant is averaged as well. For example, the allylic coupling constants in 2-bromopropene are −1.4 and −0.8 Hz, respectively, between the methyl group protons and the olefinic protons *cis* (c) and *trans* (t) to this group.

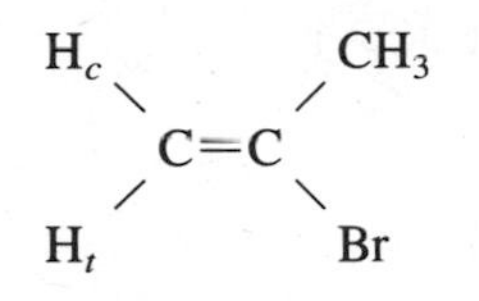

The allylic coupling constant is composed of the σ and π electronic contributions:

$$J_{\text{allylic}} = J_\sigma + J_\pi$$

The π contribution is significant only when the interaction angle is about 90° and follows a $\sin^2$ angle dependence. The zigzag σ coupling contribution has a maximum value near an angle of 0°, following a cosin^2 dependence.

A general expression for J_{allylic}, applicable for angles up to 90°, is $1.3 \cos^2 \theta - 3.0 \sin^2 \theta$. In the range 90–180°, the $\cos^2$ term in the expression is zero.

6.6.2 Homoallylic Coupling

The spin–spin interactions between X and Y nuclei in the structural system X—C—C=C—C—Y is referred to as "homoallylic" coupling. When X and Y are protons, these coupling constants, in contrast to the allylic ones, are positive and depend on the orientation of two C–H bonds with the π electronic framework. The *transoid* constants tend to be larger (by about 0.5 Hz) than the corresponding *cisoid* ones. The essentially planar cyclohexa-1,4 diene (**9**) demonstrates this difference, since the *cis* coupling is 9.63 and the *trans* one is 8.04. The size of these coupling constants reflects not only the nearly perfect alignment of the protons with respect to the olefinic bonds but also the presence of two identical interactive paths in this system.

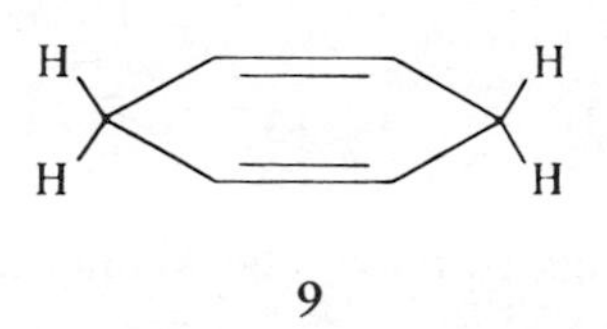

9

Where only one interactive path is possible, the allylic coupling constants are much smaller. For example, the size of the homoallylic coupling constant in 2,3-dimethylthiophene (**10**) is 0.6 Hz and increases to 3 Hz in the tetrahydropyridine (**11**).

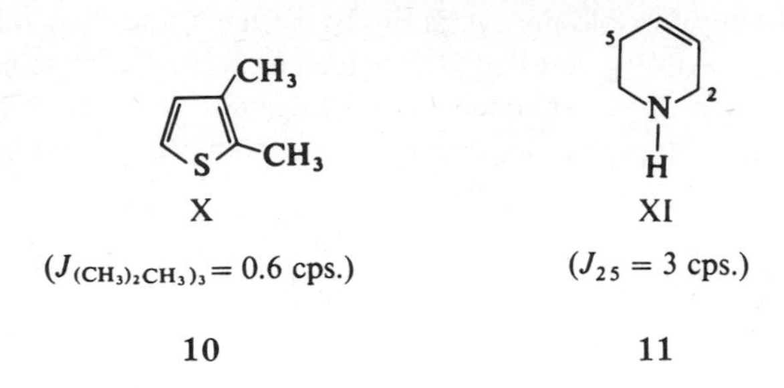

($J_{(CH_3)_2CH_3)_3}$ = 0.6 cps.) (J_{25} = 3 cps.)

10 11

6.6.3 Pseudoallylic Coupling

For the purpose of this discussion, the term quasi-allylic refers to molecules containing the general structure X–C–Z–C–Y, where Z is an sp^2-hybridized carbon or other similarly hybridized nucleus. When X and Y are protons, as in acetone, ($(CH_3)_2C{=}O$), coupling between the methyl protons on a C^{13} atom and those on the other methyl group in acetone, the C^{12} methyl group, occurs to the extent of 0.54 Hz, and represents one example of this type of coupling [34]. The coupling process operative in these instances involves overlap between the σ and π electrons, as shown in structure **12**.

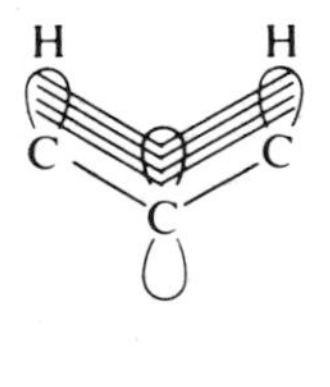

12

The coupling between the two equatorial protons (e) on the alpha carbons in *trans*-2-bromo-4-*tert*-butylcyclohexanone (**13**) is 1.7 Hz, a value enhanced by the existence of a zigzag path.

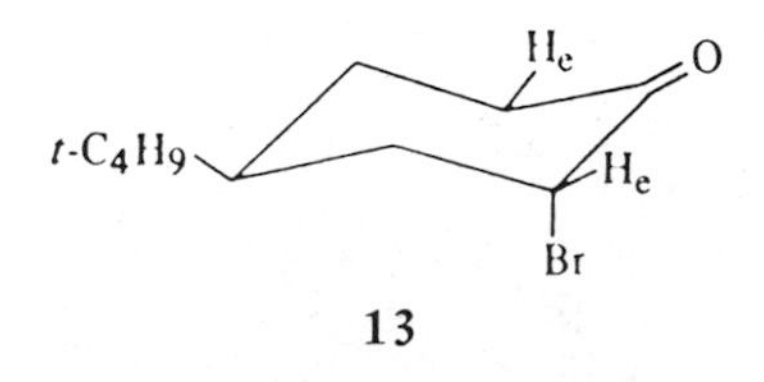

13

6.6.4 Long-Range Coupling in Polycyclic Aromatic Compounds

In addition to the already discussed *ortho*, *meta*, and para coupling constants in aromatic systems, five-bond (X—C=C—C=C—Y) and larger spin–spin interactions are frequently observed in polycyclic aromatic systems. For example, the 4,8 proton coupling constants in 1,6- (**14**) and 1,7-naphthyridine (**15**) are 0.7 and 0.8 Hz, respectively. An existing zigzag path clearly enhances the interactions in these structures.

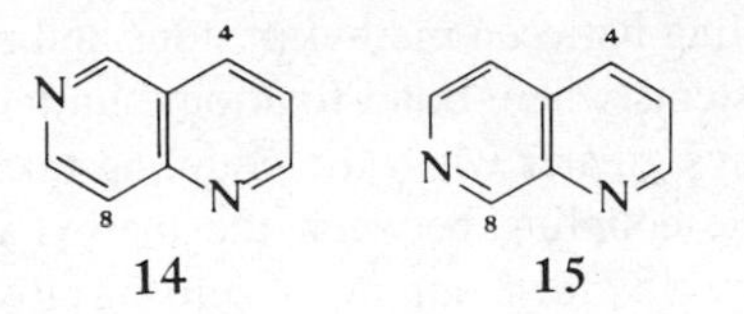

14 15

Some rather large long-range couplings have also been observed in the poly-azaindenes (**16**), where protons 1 and 3, 4 and 8, 1 and 4, and 3 and 8 are coupled to each other. In all of these instances, a π electron spin–spin coupling mechanism, augmented in some cases by the presence of a zigzag path, is almost certainly operating.

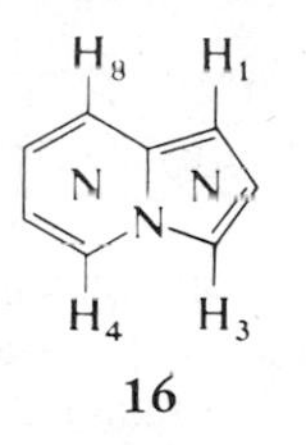

16

Interactions between aromatic ring protons and those of substituted alkyl groups are well known. The coupling constants between the *para*-protons (X in structure **17**) and the alpha protons of an alkyl substituent are of the order of 0.5 Hz.

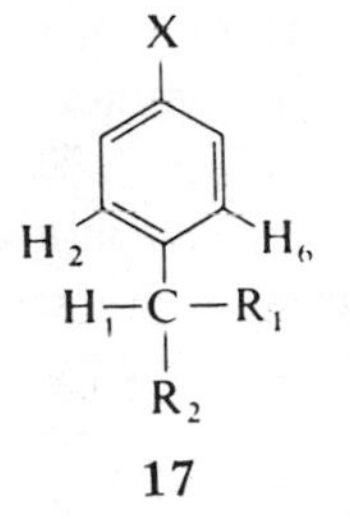

17

Similar spin–spin interactions are also observed between the methyl groups and the indicated ring protons in systems such as **18** and **19**.

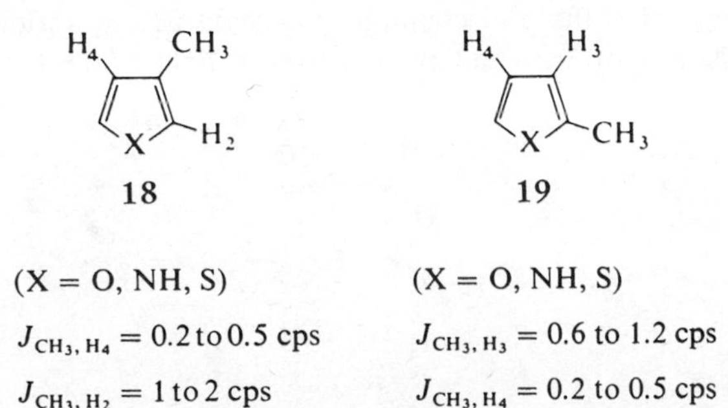

18 19

(X = O, NH, S) (X = O, NH, S)

$J_{CH_3, H_4} = 0.2$ to 0.5 cps $J_{CH_3, H_3} = 0.6$ to 1.2 cps

$J_{CH_3, H_2} = 1$ to 2 cps $J_{CH_3, H_4} = 0.2$ to 0.5 cps

The spin–spin coupling between methyl protons and ring hydrogens in aromatic and heteroaromatic systems where bond fixation, similar to that observed in naphthalene, occurs is always greater when the coupling takes place across a *localized* double bond. Thus, the coupling between the methyl group protons and the H_4 ring protons in structure **18,** for example is zero or close to it, whereas the interaction with the other protons is always approximately 1 Hz. This behavior reflects the bond order of the intervening C–C bonds, whereby in those instances where no or very small coupling occurs, the bond length is long (small bond order); the reverse exists when coupling is larger. Consequently, this coupling behavior is in accord with that already discussed for other allylic systems.

6.6.5 Summary of Long-Range Coupling Phenomena

The various long-range spin–spin interactions described and analyzed in the previous paragraphs can be summarized by the following five points:

1. Unlike *cis*- and *trans*-proton coupling constants in olefinic compounds, *cisoid* and *transoid* allylic coupling constants are of similar orders of magnitude, and the differences are usually too small to be of great predictive value.
2. The coupling constants of protons separated by an odd number of bonds are positive; they are negative when separated by an even number of bonds.
3. The size of allylic coupling constants depends on the angle that the bond joining the proton to the sp^3 carbon atom makes with the plane of the π orbitals in the unsaturated linkage.
4. The replacement of a =C—H grouping by a =C—CH_3 function alters the sign but not the size of the inter-proton coupling across the interspersed π system.
5. The largest coupling constants are observed when the bonds between the interacting nuclei have a zigzag path interrelating them.

6.7 PROBLEMS

1. The proton–proton coupling constants and proton chemical shifts of the three different types of protons in diketene are 54.78 (ppm), J = 3.92 Hz; 54.48 (ppm), J = 1.82 Hz; and 53.93 (ppm), J = 1.34 Hz. (**a**) Assign the appropriate chemical shifts and coupling constants to the various protons. (**b**) Discuss the size of the coupling constants in light of the Karplus equation.

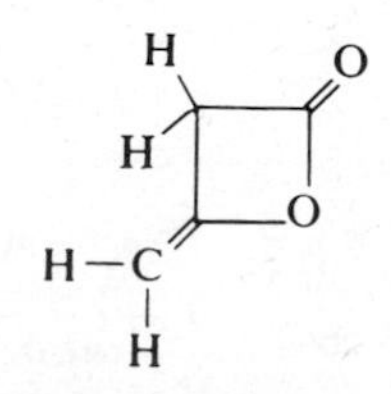

2. Estimate the π bond order of a 1,4-coupling constant in a disubstituted benzene where the constant is 0.7 Hz.
3. While long-range coupling in polycyclic heteroaromatic systems is frequently observed, it is a rarity in polycyclic carbocyclic compounds. Explain.
4. Predict the coupling pattern between the C^{13} nuclei and the protons of the methyl group in 2-methyl-propene.
5. Estimate the electronegativity of a Cl atom from the appropriate coupling constants, given in the text, for the indicated ring system.

REFERENCES

1. K. Mislow and M. Raban, *Top. Stereochem.* **1,** 1 (1966).
2. H. S. Gutowsky, *J. Chem. Phys.* **27,** 597 (1957).
3. M. Karplus, D. H. Anderson, T. C. Farrar, and H. S. Gutowsky, *J. Chem. Phys.* **27,** 597 (1957).
4. H. M. McConnell, *J. Chem. Phys.* **24,** 460 (1956).
5. J. A. Pople and D. P. Santry, *Mol. Phys.* **8,** 1 (1964).
6. J. I. Musher, *Mol. Phys.* **6,** 93 (1963).
7. W. T. Dixon, *J. Chem. Soc. (A)* 1879 (1967).
8. W. T. Dixon, *J. Chem. Soc. (A)* 1882 (1967).
9. J. A. Pople and D. P. Santry, *Mol. Phys.* **9,** 311 (1965).
10. Y. Kato and A. Saika, *J. Chem. Phys.* **46,** 1975 (1967).
11. W. McFarlane, *Q. Rev. (Lond.)* **23,** 187 (1969).
12. V. Bystrov, *Russ. Chem. Rev.* **41,** 281 (and references therein (1972)).
13. S. Sternhell, *Q. Rev. (Lond.)* **23,** 236 (1969).
14. J. B. Lambert, H. F. Shurvell, L. Verbit, R. G. Cooks, and G. H. Stout, *Organic Structural Analysis*, Macmillan, New York, 1976, p. 68 ff.
15. P. Chandra and P. T. Narasimhan, *Mol. Phys.* **12,** 523 (1967).
16. H. Gunther, *Tetrahedron Lett.* 2967 (1967).
17. H. G. Hecht, *J. Phys. Chem.* 1761 (1967).
18. M. Karplus, *J. Am. Chem. Soc.* **85,** 2870 (1963); and M. Barfield and D.M. Grant, *J. Am. Chem. Soc.* **85,** 1899 (1963).
19. R. A. Niedrich, D. M. Grant, and M. Barfield, *J. Chem. Phys.* **42,** 3733 (1965).
20. H. M. McConnell, *J. Mol. Spectrosc.* 11 (1957).
21. M. Karplus, *J. Am. Chem. Soc.* **82,** 4431 (1960).
22. J. D. Roberts, *Angew. Chem. Int. Ed.* 53 (1963).
23. C. D. Hall and L. Hough, *Proc. Chem. Soc.* 382 (1962).
24. J. Meinwald and A. Lewis, *J. Am. Chem. Soc.* **83,** 2769 (1961).
25. J. D. Roberts, D. R. Davis, and M. Takahashi, *J. Am. Chem. Soc.* **84,** 2935 (1962).
26. S. Sternhell, *Pure Appl. Chem.* **14,** 15 (1964).
27. A. D. Cross and P. W. Landis, *J. Am. Chem. Soc.* **84,** 1736 (1962).
28. L. Petrakis and C. H. Sederholm, *J. Chem. Phys.* **35,** 1243 (1961).

29. H. M. McConnell, *J. Mol. Spectrosc.* **1,** 11 (1957).
30. M. S. J. Dewar and R. C. Fahey, *J. Am. Chem. Soc.* **85,** 2706 (1963).
31. M. Karplus, *J. Am. Chem. Soc.* **82,** 4432 (1960).
32. M. Karplus, *J. Chem. Phys.* **33,** 1842 (1960).
33. P. J. Black and M. L. Heffernan, *Aust. J. Chem.* **18,** 707 (1965).
34. D. W. Moore, *J. Chem. Phys.* **34,** 1470 (1961).
35. J. R. Holmes and D. Kivelson, *J. Am. Chem. Soc.* **83,** 2959 (1961).
36. J. A. Elvidge and L. M. Jackman, *J. Chem. Soc.* 859 (1961).
37. J. P. Albrand, D. Gagnaire, and J. B. Robert, *Chem. Commun.* 1469 (1968).

7

SECOND-ORDER COUPLING PATTERNS

7.1 GENERAL CONSIDERATIONS

Although analyses of NMR splitting patterns by the $n + 1$ rule (see Chapter 3) greatly facilitate structure determination studies of unknown compounds, its application is limited to systems where the chemical shift differences between the interacting nuclei and the sizes of the coupling constants are greater than ten. When this is the case, the coupling constants and chemical shifts can be read directly from the spectral recording. However, as the chemical shift to coupling constant ratio decreases, the complexity of the spectra increases to a point where it is no longer feasible to obtain the coupling constants and/or chemical shifts directly from the recorded spectra [1–4].

For example, the typical *A* pattern of two doublets, separated by the chemical shift differences of nuclei *A* and *B*, changes as the chemical shift to coupling constant ratio decreases, to a point where the *inner* two peaks overlap and appear as a singlet. At this point, the *outer* peaks are of such low intensity as to become *invisible* over the instrument recording noise.

7.2 TWO INTERACTING NUCLEI—THE *AB* SYSTEM

The chemical shift and coupling constants of the four-line *AB* spin systems can be computed from the expressions given in Table 7.1. The relative intensities of the inner lines (lines 2 and 3 in Fig. 7.1) are always either identical to, or more intense than, the outer lines (lines 1 and 4 in Fig. 7.1). Knowledge of this behavior is of some importance when a complex spectrum is being analyzed, since it facilitates pattern identification.

While the coupling constant of an *AB* system is readily identified, it corresponds to the separation of lines 1 and 2 as well as 3 and 4, no specific peak position is identical to the chemical shift of either nucleus (*A* or *B*). Since the two *A* peaks, as well as the two *B* peaks, are not of equal intensity, the respective chemical shifts are given by the weighted average positions. The three *AB* system extremes are shown diagramatically in Figure 7.1.

As the data given in Table 7.1 show, the separation between the inner lines (lines 2 and 3) is equal to $2C - J$. Thus, the value of C can be computed. This, in turn, affords the chemical shift difference between the two nuclei (see equation in footnote in Table 7.1). Because the difference between lines 1 and 3 is the value of $2C$, a convenient check of the *AB* pattern assignment in complex multinuclear systems is available.

An analysis of the *AB* spectrum, reproduced in Figure 7.2, serves to illustrate the application of the relationships given in Table 7.1. The separation between

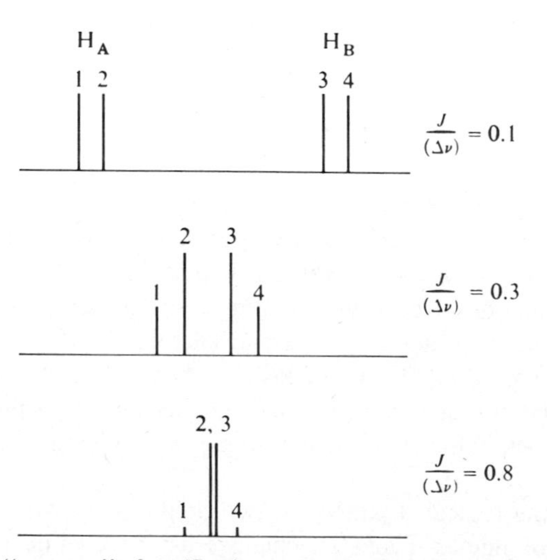

Figure 7.1. The three "extremes" of an *AB* spin system. (Assumption: J is larger than 0; ν_B is greater than ν_A.)

TABLE 7.1 Line Positions (Energies) and Intensities for Two Nuclei (AB)

Line Number	Energy[a]	Relative intensities[b]
1	$\frac{1}{2}J + C$	$1 - \sin 2\theta$
2	$-\frac{1}{2}J + C$	$1 + \sin 2\theta$
3	$\frac{1}{2}J - C$	$1 + \sin 2\theta$
4	$-\frac{1}{2}J - C$	$1 - \sin 2\theta$

SOURCE: From *High-Resolution Nuclear Magnetic Resonance* by J. A. Pople, W. G. Schneider, and H. J. Bernstein, p. 121. Copyright 1959 by McGraw-Hill Book Company. Used with permission of McGraw-Hill Book Company.

[a] $C = \frac{1}{2}[(\nu_A - \nu_B)^2 + J_{AB}^2]^{1/2}$.

[b] $\sin 2\theta = \frac{1}{2}J_{AB}/C$.

lines 1 and 2 (or 3 and 4), 5 Hz, corresponds to the *AB* coupling constant. Since the separation between lines 2 and 3, the value of $C + J_{AB}$, is 15 Hz, the constant C is 10 Hz (15–5). Application of the expression for the constant C affords the chemical shift difference, 19.4 Hz, between nuclei A and B.

With the coupling constant and $\nu_A - \nu_B$ known, the relative intensities of the spectral lines for this *AB* system can be computed by application of the expressions given in Table 7.1. The computed relative intensities are superimposed on the experimental spectrum of Figure 7.2.

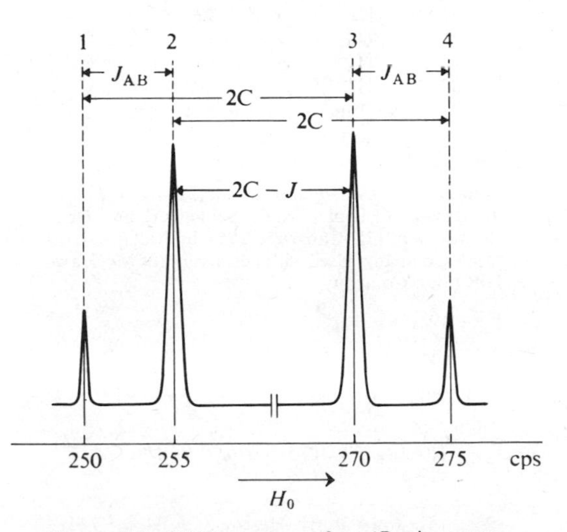

Figure 7.2. NMR spectrum of an *AB* spin system.

7.3 THREE INTERACTING NUCLEI

In homonuclear systems where the three interacting nuclei are of the same isotopic species, there exist two possible general patterns, AB_2 and ABX. In heteronuclear systems, the number of patterns is clearly much larger. However, since the chemical shift differences in the latter systems will generally be much larger than the coupling constants, normal first-order rules apply.

7.3.1 The AB_2 System

For this three-nucleus system, with two equivalent nuclei, two different coupling constants, J_{AB} and J_{BB}, can be expected. While the equivalent first-order AX_2 spectra are identified by a triplet for the A nucleus and a doublet for the X nuclei, the non-first-order AB_2 spectra can have a maximum of nine lines. Four of these are due to the A nuclei, four are caused by the B nuclei, and one line, a so-called antisymmetrical one, is caused by a combination of transitions between the A and B nuclei. Table 7.2 lists the energies (line positions) for AB_2 systems.

The shape of the spectral pattern of this system is independent of J_{BB} and depends

TABLE 7.2 Line Positions for a Three-Nucleus AB_2 Spin System

Line	Origin	Energy
1	H_A	$\nu_0(1 - E) + \frac{3}{4}J + C_+$
2	H_A	$\nu_0(1 - \sigma_B) + C_+ + C_-$
3	H_A	$\nu_0(1 - \sigma_A)$
4	H_A	$\nu_0(1 - E) - \frac{3}{4}J + C_-$
5	H_B	$\nu_0(1 - \sigma_B) + C_+ - C_-$
6	H_B	$\nu_0(1 - E) + \frac{3}{4}J - C_+$
7	H_B	$\nu_0(1 - \sigma_B) - C_+ + C_-$
8	H_B	$\nu_0(1 - E) - \frac{3}{4}J - C_-$
9	combination	$\nu_0(1 - \sigma_B) - C_+ - C_-$

SOURCE: *High-Resolution Nuclear Magnetic Resonance* by J. A. Pople, W. G. Schneider, and H. J. Bernstein, p. 125. Copyright 1959 by McGraw-Hill Book Company. Used with permission of McGraw-Hill Book Company.

$$E = \tfrac{1}{2}(\sigma_A + \sigma_B)$$

$$\nu_0 = \frac{\gamma H_0}{2\pi}$$

$$C_+ = \tfrac{1}{2}[(\nu_0(\sigma_B - \sigma_A)]^2 + [\nu_0(\sigma_B - \sigma_A)]J + \frac{9}{4}J^2)^{1/2}$$

$$C_- = \tfrac{1}{2}\{[\nu_0(\sigma_B - \sigma_A)]^2 - [\nu_0(\sigma_B - \sigma_A)]J + \frac{9}{4}J^2\}^{1/2}$$

only on the ratio of $J_{AB}/(\nu_B - \nu_A)$. Consequently, apart from a scaling factor, the shape of AB_2 spectra and the relative line intensities and line separations depend only on this dimensionless ratio. Because of this, these spectra can be analyzed without recourse to computer applications.

The data in Table 7.2 allow the following generalizations to be made:

1. The chemical shift of nucleus A is equivalent to the position of line 3.
2. The chemical shift of nucleus B is the mean position of lines 5 and 7.
3. Reduced transition energies can be calculated for various ratios of $J_{AB}/(\nu_B - \nu_A)$. These values can be used to obtain the AB coupling constant from any AB_2 spectrum, once the chemical shifts for the various nuclei have been determined.

A graphic representation of the relative changes in the various line positions, as the ratio of the AB coupling constant to the chemical shift difference between nuclei A and B changes, is given in Figure 7.3. These graphs also facilitate recognition of this type of spin system. Clearly, as the ratio increases (the chemical shift differences in the Fig. 7.3 spectra are constant), the splitting pattern becomes more and more spread out, and the intensities of lines 1, 2, 8, and 9 decrease. In many instances, if not most, the combination line (line 9) is of such low intensity as to be nonrecognizable.

The separation between lines 5 and 6 is frequently so small as to appear, at best, as a broadened singlet. The visual effect is, consequently, the erroneous appearance of a seven-line spectrum. This possibility must be kept in mind when pattern identification of an unknown spectrum is required.

If a particular spectrum represents only one AB_2 pattern and is not complicated by the presence of other peaks, J_{AB} can be computed from the following sum: $\frac{1}{3}[\nu_1 - \nu_4 + \nu_6 - \nu_8]$. However, this facile analysis is not often feasible! The

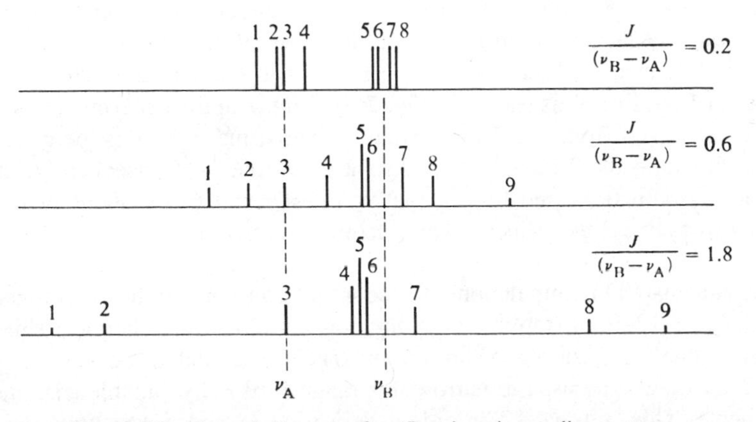

Figure 7.3. Theoretical spectra for AB_2 spin–spin coupling systems.

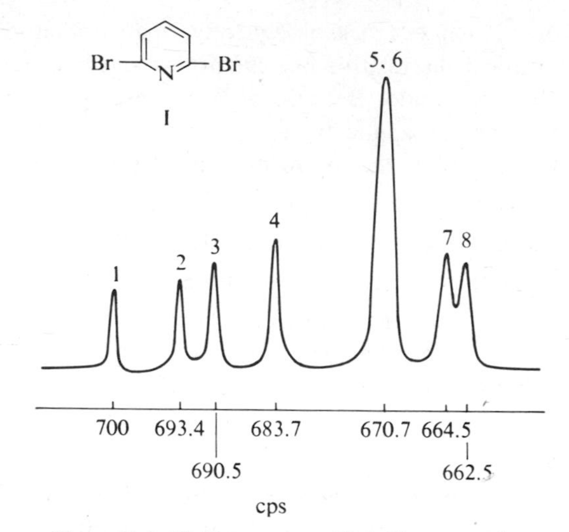

Figure 7.4. NMR spectrum of 2,6-dibromopyridine.

more usually applied process, utilized in instances where more than one splitting pattern is present, is more cumbersome. Its application to an actual example serves to demonstrate the technique.

The spectrum of 2,6-dibromopyridine given in Figure 7.4 represents a typical spectrum insofar as line 9 is not visible, and lines 5 and 6, always the most intense lines of an AB_2 pattern, are overlapping. The reduced transition energy table (Fig. 7.5) lists the relative line positions with respect to line 3 (set at 0 Hz). The peak positions of the experimental spectrum are recomputed by placing the peak at 690.5 Hz, the line 3 peak, at zero. The corresponding adjusted values for all of the peaks are listed in Table 7.3.

The relative line positions of lines 2 and 4 of the experimental spectrum, with respect to line 3, show that line 2 is closer to the reference line (line 3) than is line 4. Consequently, according to Figure 7.5, the ratio of the coupling constant to the chemical shift in this spectrum must be below 0.8 (at this ratio, the 2 lines are equidistant from line 3, as shown in Fig. 7.5). Line 1 in the experimental spectrum is further removed from line 3 than is line 4. Thus, the ratio must be greater than 0.35 (at this ratio the 2 lines are equidistant from line 3). Consequently, the ratio must lie between 0.35 and 0.80! Finally, lines 5 and 6 are about twice as far removed from line 3 as is line 1. This occurs in the range 0.35–0.40 (the use of a ruler simplifies reading this from the graph). Consequently, without taking recourse to any mathematical manipulations, the coupling constant can be determined to be in the range 8.12–9.28 (ratio = coupling constant/23.2; the chemical shift is the average of the line positions of lines 5 and 7). The tabulated reduced energies of Table 7.2 can now be used to narrow this range further by suitable trial-and-error adjustments. This, ultimately, affords a ratio of 0.38 and a coupling constant of

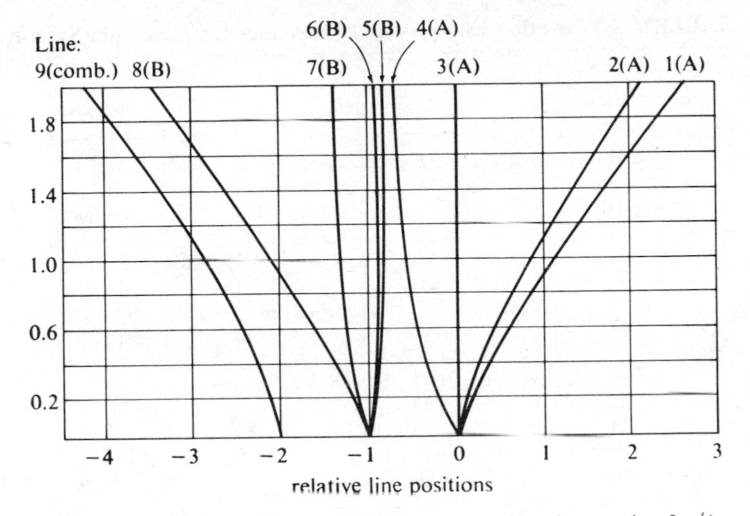

Figure 7.5. Changes in line positions of AB_2 spectra with changes in $J_{AB}/(\nu_B - \nu_A)$. Y axis $= J_{AB}/(\nu_B - \nu_A)$.

TABLE 7.3 Theoretical and Experimental Spectra for 2,6-Dibromopyridine

Line	Positions observed	Positions calculated[a]	Reduced transition energies for $J_{AB}/(\nu_B - \nu_A) = 0.38$
1	−9.5	−9.5	0.44
2	−2.9	−3.0	0.14
3	0	0	0
4	6.8	6.5	−0.30
5,6	19.8	17.8, 18.7	−0.83, −0.87
7	26.0	25.2	−1.17
8	28.0	27.3	−1.27
9	—	46.0	−2.14

[a] The scale adjustment ratio is −9.5/0.44 = −21.5.

8.82 Hz. For most purposes, this last step is not necessary, and the average of the coupling constant range can be used. This value, in the above instance, is 8.7 Hz.

7.3.2 The *ABX* System

The simplest three-spin system is one where three different magnetically active nuclei wih significantly different chemical shifts interact with each other. Under these limiting conditions, three sets of ‘‘doublets of doublets’’ are observed, one for each of the nuclei.

When two of these nuclei have similar chemical shifts and coupling constants,

TABLE 7.4 Transition Energies and Intensities for *ABX* Spin Systems

Line	Origin	Energy	Relative intensities
1	H_B	$\nu_{AB} + \frac{1}{4}(-2J_{AB} - J_{AX} - J_{BX}) - D_-$	$1 - \sin 2\phi_-$
2	H_B	$\nu_{AB} + \frac{1}{4}(-2J_{AB} + J_{AX} + J_{BX}) - D_+$	$1 - \sin 2\phi_-$
3	H_B	$\nu_{AB} + \frac{1}{4}(2J_{AB} - J_{AX} - J_{BX}) - D_-$	$1 + \sin 2\phi_-$
4	H_B	$\nu_{AB} + \frac{1}{4}(2J_{AB} + J_{AX} + J_{BX}) - D_+$	$1 + \sin 2\phi_+$
5	H_A	$\nu_{AB} + \frac{1}{4}(-2J_{AB} - J_{AX} - J_{BX}) + D_-$	$1 + \sin 2\phi_-$
6	H_A	$\nu_{AB} + \frac{1}{4}(-2J_{AB} + J_{AX} + J_{BX}) + D_+$	$1 + \sin 2\phi_+$
7	H_A	$\nu_{AB} + \frac{1}{4}(2J_{AB} - J_{AX} - J_{BX}) + D_-$	$1 - \sin 2\phi_-$
8	H_A	$\nu_{AB} + \frac{1}{4}(2J_{AB} + J_{AX} + J_{BX}) + D_+$	$1 - \sin 2\phi_+$
9	H_X	$\nu_X - \frac{1}{2}(J_{AX} + J_{BX})$	1
10	H_X	$\nu_X + D_+ - D_-$	$\cos^2(\phi_+ - \phi_-)$
·11	H_X	$\nu_X - D_+ + D_-$	$\cos^2(\phi_+ - \phi_-)$
12	H_X	$\nu_X + \frac{1}{2}(J_{AX} + J_{BX})$	1
14	comb.	$\nu_X - D_+ - D_-$	$\sin^2(\phi_+ - \phi_-)$
15	comb.	$\nu_X + D_+ + D_-$	$\sin^2(\phi_+ - \phi_-)$

SOURCE: Taken from *High-Resolution Nuclear Magnetic Resonance* by J. A. Pople, W. G. Schneider, and H. J. Bernstein, p. 134. Copyright 1959 by McGraw-Hill Book Company. Used with permission of McGraw-Hill Book Company.

$$D_\pm = \tfrac{1}{2}\{[\nu_A - \nu_B \pm \tfrac{1}{2}(J_{AX} - J_{BX})]^2 + J_{AB}{}^2\}^{1/2}$$

$$\sin 2\phi_- = \frac{\frac{1}{2}J_{AB}}{D_-} \qquad \sin 2\phi_+ = \frac{\frac{1}{2}J_{AB}}{D_+}$$

$$\cos 2\phi_- = \frac{[\frac{1}{2}(\nu_A - \nu_B) - \frac{1}{4}(J_{AX} - J_{BX})]}{D_-} \qquad \cos 2\phi_+ = \frac{[\frac{1}{2}(\nu_A - \nu_B) + \frac{1}{4}(J_{AX} - J_{BX})]}{D_+}$$

$$\nu_{AB} = \tfrac{1}{2}(\nu_A + \nu_B)$$

the complexity of the spectrum increases significantly, and the maximum number of possible absorption lines increases to 14. Of these, there are four lines each for the *A* and *B* nuclei and six lines due to the *X* nucleus. The *A* and *B* nuclei multiplets generally overlap in these systems. The chemical shift of the *X* nucleus is given by the midpoint of the six-line *X* nucleus-caused pattern.

The line numbers, origins, energies, and relative intensities of *ABX* spin systems are given in Table 7.4. Clearly, an *ABX* spectrum is determined by the chemical shifts of the three different nuclei and the three coupling constants, J_{AB}, J_{AX}, and J_{BX}. The *AB* part of an *ABX* spectrum consists of two *AB* patterns, composed,

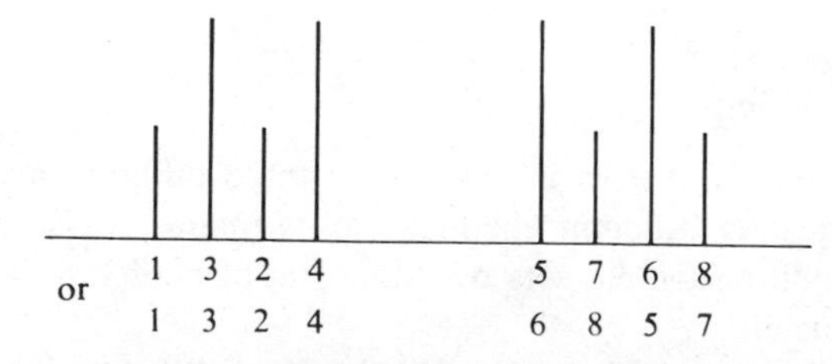

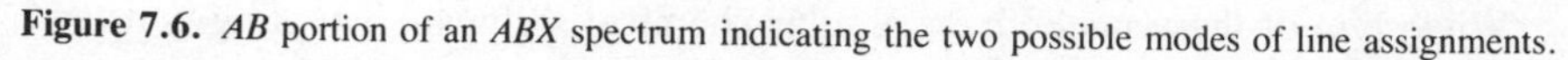

Figure 7.6. *AB* portion of an *ABX* spectrum indicating the two possible modes of line assignments.

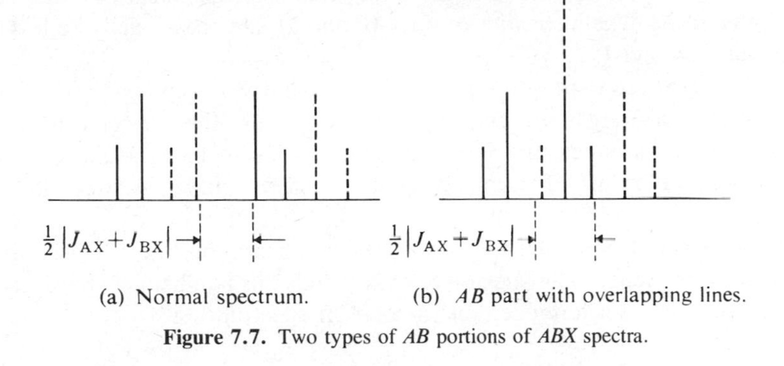

Figure 7.7. Two types of *AB* portions of *ABX* spectra.

respectively, of lines 1, 3, 5, and 7 and lines 2, 4, 6, and 8. The separation between lines 1 and 3, 5 and 7, 2 and 4, and 6 and 8 is equal and corresponds to J_{AB}. However, as Figure 7.6 demonstrates, a dilemma can arise when attempts are made to assign a given pair of doublets to a particular *AB* system. A priori, both of the line numberings shown in this figure are feasible.

However, since the separation between the centers of the two *AB* systems is equal to $\frac{1}{2}[J_{AX} + J_{BX}]$ (see Table 7.4), only one of the possible assignments is correct. Fortunately, the separation of the two most intense lines of the six-line *X* nucleus pattern corresponds to the sum of J_{AX} and J_{BX} and thus designates the correct *AB* pattern assignment. In some instances the two *AB* patterns overlap in such a fashion that a situation as shown in Figure 7.7(b) prevails. The maximum number of lines, six, that are due to the *X* nucleus appear as two symmetrically situated three-line patterns (see Fig. 7.8).

As already mentioned, the separation between the two most intense lines (lines 9 and 12) is always equal to the sum of the *AX* and *BX* coupling constants, and

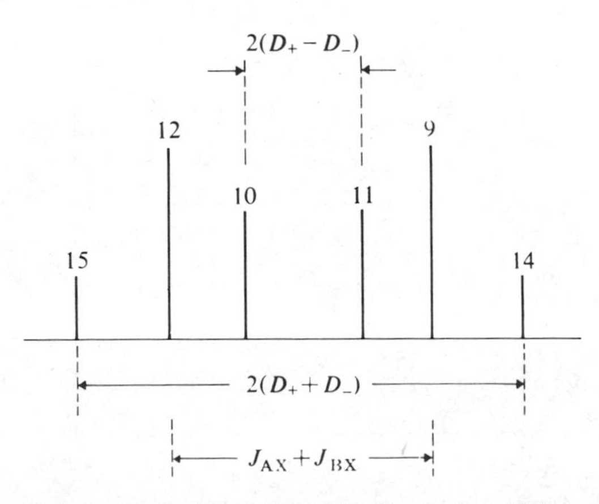

Figure 7.8. Schematic illustration of the *X* part of an *ABX* spectrum.

serves as a check for the correctness of the *AB* pattern assignments. One cautionary note must be made. The intensities of lines 10 and 11 can occasionally be less than those of lines 14 and 15!

In analyzing an *ABX* spectrum, two sets of *AX* and *BX* coupling constants with either like or unlike signs are obtained. However, only the correct combination gives the proper line intensities of the *X* pattern. The methods available for the determination of the signs of various coupling constants will be described in Chapter 9.

The reader may have noted the absence of an absorption line 13. This line is not omitted in deference to any negative connotations this number may have; rather, this absorption line is a forbidden one in an *ABX* spectrum.

7.4 FOUR INTERACTING NUCLEI

7.4.1 General Considerations

Analyses of four-spin systems are readily accomplished by first-order methods when AX_3 and A_2X_2 patterns are involved. When the complexity is increased to $AA'BB'$ systems, computer analysis is generally required. Fortunately, most of these four-nuclei systems fall into the $AA'XX'$ patterns, which can be analyzed without extensive computer utilization. These spectra are determined by the chemical shifts of the *A* and *X* nuclei and the AA', XX', AX, and AX' coupling constants.

Figure 7.9 is a graphical presentation of a general A_2B_2 system where six interactions between the four nuclei can be envisioned. If the molecule has a plane of symmetry, $J_{A_1}B_{B_2}$ and $J_{A_2}B_{B_1}$, and $J_{A_1}B_{B_1}$ and $J_{A_2}B_{B_2}$ will be, respectively, equal. Consequently, the two pairs of coupling constants can be designated by J_{AB}' and J_{AB}, respectively. Consequently, the number of coupling constants that must be considered is reduced to four for this type of system.

It has been shown that A_2B_2 spectra have a maximum of 28 lines, whereas A_2X_2 spectra have 24. One of the unique features of this type of system, which allows

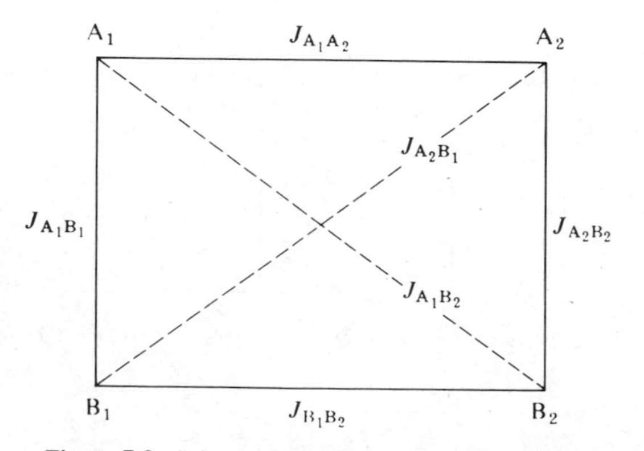

Figure 7.9. Spin-coupling constants in A_2B_2 systems.

their immediate recognition, is the total *symmetry* of these spectra. Thus, for example, in a 12-line A_2X_2 system, the maximum 12 lines caused by the A nuclei is matched by a mirror reflection of the same number of lines due to the X nuclei.

To minimize the number of possible parameters some convenient new symbols have been defined:

$$K = J_{AA} + J_{BB}$$

$$M = J_{AA} - J_{BB}$$

$$L = J_{AB} - J'_{AB}$$

$$N = J_{AB} + J'_{AB}$$

Table 7.5 gives the energies, line intensities, and line numbers for A_2X_2 coupling systems [3, 5, 6].

7.4.2 The A_2X_2 System

The energies (relative to nucleus A (or X) at 0 Hz) and the relative line intensities of the various lines are given in Table 7.5. A theoretical spectrum of one of the

TABLE 7.5 Energies and Line Intensities for A_2X_2 Spin Systems

Line	Energy relative to H_A at 0 cps	Relative intensities
1	$\frac{1}{2}N$	1
2	$\frac{1}{2}N$	1
3	$-\frac{1}{2}N$	1
4	$-\frac{1}{2}N$	1
5	$\frac{1}{2}K + \frac{1}{2}(K^2 + L^2)^{1/2}$	$\sin^2 \phi_S$
6	$-\frac{1}{2}K + \frac{1}{2}(K^2 + L^2)^{1/2}$	$\cos^2 \phi_S$
7	$\frac{1}{2}K - \frac{1}{2}(K^2 + L^2)^{1/2}$	$\cos^2 \phi_S$
8	$-\frac{1}{2}K - \frac{1}{2}(K^2 + L^2)^{1/2}$	$\sin^2 \phi_S$
9	$\frac{1}{2}M + \frac{1}{2}(M^2 + L^2)^{1/2}$	$\sin^2 \phi_A$
10	$-\frac{1}{2}M + \frac{1}{2}(M^2 + L^2)^{1/2}$	$\cos^2 \phi_A$
11	$\frac{1}{2}M - \frac{1}{2}(M^2 + L^2)^{1/2}$	$\cos^2 \phi_A$
12	$-\frac{1}{2}M - \frac{1}{2}(M^2 + L^2)^{1/2}$	$\sin^2 \phi_A$

SOURCE: Taken from *High-Resolution Nuclear Magnetic Resonance* by J. A. Pople, W. G. Schneider, and H. J. Bernstein, p. 141. Copyright 1959 by McGraw-Hill Book Company. Used with permission of McGraw-Hill Book Company.

$\cos 2\phi_S : \sin 2\phi_S : 1 = K : L : (K^2 + L^2)^{1/2}$
$\cos 2\phi_A : \sin 2\phi_A : 1 = M : L : (M^2 + L^2)^{1/2}$

nuclei in an $AA'XX'$ is given in Figure 7.12. An inspection of the relationships given in Table 7.5 reveals some diagnostically useful information:

1. The shape of these types of spectra is sensitive to the sums and differences of the coupling constants rather than to the absolute values.
2. The chemical shifts of lines 1 and 2 are identical. Thus they appear as a singlet, with relative intensity 2. Similarly, lines 3 and 4 also have the same chemical shift and, consequently, will also appear as an intense (relative intensity 2) singlet. The average position of these two intense singlets is the chemical shift of nucleus A if the A part of the symmetrical spectrum is being examined, or the chemical shift of nucleus B if the B part of the spectrum is being studied.
3. The *separation* of these two intense lines is equal to the value N which, in turn, is the sum of J_{AB} and J'_{AB}.
4. Lines 5, 6, 7, and 8 and 9, 10, 11, and 12 are two symmetrical quartets that are centered on the chemical shift of nucleus A (or B, depending on the portion of the spectrum being examined).
5. The *inner* lines of these *quartets*, lines 6 and 7 and lines 10 and 11, are always more intense than the outer lines, 5 and 8, and 9 and 12.
6. The 12 lines of the A and the B nuclei are symmetrically located about the frequencies of the A and B nuclei, respectively.

These observations also show that, in addition to the symmetrical situation existing between the A and B nuclei patterns with respect to each other, there also exists a reflection symmetry within each set of lines.

Once the various symmetry features of an $AA'XX'$ spectrum are recognized, the various coupling constants can be obtained from the values of K, L, M, and N. Since it is not feasible to distinguish between K and M, it is also not feasible to establish which value corresponds to J_{AA} or J_{BB}. Similarly, J_{AB} and J_{AA}' also cannot be directly assigned. However, since approximate ranges for different types of coupling constants are well established, the various values can be readily assigned to the appropriate spin system by a comparison of the sizes of the different ambiguous coupling constants.

The analysis of the proton portion of the NMR spectrum of *cis*-1,2,-difluoroethylene (HFC=CFH) serves to demonstrate the analysis process. The spectrum presented in Figure 7.10 is the proton portion of the NMR spectrum of this ethylene derivative. An equal pattern, although with different chemical shift for the fluorine nucleus, exists. Examination of the proton spectrum establishes that the center of the pattern is at −250 Hz from TMS. This value, 250 Hz (a positive value is assigned as described in Chapter 3), is the chemical shift of the olefinic protons in this compound. The identically patterned fluorine spectrum is centered at 660 Hz from CCl_3F as the internal fluorine standard. Thus, the fluorine chemical shift is at 660 Hz with respect to that standard.

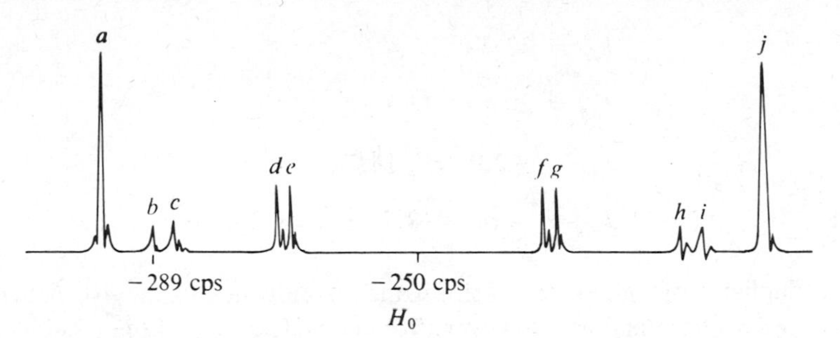

Figure 7.10. Proton NMR spectrum of *cis*-1,2-difluoroethylene. [Partially redrawn from F. G. W. Flynn, M. Matsushima, and J. D. Baldeschwieler, *J. Chem. Phys.*, **9,** 2295 (1963).]

The line positions of the proton spectrum of *cis*-1,2-difluoroethylene with respect to the center are:

Line	Cps	Relative intensity
aj	± 46.5	2.00
bi	± 38.5	0.32
ch	± 35.7	0.35
dg	± 19.1	0.65
ef	± 17.8	0.68

The strongest signals, *a* and *j*, correspond to the strong doublet whose separation of 93.0 Hz is equal to *N*.

The two possible quartets can be paired as follows: (1) lines *b*, *e* and *f*, *i* correspond to lines 5, 6 and 7, 8 (see Table 7.5), and lines *c*, *d* and *g*, *h* correspond to lines 9, 10 and 11, 12; or (2) lines *b*, *d* and *g*, *i* and lines *c*, *e* and *f*, *h* belong together.

The first pair of values yields for *K* (or *M*) (separation between lines *b*, *e*, or *f*, *i*) a value of 20.7 Hz and for *M* (or *K*) (separation between lines *c*, *d* or *g*, *h*) a value of 16.6 Hz. The second pair of assignments affords 19.4 and 17.9 Hz, respectively, for *K* and *M*.

The sum of lines 5 and 6 (or 7 and 8) is equal to $(K^2 + L^2)^{0.5}$ (see Table 7.5). This value is equal to 38.5 + 17.8 or 56.3 Hz and yields 20.7 Hz for *K* and 52.3 Hz for *L*. Repetition of this calculation employing the alternative quartet assignment shows that this alternate assignment is not correct, since no consistent value for *L* can be obtained from the two quartets.

The parameters calculated thus far (*K*, *L*, *M*, and *N*) are sufficient to compute the various coupling constants of this system:

$$J_{AB} = 72.7 \text{ or } 20.4$$

$$J'_{AB} = 20.4 \text{ or } 72.7$$

$$J_{AA} = \pm 2.0 \text{ or } \pm 18.7$$

$$J_{BB} = \pm \text{ or } \mp 18.7 \text{ or } \pm \text{ or } \mp 2.0$$

The duplicity of values results from the inability to distinguish between the absolute values of K and M. However, from considerations of the relative sizes of the various coupling constants (see Chapter 2), the following specific assignments can be made:

$$J_{HF}(gem) = J_{AB} = +72.7$$

$$J_{HF}(trans) = J'_{AB} = +20.4$$

$$J_{HH}(cis) = J_{AA} = \mp 2.0$$

$$J_{FF}(cis) = J_{BB} = \pm 18.7$$

The A_2B_2 System with $J_{AB} = J'_{AB}$

As mentioned earlier, the maximum number of lines possible for an A_2B_2 system is 28. The analysis of such a pattern without the use of a computer becomes tedious, if not impossible. However, if the AB coupling constants in an A_2B_2 are identical, the pattern is independent of J_{AA} and υ_{BB}, and its shape is controlled only by the

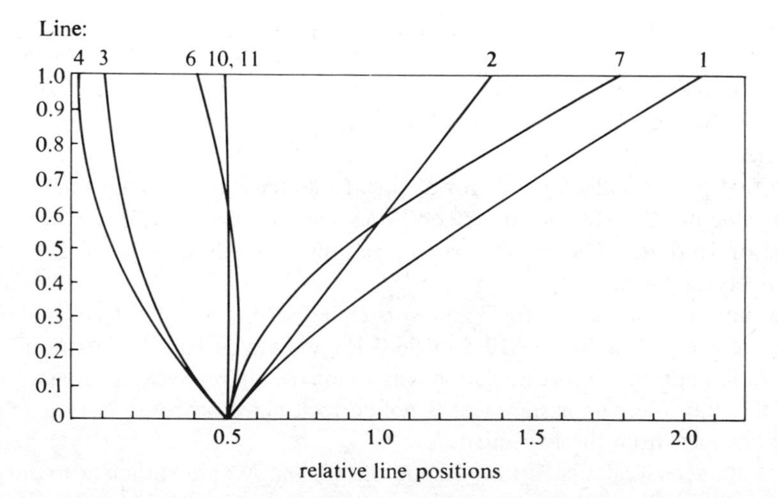

Figure 7.11. Relative line positions of A transitions for A_2B_2 spin systems with two equal J_{AB} coupling constants. Y axis $= J/(\nu_B - \nu_A)$.

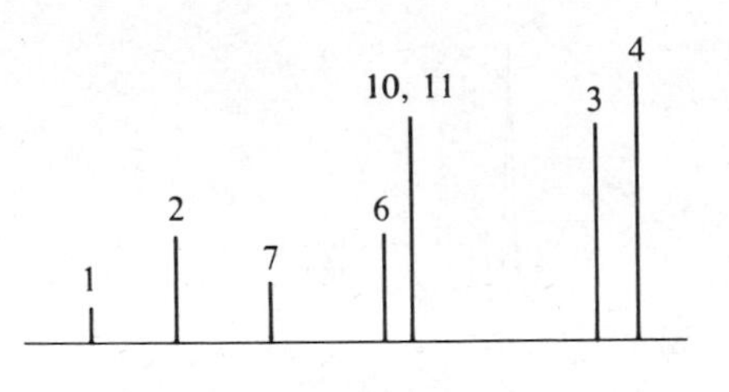

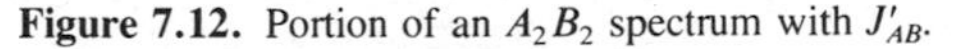

Figure 7.12. Portion of an A_2B_2 spectrum with J'_{AB}.

ratio J_{AB}/(frequency relative to the center band). The number of possible lines for each part of the symmetrical pattern is reduced to 7 (actually 8 lines, but lines 10 and 11 appear as a singlet).

A graphical representation of the changes in the chemical shifts of the various lines with changes in the J_{AB}/(frequency relative to the center band) is given by Figure 7.11. This graph can be used in a fashion identical to that earlier employed for the analysis of AB_2 spectral patterns.

The position of the line 10, 11 singlet corresponds to the chemical shift of nucleus A (or B, if the other half of the spectrum is being examined). A theoretical spectrum with $J/(\nu_B - \nu_A) = 0.3$ is presented in Figure 7.12.

Clearly, the symmetry observed for each of the nuclei patterns in the A_2X_2 instance is no longer present. However, both A and B nuclei have identical *mirror image* patterns.

7.4.4 The A_2B_2 System with $J_{AB} >> J'_{AB}$

There exist a large number of organic compounds, generally *ortho*-disubstituted phenyl derivatives, where the vicinal coupling constant (J_{AB}) is much larger than the *meta*-coupling constant (J'_{AB}) and where J_{BB} is also considerably greater than J_{AA}. The theoretical pattern of a spin system of this type is shown in Figure 7.13. Again, there is no clear symmetry for the A or B nuclei patterns. However, the A and B patterns are themselves mirror images of each other!

As the chemical shift difference between nuclei A and B increases, lines 1 and 2 on the one hand, and lines 3 and 4 on the other, coalesce into two singlets. Concomitant with this change, the intensities of lines 7, 8, 11, and 12 increase.

The experimental and theoretical H NMR spectra (exclusive of the heterocyclic ring protons) of quinoxaline and phthalazine are reproduced in Figure 7.14. These

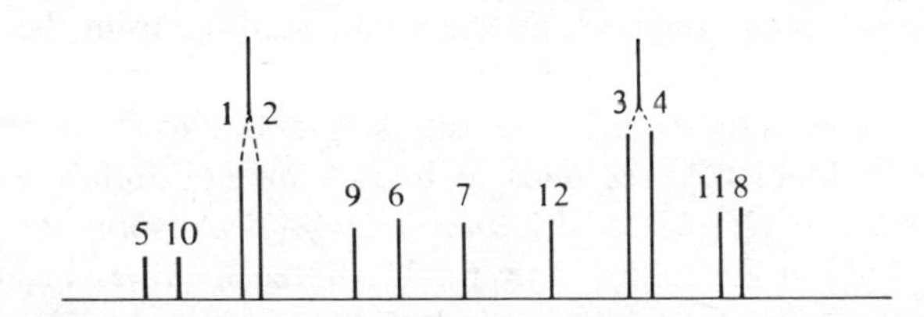

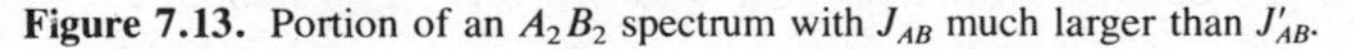

Figure 7.13. Portion of an A_2B_2 spectrum with J_{AB} much larger than J'_{AB}.

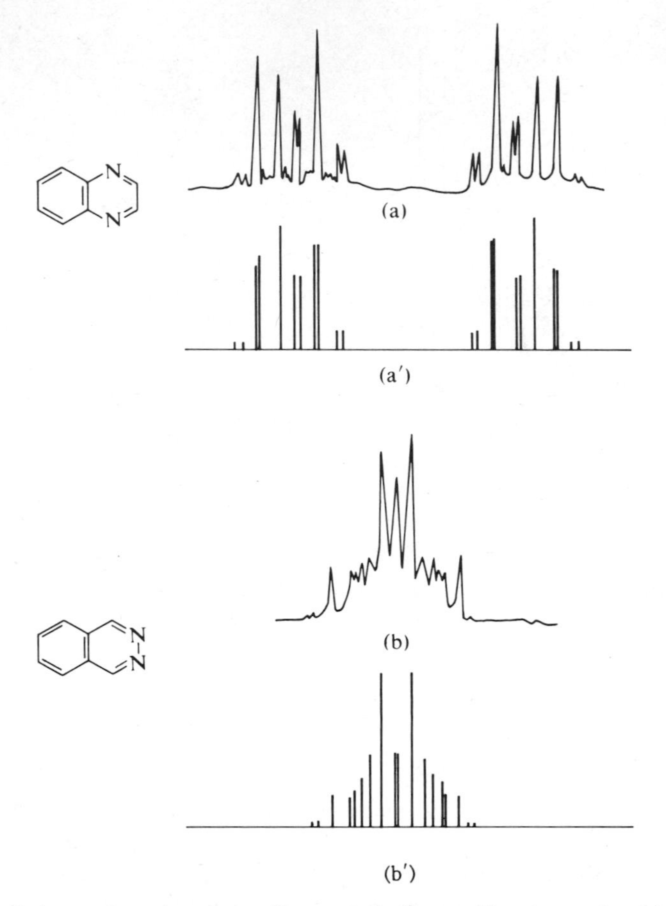

Figure 7.14. Proton–proton spin–spin coupling constants. Benzenoid proton spectra of (a) quinoxaline and (b) phthalazine. (a) Experimental spectrum. (a′) Theoretical spectrum. (b) Experimental spectrum. (b′) Theoretical spectrum. [From P. J. Black and M. L. Heffernan, *Aust. J. Chem.*, **18,** 707 (1965). Reprinted by permission of CSIRO Editorial and Publications Service.]

two spectra represent excellent examples of the pattern shift in going from an A_2X_2 to an A_2B_2 system.

The data presented in Table 7.6 give the various parameters as calculated by Black and Heffernan [5]. Figure 7.13 enables the assignment of the various lines of the two spectra. In the quinoxaline spectrum, starting from the center of the spectrum and proceeding to the right, the first peaks are due to the low-intensity lines 5 and 10, followed by the high-intensity singlet due to the overlapping 1 and 2 lines. The next doublet corresponds to lines 9 and 6, followed by the more intense overlapping singlet caused by lines 7 and 12. Absorptions 3 and 4 also appear as a singlet and are followed by the low-intensity lines 11 and 8. Similar assignments can be made for the benzenoid protons in phthalazine.

TABLE 7.6 NMR Parameters for Quinoxaline and Phthalazine

Parameter	Quinoxaline	Phthalazine
τ_1	—	0.40
τ_2	1.27	—
τ_3	1.27	—
τ_4	—	0.40
τ_5	1.94	1.87
τ_6	2.33	2.00
τ_7	2.33	2.00
τ_8	1.94	1.87
J_{56}	8.4	8.2
J_{57}	1.6	1.2
J_{58}	0.6	0.6
J_{67}	6.9	6.8
J_{68}	1.6	1.2
J_{78}	8.4	8.2

7.5 SUMMARY OF SPLITTING RULES

1. The *AB* system
   a. A doublet of doublets
   b. The coupling constant corresponds to the separation of each pair of lines
   c. The appearance of the spectrum depends on the ratio of the coupling constant to the chemical shift difference between the two nuclei
2. The AB_2 system
   a. A maximum of nine lines
   b. Line 3 appears at the chemical shift of nucleus *A*
   c. The average of lines 5 and 7 is the chemical shift of nucleus *B*
   d. The shape of the splitting pattern depends on the ratio of the coupling constant to the chemical shift difference between nuclei *A* and *B*
   e. A reduced transition energies graph is required to compute the coupling constant
3. The *ABX* system
   a. The *AB* part corresponds to two, frequently overlapping, *AB* patterns
   b. The *X* part is a six-line pattern where the separation between the most intense lines corresponds to the sum of the *AX* and *BX* coupling constants
   c. The separation of the centers of the two *AB* systems is equal to one half of the sum of the *AX* and *BX* coupling constants
   d. An energy table is required to accomplish a total spectral analysis
4. The A_2B_2 systems
   a. A maximum of 24 lines is possible
   b. The *A* and *B* spectra are symmetrical with respect to each other

5. The A_2X_2 systems
   a. The A and the X portions of the spectrum are each symmetrical within themselves
   b. The chemical shift of A is the central position of the strong doublet always observed
   c. The separation of the most intense doublet is equal to the sum of J_{AB} and J'_{AB}
   d. Two readily discernible quartets afford K, M, L, and N. These constants yield the various coupling constants.
6. The A_2B_2 systems
   a. When $J_{AB} = J'_{AB}$, the spectra are simplified, and an energy graph can be used to compute the coupling constants
   b. When J_{BB} is much larger than J_{AA}, the spectra can be analyzed by direct comparisons with theoretical spectra. (*Hint*: If J_{AA} is small, it can be set equal to zero, when $K = - M$!)

7.6 PROBLEMS

1-4. The line positions and relative intensities (in parentheses) of a series of different non-first-order spin systems are as follows: *System 1* (3 protons): 196 (0.75), 197 (1.10), 198 (1.00), 204 (2.30), 205 (1.70).

System 2 (3 protons): 244.7 (0.3), 246.3 (0.4), 247.0 (0.6), 248.6 (1.0), 252.2 (2.4), 263.9 (0.9), 254.4 (0.7).

System 3 (3 nuclei): 489.7 (0.20), 490.8 (0.22), 495.7 (0.35), 496.7 (0.70), 498.1 (0.30), 500.0 (1.00), 501.1 (0.30), 503.3 (0.70), 504.3 (0.35), 509.2 (0.22), 510.3 (0.20).

System 4 (4 nuclei): 594.4 (0.50), 595.2 (0.55), 596.0 (0.75), 596.9 (2.60), 598.6 (2.20), 601.4 (2.20), 603.1 (2.60), 604.0 (0.75), 604.8 (0.50), 605.6 (0.55). Identify the different spin systems and compute the various coupling constants and chemical shifts from the given data.

5. The following is a schematic representation of two overlapping spin systems. What are these systems? Compute the chemical shifts and coupling constants for the various nuclei.

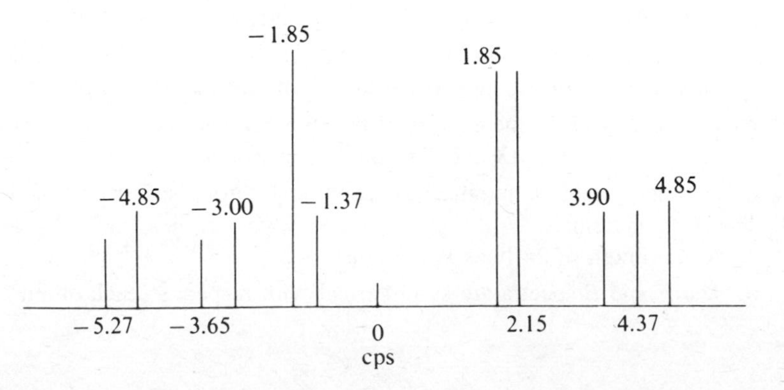

Actual Chem. Shifts

Line No.	cps
1	94.73
2	95.15
3	96.35
4	97.00
5	98.15
6	98.63
7	101.85
8	102.15
9	103.90
10	104.37
11	104.85

REFERENCES

1. H. M. McConnell, *J. Chem. Phys.* **24,** 460 (1956).
2. H. M. McConnell, A. D. McLean, and C. A. Reilly, *J. Chem. Phys.* **23,** 1152 (1955).
3. J. A. Pople, W. G. Schneider, and H. J. Bernstein, *Can. J. Chem.* **35,** 1060 (1957).
4. K. L. Williams, *J. Am. Chem. Soc.* **85,** 516 (1963).
5. P. J. Black and M. L. Heffernan, *Aust. J. Chem.* **18,** 707 (1965).
6. J. B. Lambert, H. F. Shurvell, L. Verbit, R. G. Cooks, and G. H. Stout, *Organic Structural Analysis*, Macmillan, New York, 1976, p. 68 ff.

8

CONFORMATIONAL STUDIES

8.1 GENERAL CONSIDERATIONS

Many of the discussions that have preceded this chapter have, in one way or another, dealt with the fact that changes in the screening of nuclei depend on very subtle changes of the nuclear environment. Consequently, the NMR technique can be expected to make significant contributions to the understanding of structural isomerism in its many variations.

The use of NMR spectral changes to obtain kinetic data is referred to as the "dynamic nuclear magnetic resonance" (DNMR) technique. Most generally these studies involve either C^{13} or proton NMR studies. However, several factors make C^{13} the preferred nucleus for these analyses:

1. In C^{13} NMR, the frequency differences between *exchanging* sites is frequently much larger than in the analogous proton spectra.
2. Most carbon peaks are singlets, whereas this is not generally true for proton tterns.
3. The general simplicity of carbon NMR spectra facilitates the interpretation of the spectra.
4. Since the practically obtainable temperature range for NMR studies is -180 to $+200°C$, barriers of about 5–27 kcal mole^{-1} can be studied with relative ease.

   The barrier to carbon–carbon rotation in ethanes is about 3 kcal mole^{-1}, a value significantly below the range of the DNMR technique. The barrier to C=C rotation, on the other hand, lies significantly above the 27 kcal mole^{-1} upper limit of this technique.

There exist, however, many compounds with C–C bond barriers where the DNMR technique is superbly applicable. Additionally, when two atoms that are bonded through a single bond have nonbonding electron pairs, the barriers to rotations also frequently lie within the DNMR temperature range. Among these are the N–N bond in hydrazines with a rotational barrier of 6 kcal mole^{-1}, the S–S bond in disulfides with one of about 7 kcal mole^{-1}, the N–P bond in aminophosphines at 11 kcal mole^{-1}, and the N–S bond in sulfenamides, the most hindered among these single bonds, at 15 kcal mole^{-1}.

When the barrier to internal rotation (e.g., the *cis*- and *trans*-isomers in olefinic systems) is very high, the isomers can be isolated and have different NMR spectra. In other instances (e.g., in some substituted ethylenes), isolation may not be possible, and the observed spectrum will be a composite of the various rotamers present in the mixture.

An almost classical example of the latter type of instance is observed in dimethylformamide. It is well known that, according to resonance theory, the carbon–nitrogen bond in dimethyl formamide ($O{=}CH{-}N{-}(CH_3)_2$) has a certain degree of double-bond character. Thus, a planar configuration can be anticipated for this molecule, and, if the barrier of rotation of the C—N bond is large enough, a *quasi cis*- and *trans*-isomeric mixture might exist. In fact, the two sets of methyl group protons are not equivalent at room temperature and appear as two singlets. These two singlets collapse into one when the spectrum is obtained at elevated temperatures.

Similarly, the methyl carbon atoms in N,N′-dimethyl-trichloroacetamide ($(CH_3)_2{-}N{-}CO{-}CCl_3$), appear as two singlets at lower temperatures and collapse into one singlet at elevated ones.

8.2 ROTATION ABOUT SINGLE BONDS

8.2.1 C–C, C–N, and N–N Bonds

The hypothetical compounds **1** and **2** serve to initiate the discussion delineating the application of NMR spectroscopy to conformational studies pertaining to ethanelike compounds.

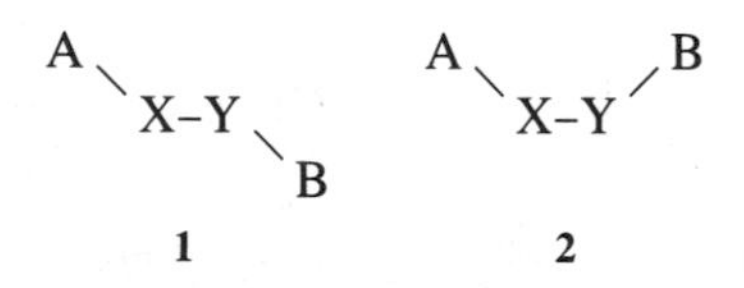

If on the one hand there is no freedom of rotation about the *X–Y* axis, the NMR spectra of the two compounds will be different. On the other hand, if there is complete freedom of rotation about the *X–Y* axis, rotamers **1** and **2** will not be two different structures but will exist as one entity only. The chemical shifts of the *A* and *B* nuclei, whatever magnetically active species they might be, would be the average of those observed for the hypothetical entities **1** and **2.**

The third possible instance is one where the rotation about the *X–Y* bond is rapid enough to give a spectral pattern caused by unequal contributions of the two rotamers, **1** and **2.** When this situation prevails, the chemical shifts and coupling constants are the result of population-weighted averages that depend on the molar populations p_1 and p_2, of the contributing species **1** and **2,** in this particular instance.

Equations 8.1 and 8.2 are the mathematical representations of this relationship. The analyses become considerably more complex when *tetra*-substituted ethylenes are involved such as

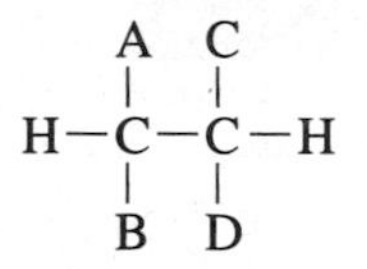

In these instances, three *low-energy* rotamers can exist.

When the rotation about the C–C bond is sufficiently slow, the proton NMR pattern of this type of molecule (assuming that the substituents *A*, *B*, *C*, and *D* are not protons) will show three different *AB* splitting patterns. If, however, rotation

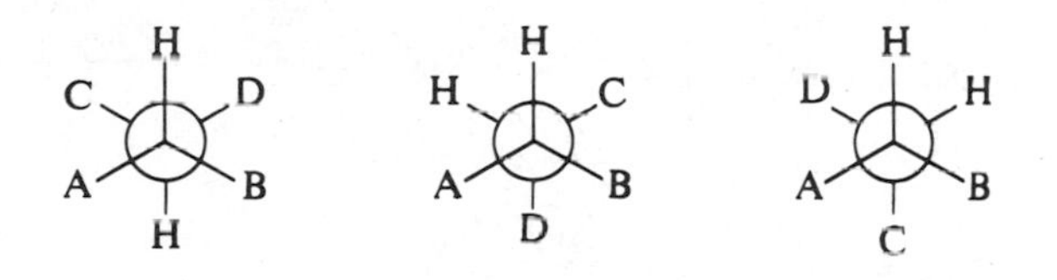

about the C–C axis is very rapid so that all three rotamers contribute equally to the averaging process, than only one *AB* pattern will be observed. A single *AB* pattern, although different from the previous one, will also be observed in the instance where rapid rotation occurs and the three rotamers contribute unequally to the averaging process. Thus, it is not possible to differentiate between the latter two processes simply by observing the proton splitting pattern. Table 8.1 lists some of the other possible spectral types resulting from different substitution patterns on ethylene [1, 2].

$$\nu_A(\text{average}) = \nu_A(\text{rotamer I}) \times p_1 + \nu_A(\text{rotamer II}) \times p_2 \quad (8.1)$$

$$J(\text{average}) = J(\text{rotamer I}) \times p_1 + J(\text{rotamer II}) \times p_2 \quad (8.2)$$

In principle, any system that exhibits a nonstatistical rotamer distribution at room temperature can be converted to one with statistical distribution at sufficiently high temperatures. Unfortunately, it is frequently not possible to attain these temperatures in NMR spectroscopy.

Nevertheless, a study of the NMR spectra of these compounds at different temperatures permits the calculation of the population distribution of the various rotamers through analyses of the different chemical shifts and coupling constants, since various mathematical treatments [3–5] have established that a limiting value of these constants will be reached as the temperature is increased. For example, the *gauge* configuration proton–proton coupling constants are about 2 Hz, whereas the *trans* ones are about 16 hz. Thus, knowledge of these *extreme* values permits computation of the populations of the various rotamers by an evaluation of the observed coupling constants at different temperatures.

TABLE 8.1 Spectral Types for Differently Substituted Ethanes

Structure type	Slow rotation spectrum	Fast rotation but unequal population (T-dependent)	Fast rotation but equal population (T-independent)
H–C(A)(D)–C(B)(E)–H (one isomer)	3 × AB	AB	AB
H–C(H)(D)–C(B)(E)–H	3 × AB	AB	AB
H–C(H)(D)–C(B)(B)–F	A_2 and AB	A_2	A_2
H–C(H)(D)–C(H)(E)–B	3 × ABC	ABC	ABC
H–C(A)(B)–C(B)(A)–H	2 × AB	AB	AB

(*Note*: It is assumed that the substituents A, B, D, E, and F do not couple with the ethane protons.)

An analysis of two specific examples will serve to demonstrate the method. The fluorine NMR spectra of $CF_2BrCHBrC_6H_5$ and $CF_2BrCBrCl_2$ are reported to give an ABX and an A_2 spectrum, respectively. The latter compound affords the fluorine singlet down to −30°C [6–8]. The nonequivalence of the fluorines in the phenyl derivative could be rationalized by either of the following two arguments: (1) The compound simply exists as rotamer **3,** the most stable of the possible rotamers, and the two fluorines are not equivalent. (2) In none of the possible rotamers (**3, 4,** or **5**) is any one fluorine subject to the same environmental influences. This is true even if no barrier to rotation exists! The *average* environment can *never* be exactly alike. Thus, the two fluorines are never equivalent! Consequently, the compound is expected to have an ABX rather than an A_2 spectrum, even when there is no restriction to the free rotation about the central bond.

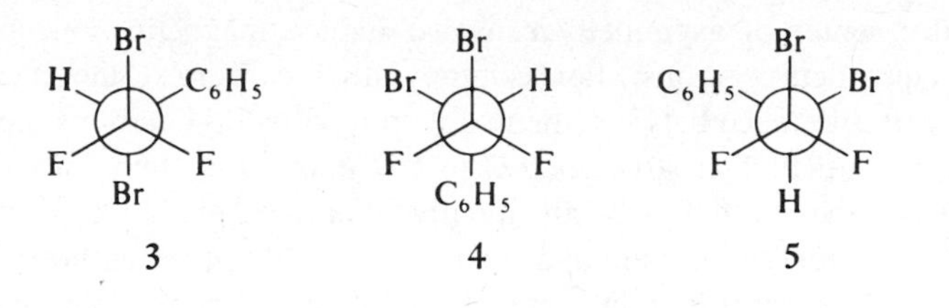

A situation such as this represents an example of *intrinsic* symmetry [9] and must always be considered as a possibility before concluding that in any particular system restricted rotation is operative to account for the NMR spectral pattern.

The matter of the fluorine equivalence in $CF_2BrCBrCl_2$ represents another interesting rotamer analysis problem. Again, two possibilities exist to account for the single nature of the fluorine atoms. The compound might exist exclusively in the symmetrical rotamer **6** configuration or it could be a rapidly equilibrating mixture of rotamers **6, 7,** and **8.**

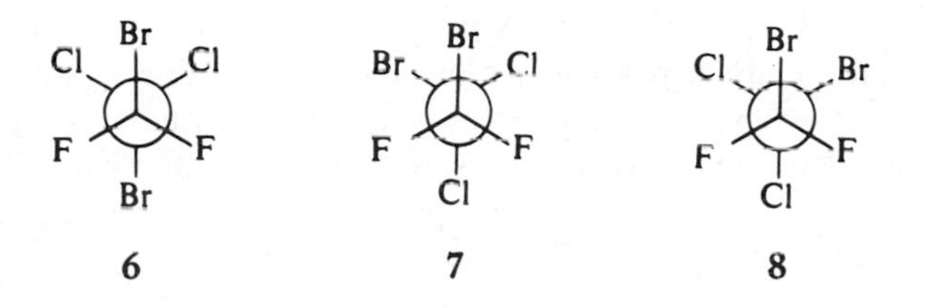

To differentiate between these two possibilities, recourse can be taken to the variable temperature analysis of the fluorine spectrum of the sterically more hindered CF_2BrCBr_2Cl. In this compound the fluorine resonance remains as a singlet down to −30°C and reaches a 1.4 : 1 mixture of a fluorine singlet and an *AB* pattern [7].

Among the possible rotamers, only structure **9** would afford a singlet for the two fluorine atoms, while structure **10** will show an *AB* pattern. Thus, rotamer **9** is more stable than rotamer **10.** As a result of this analysis, it can be concluded that the sterically less hindered $CF_2BrCBrCl_2$ compound exists as a rotamer mixture composed of rotamers **6, 7,** and **8,** with free rotation about the C–C bond. Additionally, it can also be concluded that the ABX pattern of the phenyl ethylene derivative (**3**) is due to the intrinsic symmetry of the molecule rather than any restricted rotation about the C–C bond.

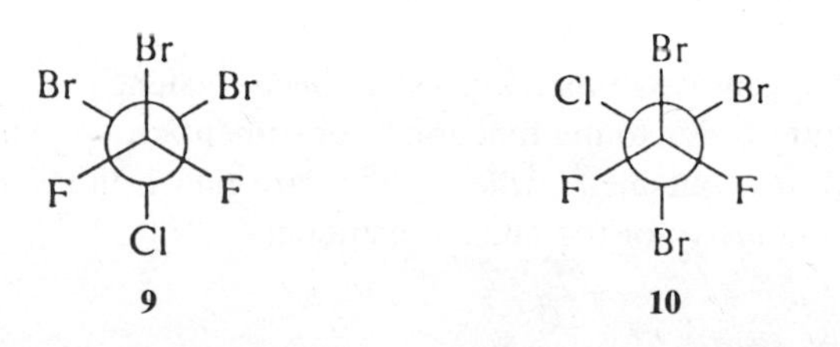

A nearby center of asymmetry can also induce magnetic nonequivalence of *normally* equivalent carbons. For isopropylalkyl carbinols, the methyl carbon chemical shift difference between the two isopropyl methyl carbons increases from 0.2 ppm (R = methyl in structure **12**) to 6.9 ppm (R = *tert.*-butyl in structure **13**). Although these differences are rapidly attenuated with increasing distances between the *asymmetric* carbon and the isopropyl group, separations by as many as three carbon atoms still show some chemical shift differences between the isopropyl methyl carbons [10–12].

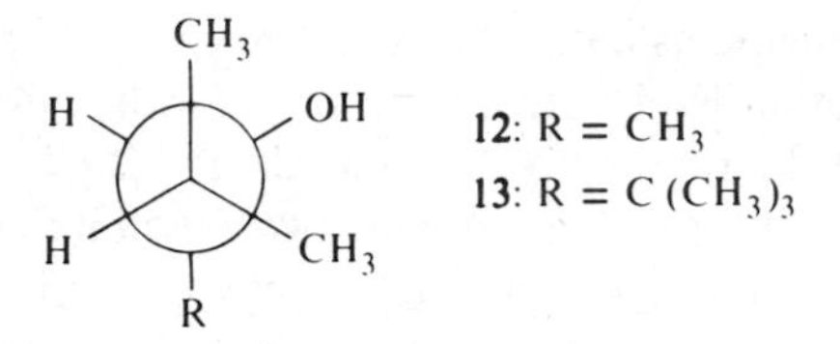

8.2.2 Nonequivalence of Vicinal Atoms in Ethanes

The various symmetry considerations are, of course, also operative in appropriately substituted *vicinally* situated nuclei as in *dl*-1,2-dibromo-1,2-dichloro-1,2-difluoroethane, where all three possible rotamers are observable at low temperatures [11]. The methyl group proton resonances in 2,2,3,3-tetrachlorobutane are a singlet at room temperature. At −44°C, the singlet is split into two singlets with an approximate intensity ratio of 2 : 1, corresponding to the trans and gauge rotamers, respectively [10].

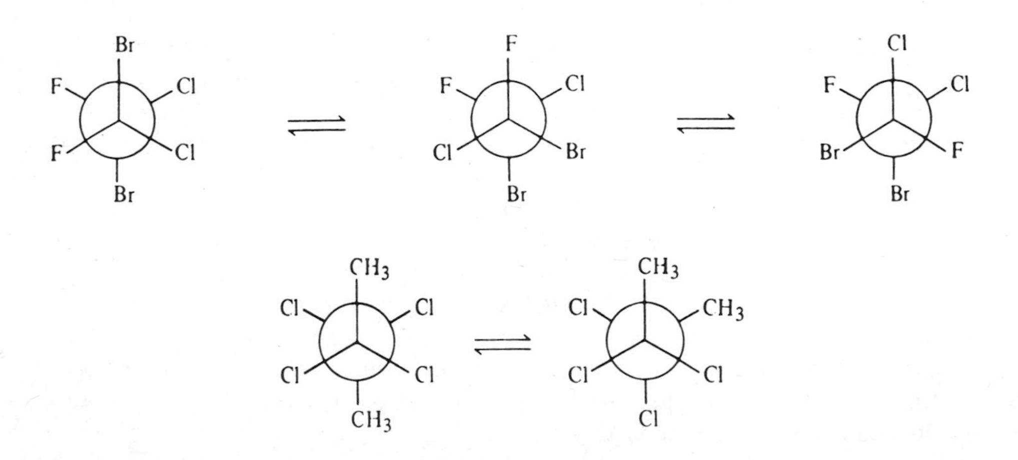

The two ester derivatives **14** and **15** represent excellent examples of two compounds were the intrinsic symmetry causing nonequivalence of the *A* and *B* protons in the ring structure is due to the functional substituent group. The *A* and *B* protons in compound **15** are equivalent, whereas they are not in the ester **14,** a reflection of the intrinsic symmetry of the latter substituent.

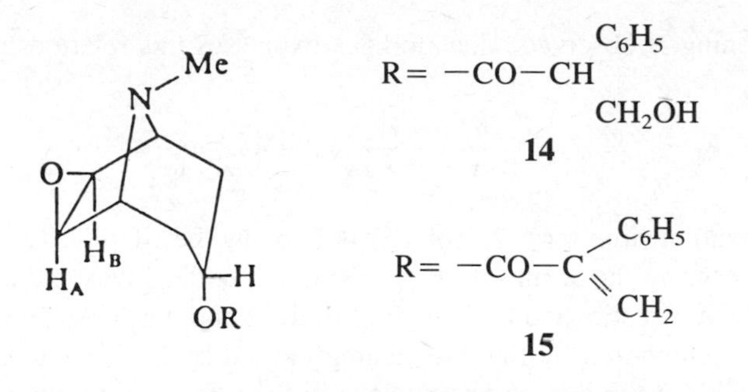

The nonequivalence of the methylene protons in diethylsulfoxide (**11**) and diethylsulfite (**12**) [11] may actually be due to electronic differences in the C–H bonds concerned rather than any intrinsic asymmetry, although magnetic nonequivalence could have been applied to account for this difference. Thus, great care must be taken in ascribing restricted rotation effects to observed nonequivalence of various nuclei.

$$CH_3CH_2-\overset{\displaystyle O}{\overset{|}{S}}-CH_2CH_3 \qquad CH_3CH_2O-\overset{\displaystyle O}{\overset{|}{S}}-OCH_2CH_3$$

11   **12**

8.3 CALCULATION OF REACTION PARAMETERS

The theory for DNMR processes has been developed by the classical procedures involving the Bloch equations and the quantum mechanical approaches utilizing density matrices [13, 14]. For the purposes of this book, it is convenient to discuss the DNMR computational methods separately for slow, fast, and intermediate exchange rates.

8.3.1 Slow Exchange Rate Processes

A given nucleus will exist in its various possible spin states (2 for a nucleus with $I = \frac{1}{2}$) for a finite time interval. This time interval is reflected by the shape of the recorded resonance curve when it is obtained under sufficiently slow passage conditions to prevent any interfering phenomena.

In principle, it is this relaxation time for exchanges between the various possible spin states that affects the signal shape most drastically. In practice, however, it is the homogeneity of the magnetic field that controls the signal shape. It is this problem that makes it necessary to define a new relaxation time, T_{exp}, which includes the line broadening caused by the magnetic field inhomogeneity.

When the separation between two proton signals, for example, is considerably smaller than the lifetimes $(\tau)_A$ and $(\tau)_B$ of the two interacting protons, additional signal broadening is observed. Equation 8.3 expresses this relationship mathematically.

$$\tau_{\mathrm{A}}^{-1} = T_{2\mathrm{A}}'^{-1} - T_{2\mathrm{A}}^{-1}, \tag{8.3}$$

If the relaxation time term T (not T') is known, the lifetime (τ) can be determined by measuring the width of the NMR signal under investigation. The relaxation time T can be estimated by including in the sample a substance whose signal width is not encumbered by line broadening caused by a slow exchange process. The signal width of the appropriate nucleus in the reference compound is taken as the value of T in equation 8.3 [15, 16].

8.3.2 Fast Exchange Rate Processes

When the lifetimes of individual rotamers are very small, a rapid exchange among the various possible forms occurs, and the contributing absorption peaks collapse into one, with the absorption frequency centered on a mean value as computed by equation 8.3 and a line width determined by

$$T'^{-1} = p_A/T_{2A} + p_B/T_{2B}$$

If the experimentally determined line width is significantly larger than computed by this expression, the exchange processes can be judged not to be rapid enough for this expression to be applicable under the particular conditions of temperature etc..

8.3.3 Intermediate Exchange Rate Processes

The remaining possibility describing exchange processes is one where the lifetimes of the participating species are of the same order of magnitude as the square root of the frequency difference between the nuclei participating in the exchange process.

A general expression that correlates the signal shapes with the lifetimes of various exchange processes has been developed by Gutowsky and Holm [15]. The interested reader may wish to examine the original literature for a detailed analysis.

For the purposes of this treatment, this expression will be simplified to deal with intermediate exchange rates with equal populations and lifetimes so that $p_A = p_B = \frac{1}{2}$ and $\tau_A = \tau_B = 2(\tau)$, and large transverse relaxation times so that $T_{2A}^{-1} = T_{2B}^{-1} = 0$.

Thus the signals will have widths that are small in the absence of exchange in comparison to their separations. Equation 8.4 is applicable to all exchange phenomena that fulfill these fundamental assumptions.

$$g(v) = \frac{K\tau(\nu_{\mathrm{A}} - \nu_{\mathrm{B}})^2}{[\frac{1}{2}(\nu_{\mathrm{A}} + \nu_{\mathrm{B}}) - \nu]^2 + 4\pi^2 r^2(\nu_{\mathrm{A}} - \nu)^2(\nu_{\mathrm{B}} - \nu)^2}, \tag{8.4}$$

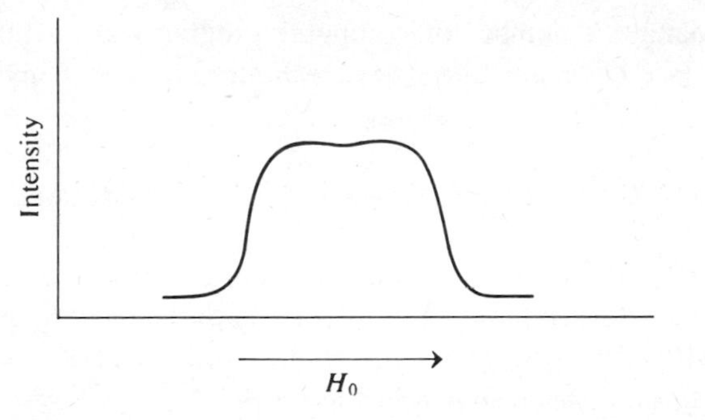

Figure 8.1. Line shape for exchange rate between two positions with equal population at an exchange rate of $\sqrt{2}/(2\pi(\nu_A - \nu_B)$.

In this expression K is a normalizing constant and ν is the applied radiofrequency. When τ is large, two lines, at ν_A and ν_B, will be observed. When τ is small, only one line, intermediate between ν_A and ν_B, will be seen.

As a particular spectrum changes from one with two separated absorption peaks to one with a singlet (i.e., NMR spectra of an appropriate compound are being recorded at ever increasing temperatures), an intermediate stage will be reached where the pattern appears as shown in Figure 8.1. When this shape exists, at the temperature just before the pattern becomes a clear singlet, equation 8.4 is simplified and equation 8.5 becomes applicable.

$$\tau = \frac{\sqrt{2}}{2\pi(\nu_A - \nu_B)}. \qquad (8.5)$$

Experimentally, the process involved in determining an exchange rate requires that the spectrum is obtained under conditions where no exchange occurs. This is generally accomplished by recording the spectrum at an appropriately low temperature. As the temperature is increased, the two singlets will come closer and closer together, and the following expression can be employed to determine the exchange rates at the different temperatures.

At the instance where the shape shown in Figure 8.1 is generated, the following expression also applies:

$$\frac{\text{Separation of peaks at } T \text{ of interest}}{\text{Separation of peaks under conditions of large } \tau} = \left[1 - \frac{1}{2\pi\tau^2(\nu_A - \nu_B)^2}\right].$$

This expression does not apply to instances where τ is small enough to generate a single line only or where the widths of the signals are not small in comparison to the chemical shift differences of the *collapsing* singlets. When either of these situations prevails, the complete expression given in reference 15 must be employed. Fortunately, a number of computer programs are available to facilitate these analyses (see *Quantum Chemistry Exchange*, Indiana University).

8.4 COMPUTATION OF THERMODYNAMIC PARAMETERS

The determination of the rate constant for exchange processes at the temperature (T_c) where two singlets collapse into one is perhaps the most frequently determined value by DNMR technique. This temperature corresponds to the situation exemplified by the spectral shape shown in Figure 8.1.

The unimolecular rate constant for these exchange processes at T_c is given by $k_c = (\tau) \times$ (distance between the two peaks at slow exchange)$/2^{1/2}$. It must be kept in mind that this expression applies only to situations where the nuclei exist in equally populated sites without any mutual spin–spin interactions.

Determination of the rate constants at several different temperatures generates the necessary information for the application of the Eyring or Arrhenius equations, which gives the free energy, the enthalpy, and, although less reliably, the entropy of activation. The rate constant k, for a zero-order process, is given by equation 8.6.

$$k = K^0 e^{-E/RT}, \tag{8.6}$$

Since the rate constant, in terms of the lifetime of a species, is $\frac{1}{2}\tau$, equation 8.6 can be rewritten as

$$\log_{10} \frac{1}{2\pi\tau(\nu_A - \nu_B)} = \log_{10} \frac{K^0}{(\nu_A - \nu_B)\pi} - \frac{E_a}{2.3RT}. \tag{8.7}$$

When the left side of the equation is plotted versus $1/T$, the slope of the resulting line is $E_a/2.3R$, and the intercept is

$$\log_{10} K^0/[\pi(\nu_A - \nu_B)]. \tag{8.8}$$

Thus, the activation energies and, in the case of restricted rotation, the frequency factor K^0 are readily obtained by application of the DNMR technique.

When these calculations are performed for the rotamers **16** and **17** of dimethyl nitrosamine [17], an energy of activation of 23 kcal/mole and a frequency factor of 7×10^{12} are obtained.

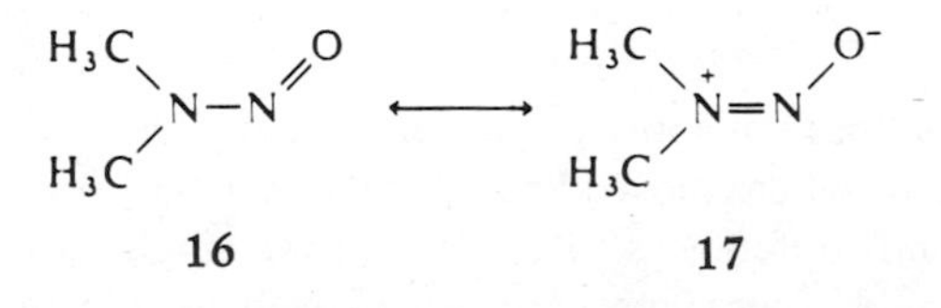

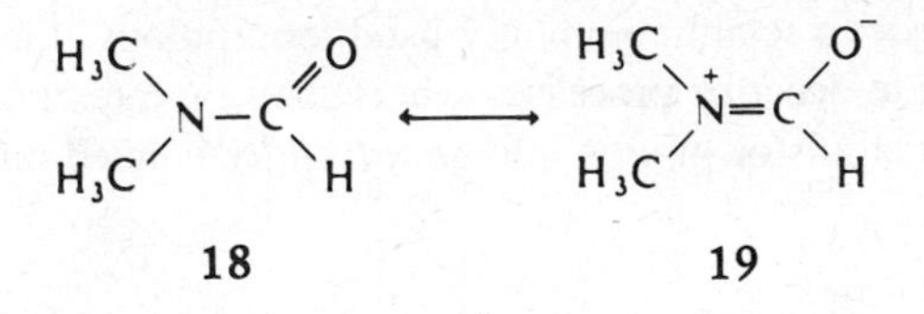

When the single bond is an N–C rather than an N–N one, as is the case in dimethyl formamide (**18** vs. **19**), the activation energy (7 ± kcal/mole) and the frequency factor (10^3–10^7) are smaller, in accordance with qualitative predictions based on resonance theory considerations.

A convenient means for estimating rotational energy barriers, when the coalescence temperature is known, is through application of the graph given in Figure 8.2. The application of the more accurate *spin echo* technique will be described in Chapter 10 [17, 18].

8.5 RING INVERSION STUDIES

The term "inversion," strictly speaking, should relate only to a configurational change as observed in, for example, pyramidal atomic inversions, rather than ring

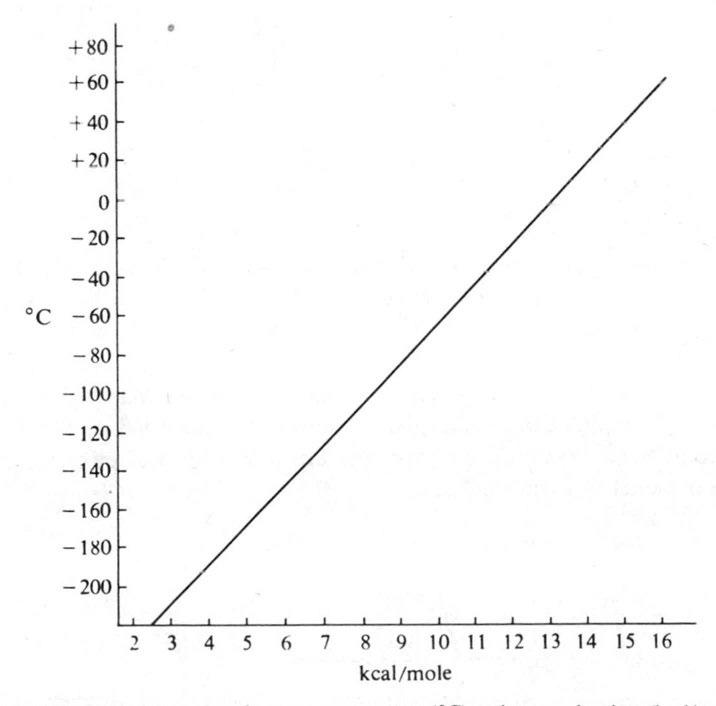

Figure 8.2. Correlation of coalescent temperatures (°C) and energy barriers (kcal/mole).

inversion processes. To retain commonly used terminology, the term "ring inversion" will be used to describe processes where the ring interconversion transforms all *axially* situated atoms or groups into *equatorially* bonded ones.

8.5.1 Six-Membered Rings

8.5.1.1 Cyclohexane

One of the most important initial applications of the DNMR technique to a cyclic compound dealt with a variable temperature study of cyclohexane [19]. The proton NMR spectrum of this compound, obtained at room temperature, shows only one peak for the 12 protons, as can be expected if there exists a rapid interconversion process between different *chair* forms of cyclohexane. When the cyclohexane proton spectrum is examined at temperatures of $-65°C$ and below, two sets of equal area peaks are obtained. These two sets of peaks correspond, respectively, to the axial and equatorial protons. These *quasi* singlets are precisely that, since they do not show the multiplicity required from the large number of spin–spin interactions among the different types of protons (several hundred lines for each type of proton can be expected) [20–22].

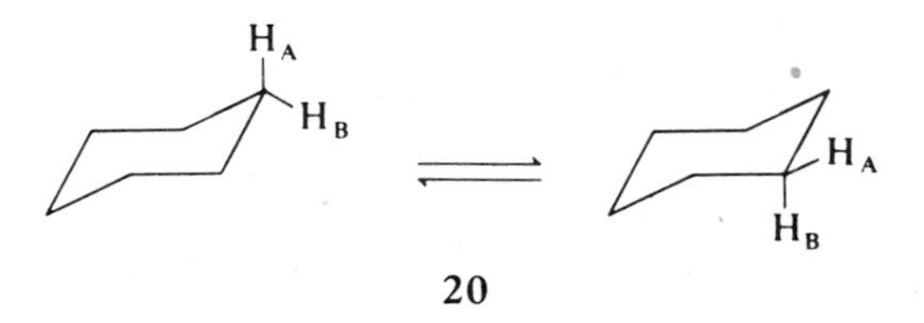

The room temperature spectrum of undecadeuterocyclohexane ($C_{12}D_{11}H$) obtained under deuterium decoupling conditions affords two narrow lines corresponding to an equatorial and axial proton, respectively.

These studies [23–25] have established that the free energy of activation for ring inversion in cyclohexane is 10.4 kcal/mole and that the enthalpy of activation is 10.5 kcal/mole (spin-echo results give a somewhat lower value, 9.1 kcal/mole [22]). The interconversion of one particular equatorial to axial proton is diagrammatically depicted in Figure 8.3 [26].

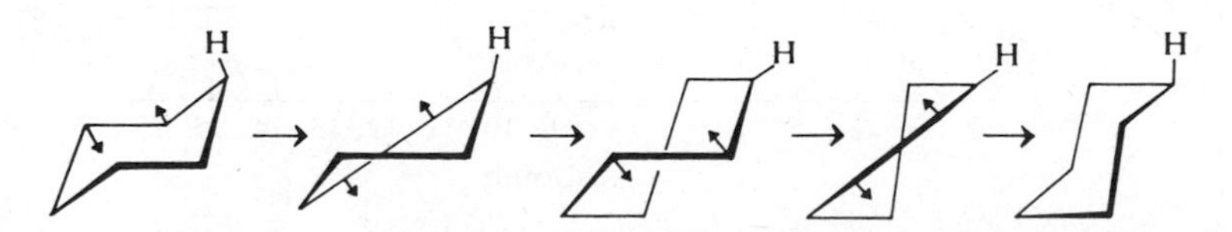

Figure 8.3. Diagrammatic path for cyclohexane ring inversion process.

8.5.1.2 *Perfluorocyclohexane*

The ring inversion rate, as determined from the fluorine spectrum of perfluorocyclohexane [27], is 69 s^{-1}, a value essentially identical to that of cyclohexane (61 s^{-1}). The existing difference between the entropies of activation of perfluorocyclohexane (−10 eV) and cyclohexane (1 eV) is probably due to F–F interactions which raise the energy of the ground-state chair conformation in the former compound.

8.5.1.3 *Methylcyclohexane*

The chemical shift difference between the axial and equatorial carbons of the methyl group in methylcyclohexane is about 5 ppm, in conformance with the expectation that carbons that are sterically perturbed (*steric compression shift*) will appear at higher field than those that are not (axial vs. equatorial substituents). At −110°C the equilibrium constant for the equatorial versus axial conformers is approximately 100 [28]

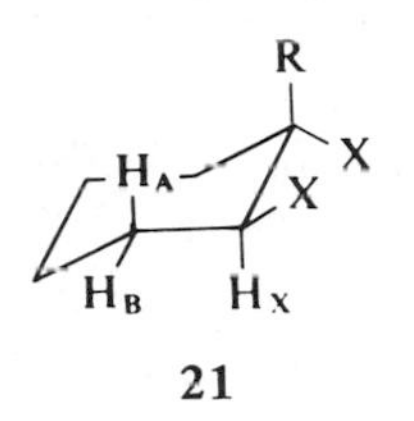

21

8.5.1.4 *Functionally Substituted Cyclohexanes*

Substituents such as hydroxyl, methyl, or halogen do not drastically alter the coalescence temperature of the cyclohexane ring system. The barrier to ring inversion in cyclohexanes is generally about 10 kcal/mole.

8.5.1.5 *Tetrahydropyran*

The replacement of a methylene group in cyclohexane by an oxygen atom, to form tetrahydropyran, also does not significantly alter the ring inversion barrier (10.7 kcal/mole).

8.5.1.6 *1,3-Dioxan*

Since the barrier to rotation about an oxygen–oxygen bond is higher than about a carbon–carbon bond, it is not surprising that the inversion barrier in 1,2-dioxane (16.1 kcal/mole) is also greater.

8.5.1.7 *Cyclohexene*

Ring inversion or, more appropriately, ring reversal—in cyclohexene occurs between two half-chair forms. The proton NMR spectrum of the deuterium-decoupled hexadeuterocyclohexene **22** shows coalescence of the two protons at −146°C, and, consequently, a barrier to ring inversion of 5.3 kcal/mole. The inversion process is envisioned to occur via a transition state where five of the six carbon atoms are in a plane [28].

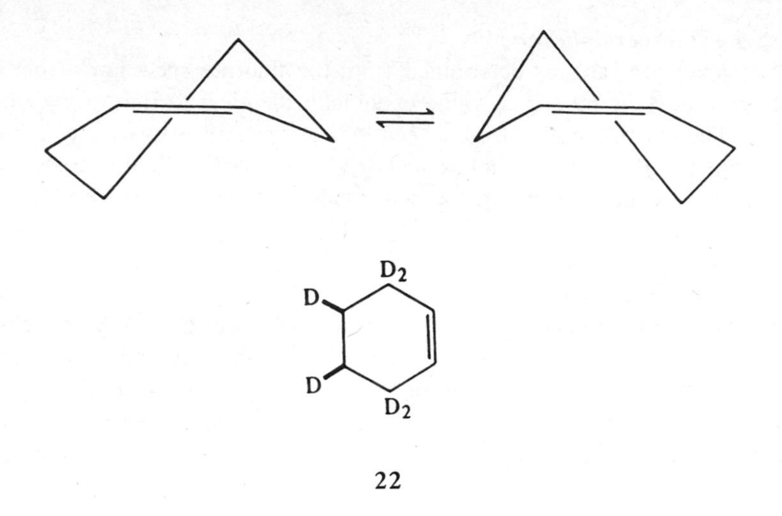

22

8.5.2 Seven-Membered Rings

The conformational interconversions in rings with seven and more members are considerably more intricate than is the case for smaller ring sizes. In these rings, the number of possible conformers increases beyond the chair and boat forms and includes crown, twist chair, boat chair, and others. The existence of these added conformational forms brings with it the existence of pseudorotation interconversions along with the ring inversion.

8.5.2.1 Cycloheptane

The various possible flexible interconversions find proof in the observation that the proton NMR spectrum of cycloheptane is insensitive to temperatures down to −180°C.

8.5.2.2 Cycloheptene

The presence of a double bond in a cycloheptane molecule prevents pseudorotation, and consequently it is not surprising that a chair–chair interconversion (7.4 kcal/mole) in 5,5-difluorocycloheptene has been observed.

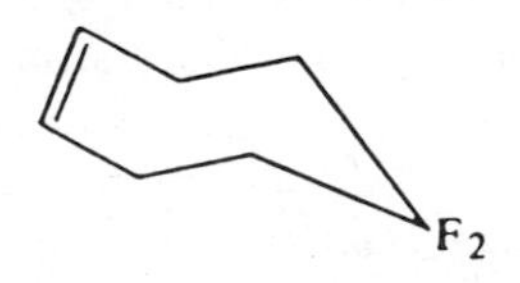

8.5.2.3 Cycloheptatriene

The preferred conformation of cycloheptatriene is known to be boat-shaped (**23** or **24**), with the two possible conformers interconverting rapidly at room temperature. The proton NMR spectrum of this compound shows split methylene protons at −143°C, which a corresponding activation energy of 7.7 kcal/mole [29, 30].

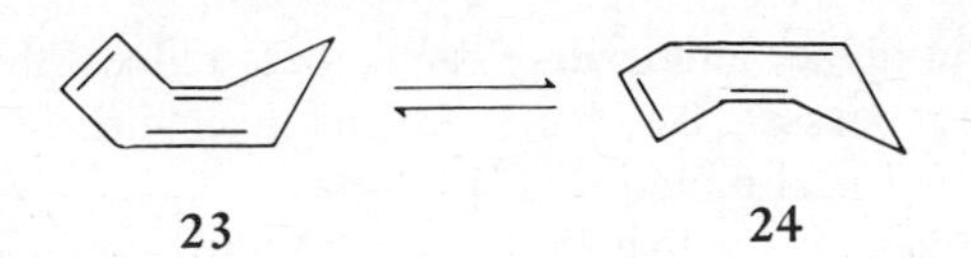

8.5.3 Eight-Membered Rings

8.5.3.1 Cyclooctane

The stable conformation of this molecule appears to be the boat-chair form, although the crown form is still considered to be the preferred conformer by some. According to models, the transformation of this flexible conformer undergoes some drastic transannular interactions when it is converted from one crown form to another. The pentadecadeuterocyclooctane derivative exhibits a rate process below −100°C with an activation energy of 7.7 kcal/mole, a barrier that is somewhat less than that observed for cyclohexane.

8.5.3.2 Cyclooctatetraene

This compound exists in the "tub" form, which at the proper temperature can undergo a bond shift (path 1) and/or a ring inversion process (path 2).

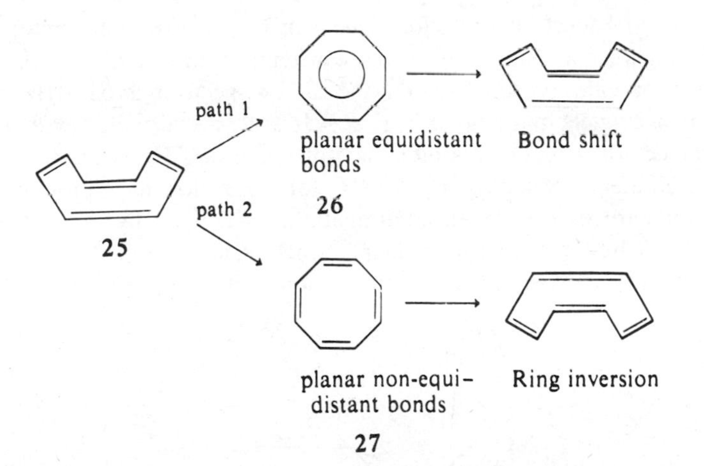

An examination of the temperature dependence of the carbon–hydrogen coupling pattern in cyclooctatetraene, which, at low temperature, is a doublet of doublets (a result of the coupling of a proton on carbons bonded to the carbon in question by a single and double bond, respectively), permitted a differentiation between the two possible paths. The coupling to the single-bonded carbon (12 Hz) is rather strong, whereas that to the double bonded carbon is rather weak. This doublet of doublets collapses at higher temperatures and, thus, reflects the increasing rate of the bond exchange processes. The barrier to this exchange has been determined, by DNMR, to be 13.7 kcal/mole [31–33].

The cyclooctatetraene derivative **28** has been used to examine the ring inversion and bond shift processes (34).

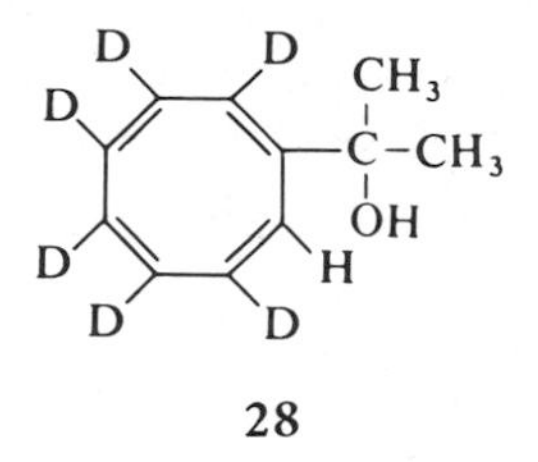

28

The changes in the methyl and ring proton chemical shifts permit the facile differentiation between the two possible inversion processes, and their rates can be found simultaneously. This study showed that conformer **26,** the planar species with equidistant bonds, is more stable than the planar nonequidistant bond species **27** by 14.7 kcal/mole.

8.5.4 Nine-Membered Rings

8.5.4.1 Cyclononane

The carbon NMR spectrum of cyclononane is much superior to the proton spectrum in recognizing the nonequivalent nuclei at low temperatures. Although proton NMR at 251 MHz provides some resolution, the carbon spectrum at 63 MHz shows two well-resolved singlets in a ratio of 2 : 1, at −162°C (6 kcal/mole), with a chemical shift difference of 9 ppm. This difference of about 500 Hz is clearly superior to the 20-Hz difference observed at 251 MHz for the cyclononane protons.

The symmetry of the frozen conformation, as defined by the C^{13} spectrum, requires that it be the "twist-boat-chair" structure that exists at the temperature. In all probability, reversal of this conformation occurs through a boat-chair intermediate.

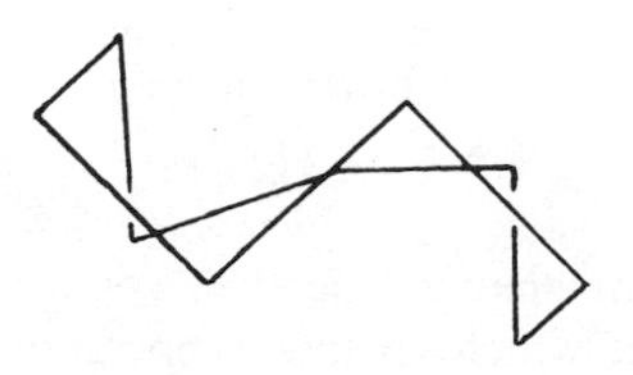

8.5.4.2 Cis,cis,cis-cyclonona-1,4,7-triene

Since the allylic protons in this structure appear as two distinct signals at −4°C, the compound exists in conformation **29** [35, 36]. These signals coalesce into one at 31°C, and thus the interconversion barrier is 14.5 kcal/mole.

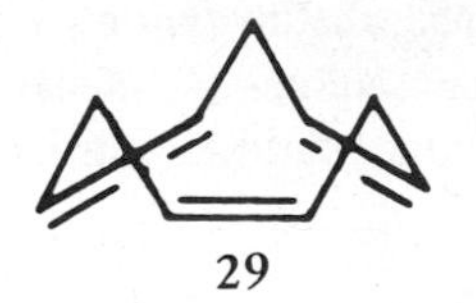

29

8.5.5 [14] Annulene

When observed at low temperature, the [14]annulene (**30**) exhibits two proton signals, one corresponding to the *inner A* and one to the *outer B* protons [37, 38]. Since these signals coalesce into one at higher temperatures, and because of the deshielded nature of these protons, this compound has been defined as "aromatic."

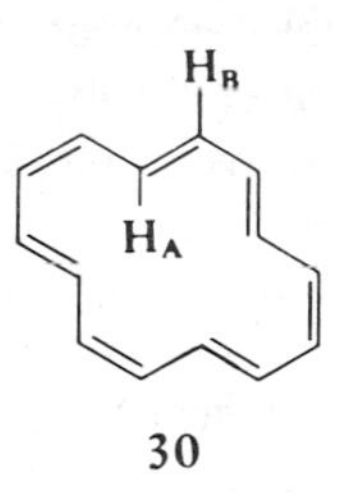

30

8.6 ATOMIC INVERSION PROCESSES

Atoms with a nonbonding pair of electrons, such as nitrogen, phosphorus, sulfur, arsenic, certain carbanions, and so on, and are bonded to three other groups in a pyramidal fashion can undergo unimolecular configuration inversion. The process occurs through an "umbrella-like" inversion, where at the transition state, the central atom is trigonally hybridized and the lone pair of electrons is in a *p* orbital. This inversion process can be pictorialized as follows.

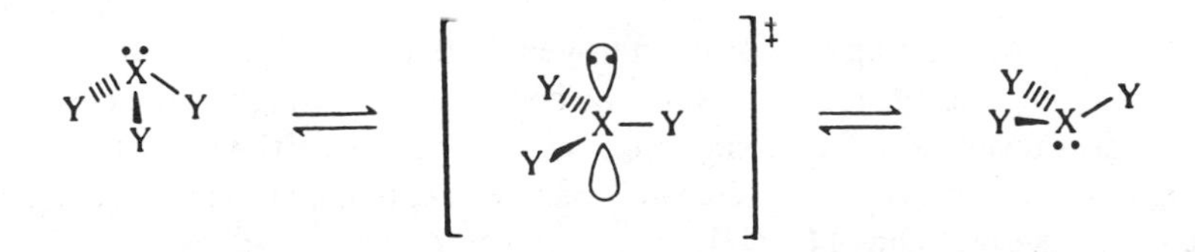

Some of the first- and second-row elements of the periodic table offer interesting contrasts. While the nitrogen and oxygen compounds normally invert too rapidly for the DNMR technique, the phosphorus, sulfur, and arsenic compounds invert too slowly for this process to be applied at room temperature [39–40].

8.6.1 Nitrogen Inversions

8.6.1.1 Cyclic Compounds

AZIRIDINES. When a nitrogen atom is part of a strained ring, as in aziridines, the angle strain increases during the inversion in going from the ground state to the transition state. As a consequence, the inversion barrier is considerably higher (18–20 kcal/mole) than in open-chain amines (4–7 kcal/mole) [41, 42].

The equilibrium involved in the aziridine-nitrogen inversion is

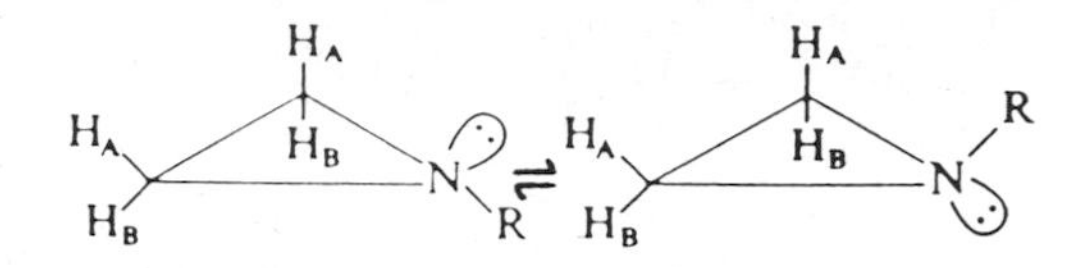

While, rapid inversion will cause all four of the ring protons to appear as being magnetically equivalent, a slow inversion rate will generate an A_2B_2 pattern. The H NMR spectra of aziridines, obtained at low temperatures, show an A_2B_2 pattern. This multiplet collapses into a singlet at about 100°C.

The barrier to nitrogen inversion in N-substituted aziridines decreases with increasing size of the substituent (methyl > ethyl > *iso*-propyl > *tert.*-butyl).

AZETIDINES. The effect of ring strain on the nitrogen inversion rate has also been observed in the four-membered ring, N-substituted azetidines. The 1,3,3-trimethyl azetidine has an inversion energy of 9 kcal/mole, considerably lower than the more strained azetidines.

PYRROLIDINES. The barrier to inversion rapidly becomes too low for normal DNMR approaches as the ring size is increased from four- to five-membered. However, the barrier can be lowered by replacing a carbon substituent with a more electronegative one. This has the effect of increasing the *s* character of the ground-state lone pair of electrons and, although the lone pair must still be *p*-hybridized, increases the barrier. For example, the inversion energies for the nitrogen in N-chloropyrrolidine and N-methyloxazolidine are 12 and 16 kcal/mole, respectively, while the barrier for pyrrolidine itself is outside the DNMR technique range.

PIPERIDINES. As the ring size of the nitrogen-containing ring increases, the problem of analyzing the nitrogen inversion rate becomes increasingly more complicated not only as a result of its increasing rate, but also because of the problems created by ring inversion processes. Fortunately, some means have been developed that overcome these problems [43, 44].

The NMR spectra of the quinuclidine ring system have been studied in great detail, and their analyses have served superbly in overcoming the complexities of inversion etc..

The known stereochemistry of the ring fusion places H_C *trans* to the lone pair of electrons in structure **31** [44]. The proton H NMR spectrum shows that the chemical shift difference between the methylene protons, H_A and H_B, is 0.93 ppm. The proton (H_A) that is anticoplanar to the nitrogen lone pair of electrons is upfield from the other methylene proton (H_B).

When the lone pair of electrons is removed by protonation (structure **32**), the chemical shift difference between H_A and H_B is decreased to 0.4–0.5 ppm., a difference that is normal for *axial* versus *equatorial* protons. This lone-pair effect on the chemical shift differences of methylene protons adjacent to an amine nitrogen can be used for nitrogen and ring inversion studies.

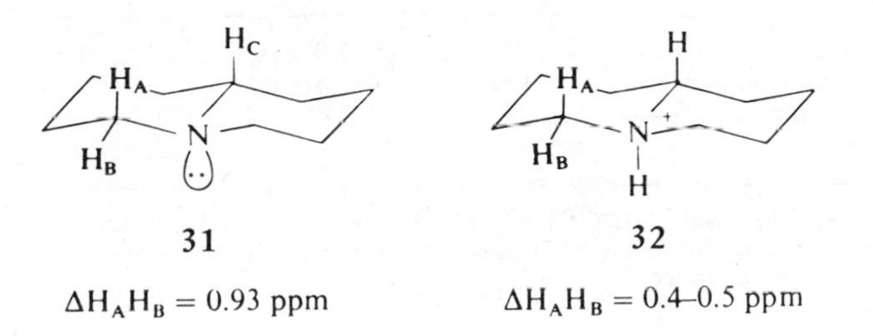

Lambert and Keske [45] have applied this technique to the analyses of a number of deuterated piperidine derivatives (structures **33** and **34**), in order to establish whether the lone pair of electrons on the nitrogen are *axially* or *equatorially* located or whether a rapid equilibrium between the two sites exists.

The chemical shift difference between the methylene protons *alpha* to the nitrogen atom in the piperidine **34** is 0.436 in deuterated methanol and 0.458 ppm in cyclopropane as solvent. Since addition of acid does not alter these chemical shift differences (they are typical of normal *axial* versus *equatorial* proton chemical

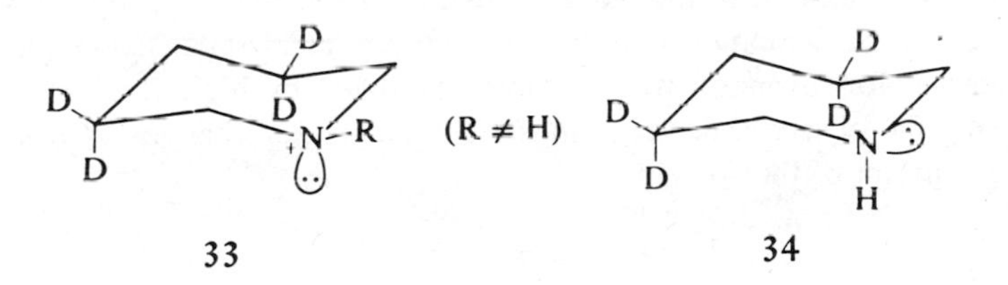

shifts), it can be concluded that, under the conditions of these experiments, the lone pair of electrons in these piperidines exists in the equatorial configuration.

When *R* in structure **33** is either a methyl or *tert.*-butyl group, the chemical shift difference between the methylene protons is 0.943 and 0.997 ppm, respectively, in perdeuteromethanol solution. In nonpolar cyclopropane, this difference becomes 1.057 ppm for the *tert.*-butyl derivative. Contrary to the behavior of the non-N-alkylated compound, protonation of these compounds decreases the chemical shift difference to 0.40 ppm between the methlene protons that are *alpha* to the nitrogen. It is of some interest to note that the methylene protons *gamma* to the ring nitrogen have chemical shift differences of 0.52 ppm regardless of the solvent or acidity of the medium.

These data establish that under the experimental conditions cited, the N-methyl and N-*tert.*-butyl derivatives **33** exist largely with the lone pair of electrons in the *axial* position [46–48].

TABLE 8.2 Correlation between Solvent and Energy of Activation of the Nitrogen Inversion in Benzyl Methoxy Methylamine

Solvent	T_c	E_a(kcal/mole)	Dielectric constant of the solvent
n-hexane	−16	12.9 ± 0.3	1.89
carbon disulfide	−27	12.4 ± 0.5	2.64
methylene chloride	−34	9.4 ± 0.4	8.3
acetone	below −70	?	20.7

SOURCE: Taken from D. L. Griffith and J. D. Roberts, *J. Am. Chem. Soc.*, **87** (1965), 4089, copyright 1965 by the American Chemical Society. Reproduced by permission of the copyright owner.

8.6.1.2 Acyclic Compounds

Although nitrogen inversion in acyclic amines is not amenable to study by DNMR, some N-methoxy derivatives, $X{=}CH_3O$ in structure **39** are [49]. The NMR spectrum of the methylene protons in compound **39** becomes an *AB* pattern at −60°C, with the coalescence temperature, T_C, the temperature at which maximum line broadening is observed, depending on the solvent (see Table 8.2). These data allow the conclusion to be drawn that there exists a linear correlation between the energy of activation and the dielectric constant of the solvent [50, 51]. Griffith and Roberts [49] have suggested that in the planar transition state, the C–N dipoles will tend to reinforce the N–O dipoles more than in the pyramidal state. Consequently, the transition state for inversion will be more stabilized by high-dielectric solvents, thus causing the observed decrease in the energy barrier changes with increasing dielectric constant of the solvent.

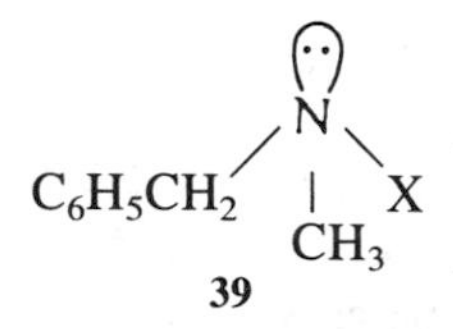

39

Rotational barriers of nitrogen-containing compounds, where the inversion rates cannot be measured by application of the DNMR technique, can be studied through N^{15} NMR. This approach is based on the fact that structural influences that effect delocalization in the ground state are also those that dominate the N^{15} chemical shifts. Among a number of relationships that have been developed, the equation that relates the chemical shift differences between an enamine and the corresponding saturated amine provides a good example of this approach [52]:

$$\Delta G = 2.91 - 0.19\ \Delta(N^{15}\text{enamine} - N^{15}\text{amine})$$

8.6.1.3 Trigonally Hybridized Nitrogen Compounds

When the nitrogen is trigonally hybridized, as in imines, a planar inversion process causes the conversion between the *syn* and *anti* forms.

In the N-phenylimine of acetone, the DNMR-measured energy barrier to inversion is 20 kcal/mole; the N-phenylimine derivative of tetramethylurea inverts somewhat more readily (18 kcal/mole).

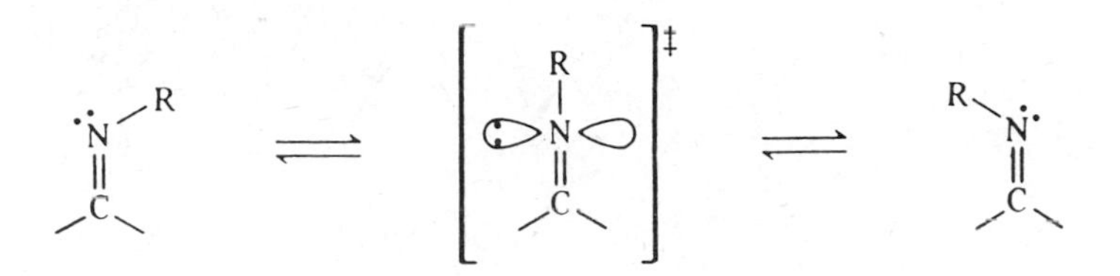

8.6.2 Phosphorus Inversions

8.6.2.1 Acyclic Phosphorus Compounds

With elements found in the second row of the periodic table, such as phosphorus, structural modifications must be made in order to lower the inversion barrier to make the process amenable to study by the DNMR technique.

While electronegative substituents are used to decrease the inversion barriers in nitrogen compounds (e.g., methoxyl groups), it takes electropositive substituents to accomplish the reverse—that is, to lower the inversion barrier rate. For example, the 32 kcal/mole rate in a particular phophine is reduced to 19 kcal/mole when one of the alkyl groups is replaced by a $Si(R_3)$ function.

$$P(CH_2C_6H_5)(CH_3)(C_6H_5) \quad \textbf{vs} \quad P(C_6H_5)[CH(CH_3)_2]\,[Si(CH_3)_3]$$

8.6.2.2 Cyclic Phosphorus Compounds

Since the lone pair of electrons becomes *p*-hybridized in the interconversion process, any substituent that can interact through direct conjugation with the phosphorus atom will lower the inversion barrier energy. Although this effect is not as strong in phosphorus as in nitrogen compounds, the difference in the DNMR-detected inversion barriers between the aromatic phosphole (16 kcal/mole) and a closely related tetrahydro derivative (36 kcal/mole) can be accounted for by application of this argument.

P

$CH(CH_3)_2$

8.6.3 Pentacovalent Structures

8.6.3.1 General Considerations

In pentacovalently bonded atoms where a bipyramidal structure exists, the *axial* and *equatorial* positions can be interchanged through an intermediate *tetragonal* pyramid.

This type of interconversion can occur when the central atom is sulfur, phosphorus, etc..

8.6.3.2 Sulfur Compounds

The room temperature fluorine spectrum of sulfur tetrafluoride shows only one fluorine signal. This signal is converted into two triplets at $-100°C$, a pattern that is consistent with the ''frozen'' structure.

The inversion barrier for this process is approximately 10 kcal/mole.

8.6.3.3 Phosphorus Compounds

The phosphorane has only one methoxyl group proton signal at room temperature (excluding the phosphorus coupling). This singlet is transformed into three different singlets at $-70°C$, an observation that is consistent with the indicated frozen structure. The barrier to this interconversion has been estimated to be 10 kcal/mole.

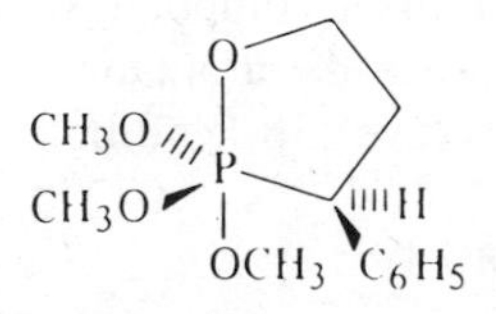

8.7 STRUCTURAL REORGANIZATION PROCESSES

Structural reorganization refers to chemical processes such as are typified by valence bond tautomeric interconversions and to processes involving organometallic entities where the spatial relationship between the organic ligand and the metal ion change unimolecularly. The classical example of the application of the DNMR technique to a valence bond tautomeric process is the interconversion among the possible structures of 3,4-homotropilidene.

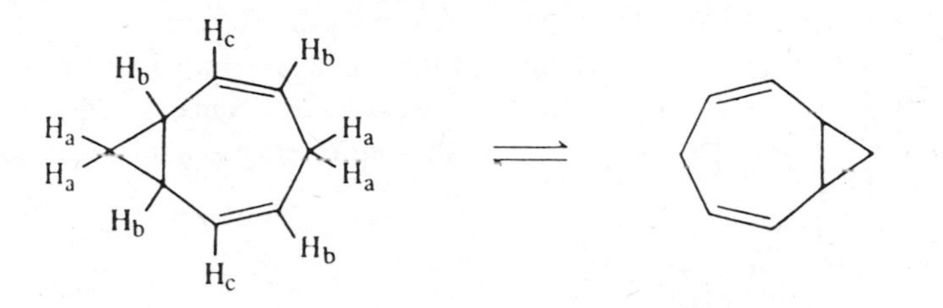

The ground-state structure of this molecule has *five* different protons, and the low-temperature H NMR spectrum of this compound is consistent with this. However, above room temperature, the indicated Cope rearrangement occurs, and the proton H NMR spectrum contains only *three* different types of protons (H_A, H_B, and H_C). The activation energy of this rearrangement is estimated to be about 14 kcal/mole.

The very complex proton NMR spectrum of the amazing bullvalene simplifies to a singlet at 180°C, since all of the protons become equivalent through a multiple sequence of Cope rearrangements. The activation energy for this unique process is 12.8 kcal/mole.

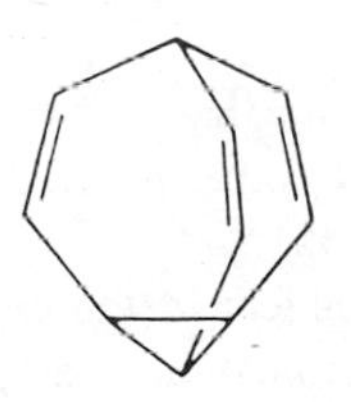

The earlier-described cyclooctatetraene interconversion process is another of the many examples of DNMR application to fluxionally interconverting systems.

An interesting DNMR study of an organometallic compound is the variable temperature proton behavior of tetramethylalleneiron tetracarbonyl, where only one methyl group peak appears at room temperature.

$$(H_3C)_2C{=}C{=}C(CH_3)_2 \cdot Fe(CO)_4$$

At −60°C, the H NMR spectrum shows three different methyl groups in the ratio of 1:1:2, in agreement with the above structure. The interconversion barrier of the process is about 9 kcal/mole, and the iron tetracarbonyl unit is envisioned to circulate about the allenic π cloud by moving *orthogonally* from one alkenic unit to the other and to the reverse side of the first, and finally to the opposite side of the second unit, and so on. This sequence will average the chemical shifts of the four methyl groups!

Among the many cyclopentadienyl complexes that have been studied, compound **39** is somewhat unique, since the iron is bonded to a π as well as to a σ bond. At $-80°C$ the spectrum contains a singlet for the π-bonded ring and an $AA'BB'X$ pattern for the σ-bonded one. The latter pattern coalesces into a singlet above room temperature (interconversion barrier = 10 kcal/mole). Since a sequence of $1 \leftrightarrow 2$ shifts involves a *least-motion* path, it is the preferred mechanism to account for this process [53].

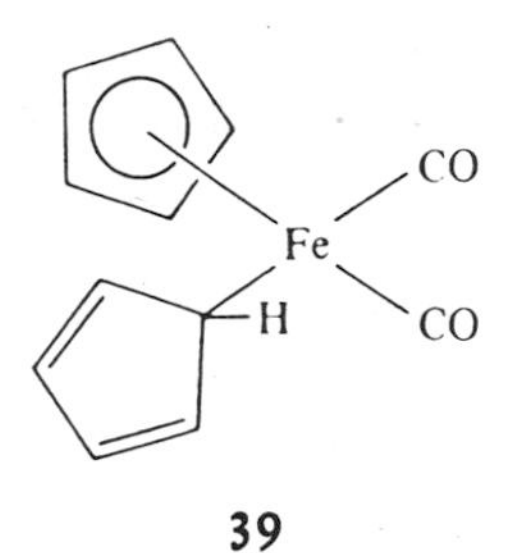

39

8.8 ACID–BASE EQUILIBRIA

Because atom inversion rates greater than 50 s^{-1} do not lend themselves to study by means of the DNMR technique, other methods have been developed to accomplish these analyses [50, 51, 54, 55].

One of these non-DNMR-dependent methods, developed by Saunders and Yamada [55] is best described by application to a specific example.

The H NMR spectrum of amine **40** in concentrated hydrochloric acid shows coupling of the methyl as well as methylene protons to the proton on the nitrogen. As the acid concentration is decreased, the methylene proton doublet coalesces into a singlet as the exchange rate between the N-protonated form (**41**), and the free amine (**40**) becomes rapid. This exchange rate can readily be measured from line-broadening data [54]. When this is done, a first-order rate constant of 5 s^{-1} is obtained for a solution with a salt concentration of 0.35 M in a 2 N hydrochloric acid solvent.

$$\underset{\mathbf{40}}{C_6H_5{-}CH_2{-}N(CH_2C_6H_5){-}CH_3} + H^+ \rightleftharpoons \underset{\mathbf{41}}{C_6H_5{-}CH_2{-}\overset{+}{N}H(CH_2C_6H_5){-}CH_3}$$

The pK_A of the dibenzylmethylamine (**40**) is 7.5, and at that pH the rate constant for hydrogen transfer is $6 \pm 3 \times 10^8$ L/mole s. Since at weaker acid concentrations the N–H exchange rate increases while the methylene protons remain nonequivalent, it appears that the protonated species can be deprotonated and reprotonated without the nitrogen undergoing an inversion. Finally, at a pH of 2.5, the methylene proton doublet coalesces into a singlet. Clearly, the inversion can only occur during the time interval that the nitrogen exists as a free amine, and because the interconversion rate between the amine and the salt is rapid, the following relationship must prevail.

$$K_{\text{interchange}} = K_{\text{inversion}} \times \frac{[\text{amine}]}{[\text{salt}]} + [\text{amine}].$$

Since the pK_A is known, the concentration can be calculated as a function of the pH, and the *interchange* constant is obtained from the line shapes of the *AB* system by means of Alexander's equation [50]. At a pH of 3.5, $K_{\text{interchange}}$ is 21 s^{-1}, and the inversion constant, $K_{\text{inversion}}$, for dibenzylmethylamine is $2 \pm 1 \times 10^5$ s^{-1} [56, 57].

8.9 PROBLEMS

1. The room temperature H NMR spectrum of compound **42** has a narrow singlet for the methyl group protons, a quintuplet for the bridgehead, and a triplet for the vinylic hydrogens. The methyl group proton singlet becomes three peaks of different intensity, and the vinylic protons signal is changed to a featureless broad multiplet when the spectrum is obtained at −58°C [58]. Explain!

N—CO_2CH_3
N—CO_2CH_3

42

2. Explain how NMR spectroscopy can be used to differentiate between the *cis* and *trans* isomers of the piperidine **43** [59].

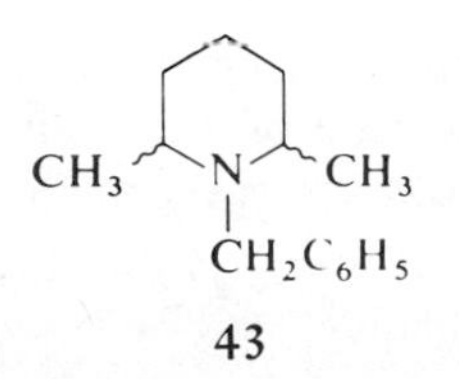

43

3. The *iso*-propyl methyl group protons of the cyclopentenone **44** appear as two doublets. Explain [60]!

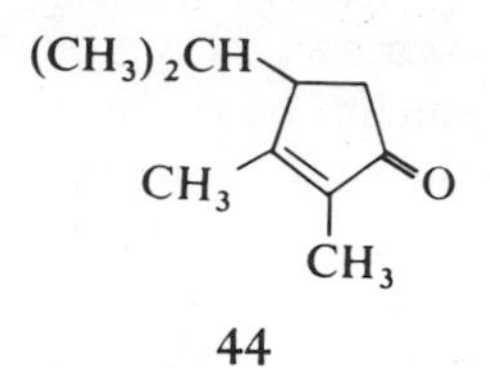

44

4. The geminal proton–nitrogen coupling constant in the iminophenol **45** is 21 Hz at 0°C and 32.4 Hz at 50°C when measured in deuteriochloroform solution. At −46°C, the same coupling constant is 68.2 Hz when the solvent is methanol. Explain the significance of this difference [61].

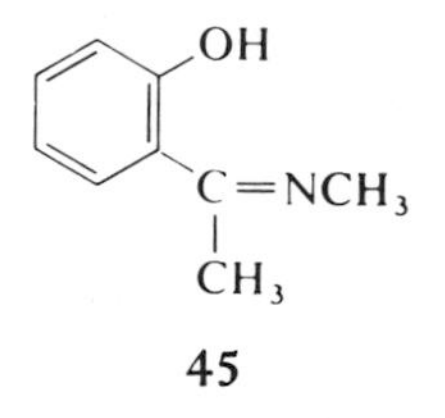

45

5. The room temperature H NMR spectra of the *cis* and *trans* decalins are significantly different. One of the isomers gives a somewhat line-broadened singlet, while the other has a proton pattern best described as consisting of three broad, partially overlapping singlets. Which of these patterns belongs to what isomer?

6. Two of the many possible hexachlorocyclohexane isomers give only one single proton resonance line. The only difference is the chemical shift, where one of the isomer singlets is more deshielded by 0.67 ppm than the other. Identify the isomers.

7. 2,4-Pentanediol has been separated into a low-melting (48–49°C) and a noncrystallizable isomer [62]. The methylene protons of the low-melting isomer appear as a triplet, whereas those of the other isomer have each member of the triplet split into a doublet. Which is the *dl* and which is the *meso* isomer?

8. When trichloroactaldehyde is trimerized, two isomeric substances, m.p. 116 and 152°C, respectively, are obtained. The H NMR spectrum of the higher melting isomer gives only one proton signal (delta 5.73), whereas the other has two singlets in the ratio of 1:2 (delta 4.98 and 5.38, respectively). Deduce the structures of these two compounds.

9. Suggest how the *cis*- and *trans*-2,4-dimethylazetidines (**46**) could be identified by the NMR technqiue.

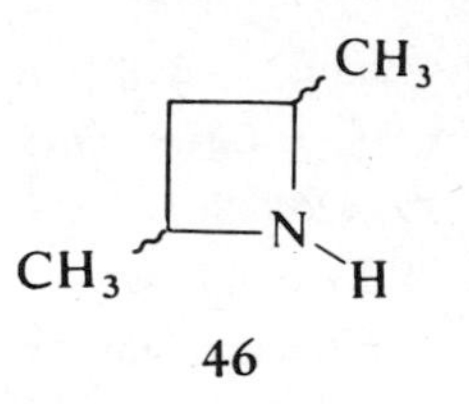

46

10. How could NMR spectroscopy be used to establish whether the lone pair of electrons on the nitrogen in the azetidine **47** is *cis* or *trans* to the *cis*-methyl groups?

11. 1,3,5-Trimethyl-1,3,5-triazacyclohexane **47** shows only one methyl group signal (Δ 6.85) down to at least −60°C. Analyze this result in terms of the nitrogen atomic and ring inversion processes.

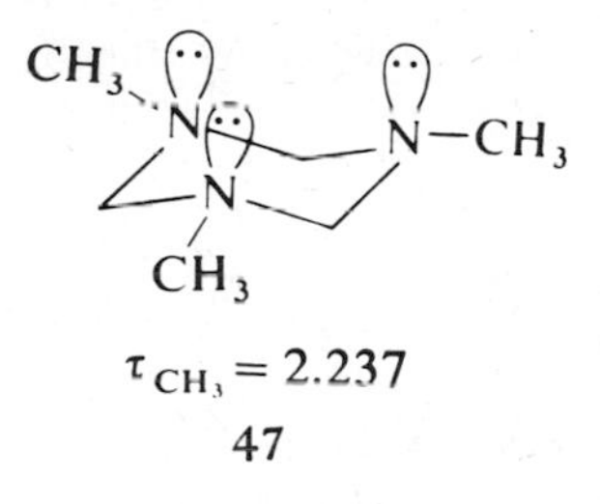

$\tau_{CH_3} = 2.237$

47

REFERENCES

1. J. A. Pople, W. G. Schneider, and H. J. Bernstein, *High-Resolution Nuclear Magnetic Resonance*, New York, McGraw-Hill, 1959.
2. M. van Gorkom and G. E. Hall, *Q. Rev.* 14 (1968).
3. J. A. Pople, *Mol. Phys.* **1,** 3 (1958).
4. N. Sheppard and J. J. Turner, *Proc. R. Soc.* **252A,** 506 (1959).
5. H. S. Gutowsky, *J. Chem. Phys.* **37,** 2196 (1962).
6. J. J. Drysdale and J. D. Roberts, *J. Am. Chem. Soc.* **79,** 319 (1957).
7. P. M. Nair and J. D. Roberts, *J. Am. Chem. Soc.* **79,** 4567 (1957).
8. E. O. Bishop, *Ann. Rep.* **58,** 67 (1961).
9. G. M. Whitesides, F. Kaplan, K. Nagarajan, and J. D. Roberts, *Proc. Natl. Acad. Sci. USA* **48,** 1112 (1962).
10. Reference 6, Chapter 7.
11. F. Kaplan and J. D. Roberts, *J. Am. Chem. Soc.* **83,** 4666 (1961).
12. J. I. Kroschwitz, M. Winokur, H. J. Reich, and J. D. Roberts, *J. Am. Chem. Soc.* **91,** 5927 (1969); and T. H. Siddall, *J. Phys. Chem.* **70,** 2249 (1966).
13. H. S. Gutowsky, D. W. McCall, and C. P. Slichter, *J. Chem. Phys.* **21,** 279 (1953).
14. H. M. McConnell, *J. Chem. Phys.* **28,** 430 (1958).
15. H. S. Gutowsky and C. H. Holm, *J. Chem. Phys.* **25,** 1228 (1956).
16. L. H. Piette and W. A. Anderson, *J. Chem. Soc.* 899 (1959).

17. C. E. Looney, W. D. Phillips, and E. L. Reilly, *J. Am. Chem. Soc.* **79,** 6136 (1957).
18. A. Allerhand and H. S. Gutowsky, *J. Chem. Phys.* **42,** 1587, 3040, 4203 (1965).
19. J. E. Anderson, *Quart. Rev.* **19,** 426 (1965).
20. J. I. Musher, *J. Chem. Phys.* **35,** 1159 (1963).
21. F. R. Jensen, D. S. Noyce, C. H. Sederholm, and A. J. Berlin, *J. Am. Chem. Soc.* **84,** 386 (1962).
22. A. Allerhand, F.-M. Chen, and H. S. Gutowsky, *J. Chem. Phys.* **42,** 3040 (1965).
23. F. A. L. Anet, M. Ahmed, and L. D. Hall, *Proc. Chem. Soc.* 145 (1964).
24. F. A. L. Anet and M. Z. Haq, *J. Am. Chem. Soc.* **87,** 3147 (1965).
25. F. A. Bovey, F. P. Hood III, E. W. Anderson, and R. L. Kornegay, *Proc. Chem. Soc.* 146 (1964).
26. J. B. Hendrickson, *J. Am. Chem. Soc.* **83,** 4537 (1961).
27. G. V. D. Tiers, *Proc. Chem. Soc.* 389 (1960).
28. F. A. L. Anet, C. H. Bradley, and G. W. Buchanan, *J. Am. Chem. Soc.* **93,** 258 (1971).
29. J. E. Anderson, *Q. Rev. (Lond.)* **19,** 426 (1965).
30. F. R. Jensen and L. A. Smith, *J. Am. Chem. Soc.* **86,** 956 (1964).
31. F. A. L. Anet and J. S. Hartman, *J. Am. Chem. Soc.* **85,** 1204 (1963).
32. F. A. L. Anet, *J. Am. Chem. Soc.* **84,** 671 (1962).
33. J. D. Roberts, *Angew. Chem.* **75,** 20 (1963).
34. F. A. L. Anet, A. J. R. Bourn, and Y. S. Lin, *J. Am. Chem. Soc.* **86,** 3576 (1964).
35. K. G. Untsch and R. J. Kurland, *J. Am. Chem. Soc.* **85,** 5709 (1964).
36. W. R. Roth, *Ann.* **671,** 10 (1964).
37. G. C. Levy and G. L. Nelson, *Carbon-13 Nuclear Magnetic Resonance for Organic Chemists*, Wiley-Interscience, New York, 1972, p. 200.
38. Y. Gaoni and F. Sondheimer, *Proc. Chem. Soc.* 299 (1964).
39. S. Wolfe and J. R. Campbell, *Chem. Commun.* 872 (1967).
40. C. Altona, H. R. Buys, H. J. Hageman, and E. Havinga, *Tetrahedron* **23,** 2265 (1967).
41. J. D. Roberts and A. T. Bottini, *J. Am. Chem. Soc.* **80,** 5023 (1958).
42. S. J. Brois, *J. Am. Chem. Soc.* **89,** 4242 (1967).
43. F. Bohlman, *Chem. Ber.* **91,** 2157 (1958).
44. H. P. Hamlow, S. Okuda, and N. Nakagama, *Tetrahedron Lett.*, 2553 (1964).
45. J. B. Lambert and R. G. Keske, *J. Am. Chem. Soc.* **88,** 620 (1966).
46. M. J. T. Robinson, *Tetrahedron Lett.*, 1153 (1968).
47. J. B. Lambert and R. G. Keske, *Tetrahedron Lett.*, 2023 (1969).
48. K. Brown, A. R. Katritzky, and A. J. Waring, *Proc. Chem. Soc. (Lond.)*, 257 (1964).
49. D. L. Griffith and J. D. Roberts, *J. Am. Chem. Soc.* **87,** 4089 (1965).
50. S. Alexander, *J. Chem. Phys.* 2741 (1964).
51. J. J. Delpuech and M. N. Deschamps, *Chem. Commun.*, 1188 (1967).
52. H. S. Gutowsky and P. A. Temussi, *J. Am. Chem. Soc.* **89,** 4358 (1967).
53. Bacon, 1971, ch. 8.

54. A. Lowenstein and S. Meilboom, *J. Chem. Phys.* **27,** 1067 (1957).
55. M. Saunders and F. Yamada, *J. Am. Chem. Soc.* **85,** 1882 (1963).
56. R. K. Harris and R. A. Spragg, *Chem. Commun.* 314 (1966).
57. J. B. Lambert, *J. Am. Chem. Soc.* **89,** 1836 (1967).
58. J. E. Anderson and J. M. Lehn, *Tetrahedron Lett.* **24,** 123 (1968).
59. R. K. Hill and T. H. Chan, *Tetrahedron Lett.* **21,** 2015 (1965).
60. T. S. Sorensen, *Can. J. Chem.* **45,** 1585 (1967).
61. G. O. Dudek and E. I. Dudek, *Chem. Commun.*, 464 (1965).
62. J. Pritchard and E. Vollmer, *J. Org. Chem.* **28,** 1545 (1953).

9

PULSED AND MULTIPLE RESONANCE NMR—THEORY AND APPLICATIONS

9.1 PULSED NMR SPECTROSCOPY

9.1.1 General Principles

The traditional excitation process in FTNMR uses a repetitive application of a radiofrequency pulse designed to excite all of the nuclei of the particular nuclear species being studied. The *free induction decay* (*FID*) curves obtained from each pulse are summed and ultimately analyzed by means of the Fourier transform method to generate the familiar frequency domain spectra (see Chapter 4). The use of *supercon* magnets capable of sustaining magnetic fields of 4 T and higher and the capability to use greater diameter sample tubes, along with the development of the PFT (Pulsed Fourier transform) technique, have added new dimensions to NMR spectroscopy. Fundamentally, PFT employs variations of the frequency, intensity, and application time of appropriate RF pulses, along with different time intervals between these pulses.

The determination of the appropriate time interval between application of the pulses is one of the prime requirements that must be met prior to utilizing this technique. If the interval between RF pulses is too short, significant saturation effects occur, and the signal intensity decays rapidly with time. This happens in N^{15} NMR spectroscopy, for example, when the interval between the applied pulses is less than $6T_1$. On the other hand, a time interval that is too long will decrease the effectiveness of the PFT method since the FIDs will decay.

Another electronic technique available to minimize saturation is a decrease of the pulse duration time, t_p, and/or of the pulse angle. Although these changes can cause a loss in signal intensity if not carefully controlled, this may be offset by the shorter time interval required between the application of the pulses. An example of a chemical technique, which minimizes the saturation problem, is represented by the use of T_1 reagents such as $Cr(acac)_3$ in N^{15} spectroscopy.

Pulsed FT NMR experiments are conveniently pictorialized by referring to a rotating three-dimensional frame of reference. In this frame of reference, the entire coordinate system is envisioned as rotating at the Larmor frequency corresponding to the applied magnetic field. The RF field, B_1, oscillating at the Larmor frequency, is fixed in the rotating reference frame [1, 2]. In this representation, the spin system and the RF excitation are viewed as simple vectors (see Fig. 9.1).

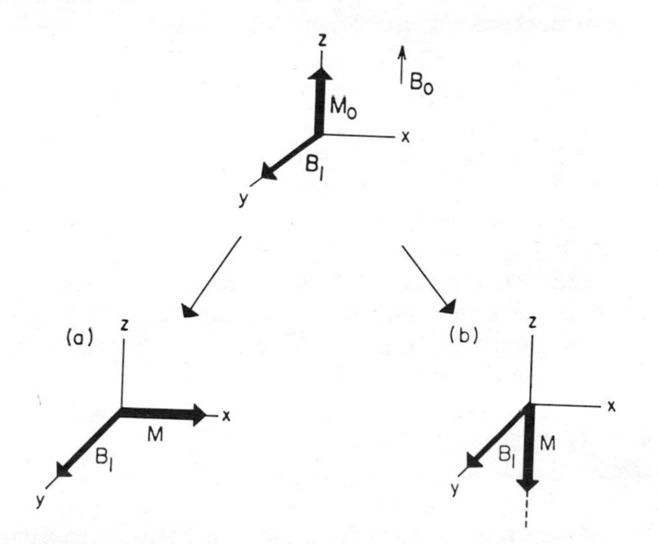

Figure 9.1. Rotating frame reference: (a) 90° pulse; (b) 180° pulse.

When a magnetic field, B_0, is applied in the direction of the Z axis, interactions between the nuclei and the lattice (spin–lattice relaxation) establish an excess of nuclei in the lower energy level according to the Boltzmann distribution law (cf. Chapter 1). As a result, the nuclei will have a net equilibrium magnetization, M_0, in the same direction as the applied magnetic field. Once this equilibrium condition has been attained, a radiofrequency field, B_1, is applied, for a time t, in the direction of the y axis. While this technique can cause saturation of the system during the on time, it is minimized if the amplitude of the applied RF field is reasonably large and the duration of the pulse is appropriately short. The application of a properly timed (t) RF pulse, B_1, causes the nuclear magnetization to rotate around the Y axis, a 90° shift (from the Z to the Y axis).

If the time of the pulse that caused the 90° shift is doubled, the rotation of M_0 will also be doubled and the magnetization vector will become aligned in the direction of the Z axis (Fig. 9.1). This type of pulse is referred to as a 180° one. These processes are graphically presented in Figure 9.1.

Since NMR instruments measure the magnitude of resonance signals in the X–Y plane only, no signal is detected if the nuclei are either exposed to the applied magnetic field alone, where the magnetization is in the direction of the Z axis, or when a 180° pulse is applied, where the magnetization vector is in the $-Z$ direction. The situation where the excess nuclei are in the higher energy level is referred to as having a "negative spin temperature."

With a 10,000 Hz frequency, a particular RF field will induce nuclear spin precessions of 10,000 revolutions per second, and the spin vector will make one revolution around B_1 in 0.0001 s. In one fourth of that time, the spin vector travels through only 90° and will be at right angles to the applied magnetic field and perpendicular to B_0. If the RF field is turned off at the point where the spin vector is at 90°, a maximum resonance signal will be observed (barring any other interferences!). A numerical example (see Figure 9.2) serves to describe this technique pictorially.

The pulse interval is primarily controlled by the time required for spin–lattice relaxation to restore equilibrium magnetization, M_0. Pulse times normally range from 10 to 100 μs, with the lowest frequency nuclei requiring longer pulse times than the higher frequency ones. Other factors contribute to the required time interval as well. For example, when operating at very low frequencies, *acoustic ringing*, resulting from RF pulses creating standing waves in the sample walls, has to be allowed to dampen before another RF pulse is applied [3–5]. This is the situation in Fe^{57} or Rh^{103} signal acquisition, where the pulse must be stopped for some milliseconds. Although in these instances the relaxation times are so long that the acoustic ringing time does not significantly worsen the spectral aquisition situation, this time delay can cause considerable signal loss for some nuclei, because of the T_2 relaxation of the nuclear spin system.

As the spectral window for any one nuclide is widened (the result of increasing the applied magnetic field), the pulse power required for broad-band excitation increases. The excitation distribution for a pulse of 25 μs, for example, is such

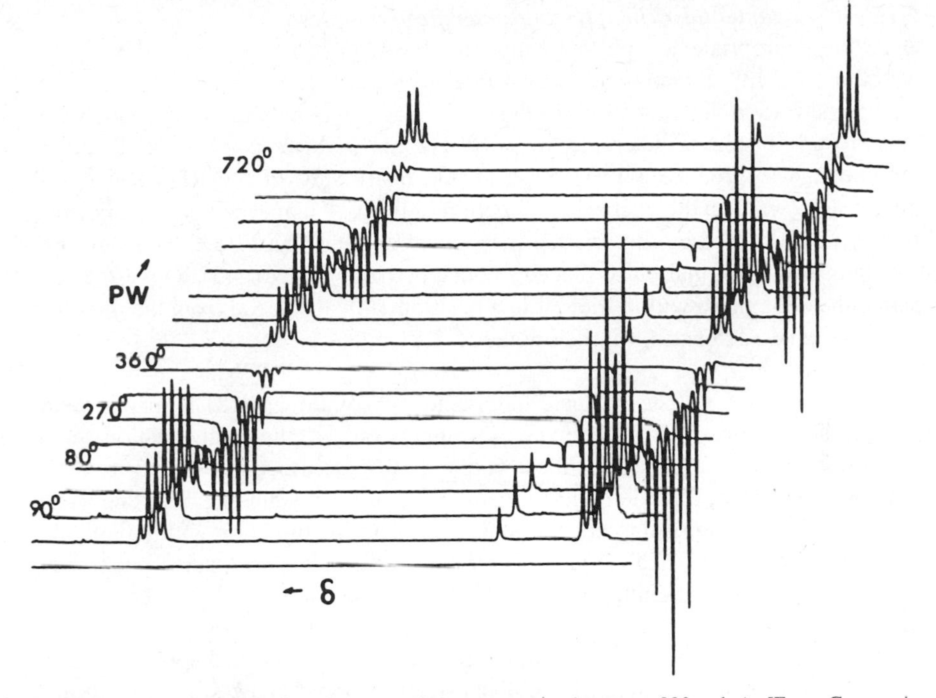

Figure 9.2. Pulse width variation spectra (3rd spectrum is close to a 90° pulse). [From George A. Gray, *Methods of 2D Nuclear Magnetic Resonance* (preprint), Varian Associates, Palo Alto, CA. Reproduced with permission.]

that at 2.3 T the excitation energy for C^{13} is only minimally decreased across the 250-ppm range, whereas at 11.7 T there is a significant energy loss across this field range. This problem is resolved by adjusting the pulse flip angle to a value considerably below the 90° maximum value.

Factors such as these determine the experimentally set pulse flip angle ($<90°$, between 20 and 45°) and the pulse interval. In some N^{15} experiments, for example, this interval is 10 s or more.

9.1.2 Multipulse Techniques

The currently available multinuclear NMR spectrometers, generally based on supercon magnets and employing FT-based analyses, have built into them programmable pulse interfaces. Thus, it is relatively easy to program a variety of different pulse sequences for different nuclides and experimental conditions. [6]

The Varian XL series of supercon FTNMR spectrometers, for example, provide a pulse sequence controller that automatically controls phases, pulses, and delays, thus allowing the acquisition of FTNMR spectral data using many different pulse sequences.

9.1.2.1 Fundamental Principles and Specific Examples

Once the appropriate RF pulse is applied, the magnetization vector flips in the direction of the X–Y plane and is then free to precess. If this pulse is a 90° one, the magnetization vector will be in the X–Y plane, and application of another 90° pulse to this system will redirect it to the $-Z$ axis. This second burst, if left on long enough to produce a 180° pulse (a 90° phase shift of the RF frequency is often used), will flip the vector in a direction in the X–Y plane, which is equidistant from the axis caused by the original magnetization pulse but of opposite sign. For example, if the original pulse placed the magnetization vector 45° above the X–Y plane, the 180° pulse will reorient it in a position 180° displaced from the direction of the original vector and 45° below the X–Y plane.

REFOCUSING PULSES. The intrinsic inhomogeneity of the applied magnetic field, among other factors, causes a breakup (a ''fanning out'' of the magnetization vector once an RF field is applied). Of these vectors, 50% are moved in the positive and 50% in the negative direction with respect to the original signal vector. The signal resulting from this spreading corresponds to a decay in amplitude with a particular time constant. An FT analysis of this FID would provide resonance peaks with wide band widths containing significant contributions caused by the vector defocusing.

When the 180° pulse is applied along the Y axis, the individual spin vectors are forced to new positions within the X–Y plane, and normal precession can continue (this 180° pulse is referred to as a refocusing pulse). At time $2t$, after the beginning of the sequence, the spins will have returned aligned along the Y axis. This type of refocusing eliminates any necessary phase correction caused by the dephasing of spins during the pulse sequence period $2t$ and corrects any dephasing caused by the inhomogeneity of the magnetic field.

If detection is initiated after time $2t$, the signal that is still present is referred to as an echo, or spin echo. The use of this signal has also found applications in a number of multiphase sequences [7].

POLARIZATION TRANSFER. When dealing with a heteronuclearly coupled system (A–X), the application of two simultaneous 180° pulses, one to each nuclide, causes an interchange of the labels (signs) on the spin vectors (polarization transfer), and instead of refocusing after $2t$, the spins will continue to diverge (a modulation of the spin echo will result). Finally, the nuclei will focus along the $-Y$ axis after $1/J$ (J = coupling constant between A and X) and along the Y axis after $2/J$. This technique is extensively used in 2D NMR and will be described in Chapter 10.

A process that identifies pairs of nuclei (A and X) as either being bonded to each other and/or as being spatially close to each other is based on the application of two 90° pulses applied, respectively, to each nucleus. The basis of this technique is that the signal of the observed nucleus is modified by the chemical shift of the other. One way to study these interactions is to transfer polarization from the A to the X nuclei. If the sign and the magnitude of the polarization are modulated at the frequency of the nucleus that is responsible for polarization transfer, the information can be transferred to the other nucleus when a proper alignment of the spin components exists [8, 9].

This concept, where magnetization transfer is achieved through spin–spin coupling, is not only very sensitive to the size of the coupling constant but also requires a very close match of the RF amplitudes of the decoupler and observing channels. Although the method has been modified [10], some experimental difficulties still exist.

INEPT (*insensitive nuclei enhanced by polarization transfer*) is another technique [11], which is designed to provide spectral sensitivity enhancement. It operates with the following pulse sequences:

$$90°_H(X) - \tau - 180°_H(X),\ 180°_I - \tau - 90°_H(Y),\ 90°_I$$
$$- \Delta - \text{DCPLE, ACQ DCPLE, ACQ}$$

In this expression, X and Y refer to the phase of the decoupler pulse, I is the observed nucleus in an I–H bonding situation, τ is a delay determined as $1/4J_{HI}$, and Δ is a preacquisition delay (2 τ for a doublet, τ for a triplet). Figure 9.3 shows a one-pulse N^{15} spectrum of 90° of formamide (without NOE) and the spectrum obtained from a one-cycle accumulation using the INEPT sequence. While the S/N ratio for the former is 26 : 1, it is 160 : 1 for the latter. [5].

INVERSION RECOVERY. A *pulse-sequence technique*, which permits measurement

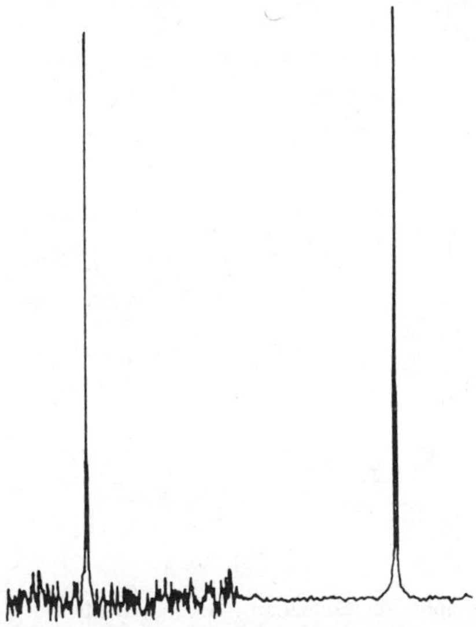

Figure 9.3. N^{15} spectrum of formamide. Left: normal; right: INEPT. [J. J. Dechter, *Progress in Inorganic Chemistry*, S. J. Lippard, Ed., Vol. 29, John Wiley, New York, p. 297 (1982). Reproduced with permission.]

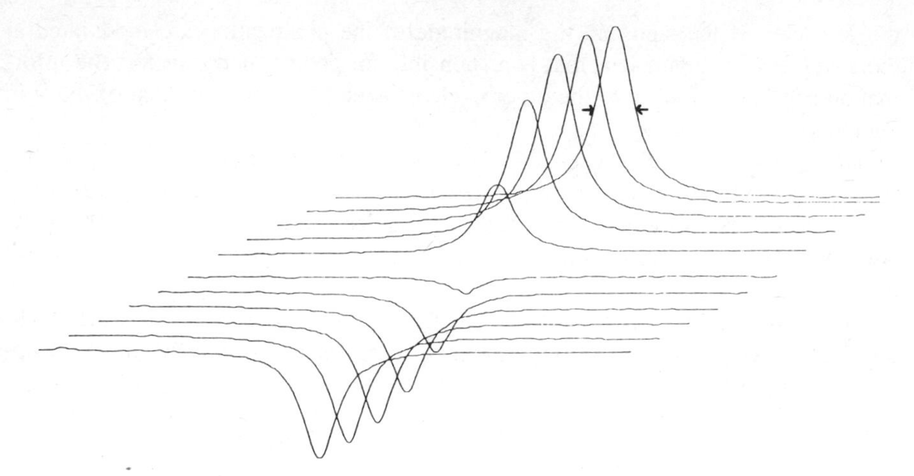

Figure 9.4. Inversion recovery experiment: Cu^{63} 53.02 MHz. [Varian Catalog, *Other Nuclei NMR on the XL-200*. Reproduced with permission.]

of T_1 and is referred to as the "inversion recovery" method, is based on the sequence: 180°-*t*-90°. The 180° pulse inverts the energy level populations producing the normal Boltzman distribution of nuclei, with the excess nuclei in the higher energy level. This distribution begins to decay rapidly and, if no additional RF pulses are applied, reestablishes itself in the normal Boltzman distribution (the lower energy spin states are in excess). The 90° pulse is applied after time *t*, and the free induction decay curve is recorded. If the time *t* is very long, a normal spectrum will be observed (since the normal distribution will have reestablished itself). If, however, *t* is of the order of magnitude of T_1, then the FID curve will provide additional information.

The spectral accumulations of some typical inversion recovery experiments on Cu^{63} and Al^{27} ions (Figs. 9.4, 9.5), along with one applied to an organic compound

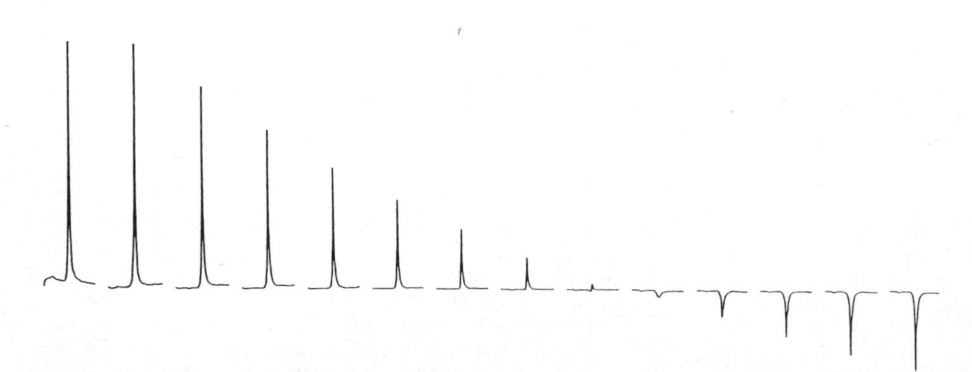

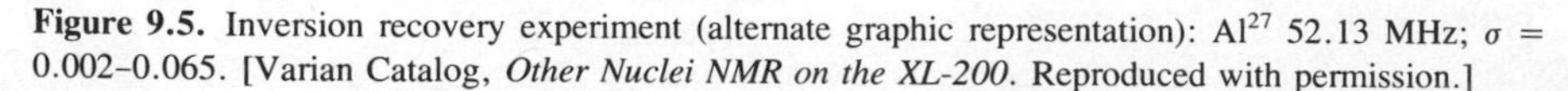

Figure 9.5. Inversion recovery experiment (alternate graphic representation): Al^{27} 52.13 MHz; σ = 0.002–0.065. [Varian Catalog, *Other Nuclei NMR on the XL-200*. Reproduced with permission.]

(Fig. 9.6), serve as excellent examples of this technique and also demonstrate two different means of presenting the data.

PRFT (PARTIALLY RELAXED FOURIER TRANSFORM). Levy et al. [12] have provided an excellent example that contains all of the theoretically possible information that can be obtained from this type of multipulsed spectral aquisition. In Figure 9.7, the A peak of the partially relaxed Fourier transform (PRFT) spectrum is inverted because $t \ll T_1$ and little relaxation has occurred after the 180° pulse was applied. The signal of the *B* nucleus is essentially zero because *t* is approximately equal to $T_1 \times \ln(2) = 0.69T_1$. The small, positive *C* nucleus signal indicates that *t* and T_1 are of the same order of magnitude. When *t* is much larger than T_1, the signal is positive and of full intensity, as shown for nucleus *D*.

A variation of this pulse sequence, which affords spectra with the absorption lines represented in the same direction, was developed by Freeman and Hill [8] and can be described by $(-T-90°_{\text{infin.}}-t-180°-t-90°_t)_x$, where *T* is the waiting time between initiation of a repetitive pulse sequence, the $90°_{\text{infin.}}$ pulse generates a relaxed system, and the FID resulting from the $90°_t$ contains the PRFT data. In this sequence, the computer alternates between collecting and adding the data from the infinite pulse and subtracts the $90°_t$ FID from the collected information. An example of the spectral plots resulting from the difference spectra ($S_{\text{infin.}}-St$) is given in Figure 9.8.

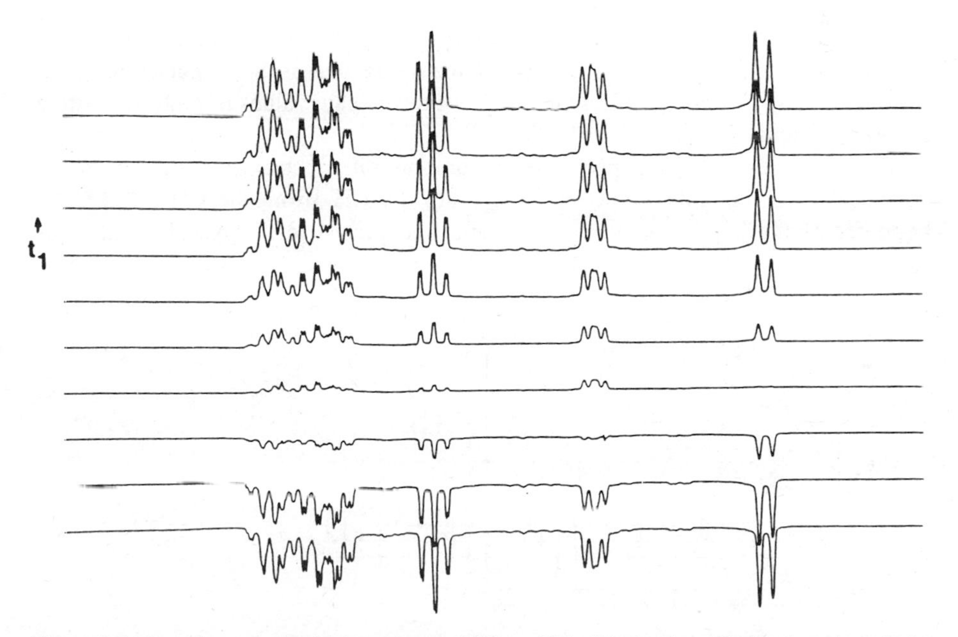

Figure 9.6. Inversion recovery T_1 experiment ($180°-t_1-90°$). [G. A. Gray, Varian Associates (Palo Alto, CA) preprint *Methods of Two-Dimensional NMR*. Reproduced with permission.]

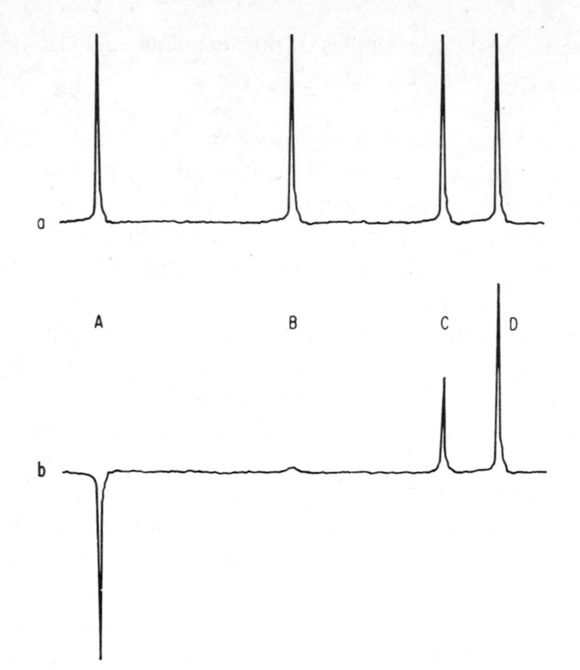

Figure 9.7. PRFT spectrum (a) Normal. (b) Partially relaxed FT spectrum. [Source: Levy and Nelson, *Carbon-13 NMR for Organic Chemists*, John Wiley, 1972, p. 183. Reproduced with permission.]

If $t >> T_1$, ($S_{infin.}-S_t$) is similar to ($S_{infin.}-S_{infin.}$), and the peak height will approach zero. If $t << T_1$, S_t will be of the same size as $S_{infin.}$ but of negative sign. ($S_{infin.}-S_t$) is thus equivalent to $25_{infin.}$. Intermediate values of t afford positive but lower-intensity peaks.

The XL series of Varian spectrometers, among other instruments, provide facile means of obtaining subspectra which separate the CH- from the CH_2- and CH_3-bonded carbons in a compound. This "attached proton test" (A.P.T.) pulse se-

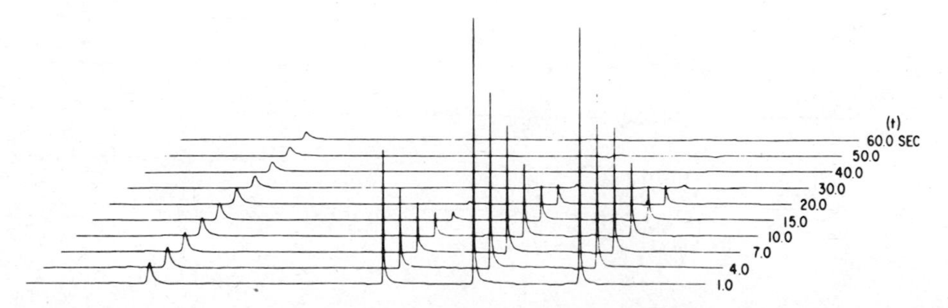

Figure 9.8. Freeman-Hill plot of PRFT spectra of nitrobenzene. [Source: Levy and Nelson, *Carbon-13 NMR for Organic Chemists*, John Wiley, 1972, p. 184. Reproduced with permission.]

quence modification of PRFT, along with use of NOE, has become the method of choice for obtaining routine C^{13} spectra. Figure 9.9 is an example of a normal proton decoupled and APT spectrum of a steroid, provided by Varian Associates.

ROUTINE PULSE SEQUENCE. The most commonly used pulse sequence for routine NMR spectroscopy utilize a 90°–t–180°–t sequence, where at time $2t$, after the beginning of the sequence, the spins will have returned to alignment along the y axis [1, 7].

9.1.2.2 Generally Useful FTNMR Pulse Sequences

The oldest of the useful pulse sequences in FTNMR was proposed in 1969 for use with slowly relaxing nuclei such as N^{15} and metal nuclides such as Cd^{113}. This "DEFT" (driven equilibrium Fourier transform) method is designed to achieve improved sensitivity by forcing nuclei (RF pulsing) to relax immediately following the data acquisition period in the pulse sequence (cf. refocusing pulses). Modification of this approach has led to the "SUPERDEFT" and "SEFT" (spin echo Fourier transform) pulse sequence modifications [1, 2].

Some of the many pulse sequences, along with pertinent descriptive information, that are applicable to the determination of T_1 and that facilitate spectral data accumulations are the following:

IRFT (inversion recovery): T–180°–τ–90°; delay = T → 4T_1 s; S/N factor increase is about 2; relatively slow sequence.

FIRFT (fast inversion recovery): T–180°–τ–90°; delay = T → 1 s; self-optimizing sequence, fast repetitions.

PSFT (progressive saturation): 90°–t; delay T → 0.3 s; accurate pulse angles

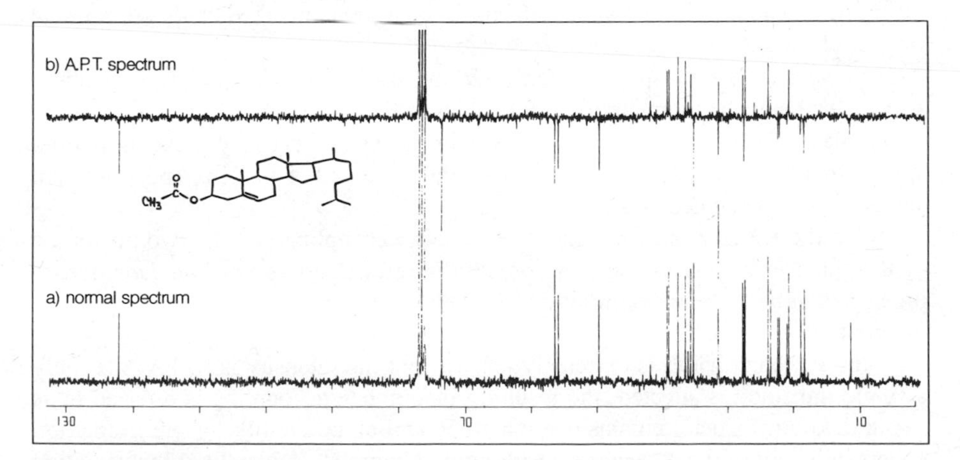

Figure 9.9. Attached Proton Test and Normal Spectra of a Steroid. [*Varian XL-300 Spectra Catalog No. 1.* Reproduced with permission.]

must be used; resolution is proportional to the pulse interval, one of the faster sequences.

SRFT (saturation recovery): Z^*–90°–Z^*–τ–90°; delay T → 1 s; pulse burst or Z-axis homospoil (Z^*) must be available; fast repetitive sequence.

DNOEFT (dynamic nuclear Overhauser effect FT): T[τ = decouple]–90°; delay T → $10T_1$ s; determines T_1 and NOE simultaneously; relatively slow sequence [13].

Other pulse sequences, along with their major applications, that have been developed in response to NMR chemical shift and coupling constant determination needs are these:

Carr–Purcell–Meilboom–Gill–T_2 sequence: Heteronuclear 2D shift correlations.

Quadrupole echo cross polarization: Homonuclear and heteronuclear NOE 2D.

Multiple-contact cross-polarization: NOE 2D.

Quadrature selective excitation: C^{13} satellite studies.

Homo- and heteronuclear 2D J Spectroscopy: Selective inversion, selective solvent suppression, T_1-rho determination.

9.2 MULTIPLE RESONANCE SPECTROSCOPY

9.2.1 General Considerations

The *multiple* resonance technique utilizes, as the name implies, more than one RF field to impact magnetically active nuclei. When *two* different RFs are employed, the process is referred to as "double resonance." It involves the application of a second RF field, B_2, appropriate to cause resonance of a nucleus in an NMR experiment where a particular A nucleus is being examined with its appropriate resonance RF, B_2.

If nucleus A is spin–spin coupled to B and the applied B_2 field is sufficiently strong, the magnetic moment of the B nucleus will become *quantized* perpendicular to the magnetic field rather than parallel with it. This effectively prevents B from spin–spin interacting with nucleus A and the observed A nucleus spectrum will appeared as noncoupled to B.

When the RF is strong enough to cause spin decoupling and the two nuclei are of the same elemental species, the process is referred to as *homonuclear* decoupling. When the two nuclei are of different elemental species, the process is called *heteronuclear* decoupling. [14–18].

If the RF band width is extremely narrow and the intensity is so low that only a single transition is affected, the resulting incomplete decoupling is referred to as "spin tickling." There remains one other decoupling possibility, where irradiation covers only part of the resonance, resulting in incomplete decoupling; this situation is called selective spin decoupling. These two processes have their own particular uses and will be described later in this chapter.

Although the effects are not particularly large, changes in the resonance positions of the decoupled nuclei do occur. This phenomenon, the Bloch–Siegert shift, must be taken into consideration when describing the chemical shifts of decoupled nuclei [6, 14].

To facilitate description of these double-resonance eperiments, standardized notations have been introduced. The symbol A–[X] designates that B_2, the strong decoupling field, is applied at or near the resonance frequency of nucleus X, and nucleus A is the species being examined. Thus, H–[F] describes a proton NMR experiment in which the fluorine nucleus is being irradiated—a heteronuclear resonance experiment. The H_1–[H_2] symbolism describes a proton NMR experiment, where H_1 is being examined, under conditions where H_2 is irradiated—a homonuclear decoupling experiment.

9.2.2 Spin Decoupling Processes

9.2.2. Noise Decoupling

An amazingly wide range of frequencies can be irradiated by using either white noise or broad-band oscillators. A single H^1 decoupling frequency is used as the center of a finite excitation band. This single frequency can be modulated by a pseudo noise generator, which yields an effective excitation throughout a preset band width. The band width can be set to cover, for example, all proton resonances in a particular sample.

This technique is generally employed in heteronuclear experiments to remove the spin coupling of a particular nuclide. Among these, natural abundance C^{13} spectra are normally noise-decoupled from protons.

The one-bond C^{13}-H coupling constants are numerically large, and the carbon resonances can be split into wide-ranging multiplets. When these protons are wide-band decoupled, >10 order of magnitude sensitivity enhancements can and do occur. This enhancement is a result of the summation effect of the various resonance lines of a given multiplet when they are coalesced into a singlet. Additionally, the NOE (see below) causes an additional peak area increase. The shortcoming of this decoupling process is, of course, that all coupling information is lost, since all carbon peaks appear as singlets (unless they are coupled to nonproton nuclides). An example of two C^{13} spectra of imidazo[1,2-a]pyridine, one without proton decoupling and one with noise-decoupling, is given in Figure 9.10. The difference in accumulation time between these two spectra is at least a factor of 1:50!

A technique known as single-frequency off-resonance decoupling permits recovery of some of the C–H coupling constants. The H^1 irradiation frequency in this technique is kept at very high power levels with the center frequency moved between 500 and 1000 Hz from the protons to be irradiated. Concurrent with this, the excitation band width generator is switched off. When this process is operating,

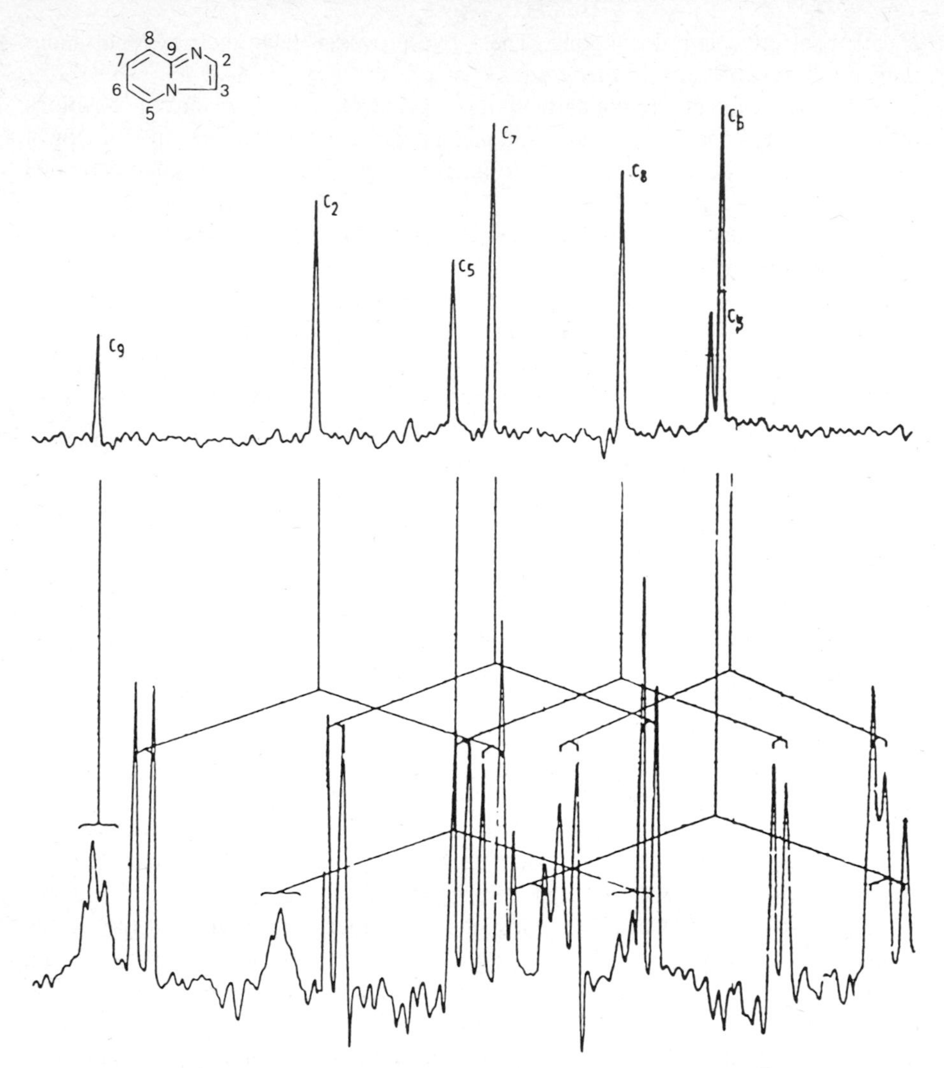

Figure 9.10. (a) Proton noise-decoupled spectrum. (b) Nondecoupled spectrum. [provided by Dr. E. S. Hand, U. of Alabama]

the one-bond C–H coupling patterns return, although the coupling values are not the same as those observed in the absence of a decoupling field. The spectral assignments can be made on the basis of the multiplicity of the resulting multiplets. The H-decoupled and off-resonance decoupled spectra of phenol serve to illustrate this modified noise-decoupling process [12].

9.2.2.2 Spin–Spin Decoupling

Although these double-resonance experiments can be carried out under either frequency or field sweep conditions, the spectra resulting from the two methods are not equivalent. In the frequency mode, the irradiating frequency is set at the resonance position of the proton to be decoupled. The spectrum is then swept in the usual fashion. Under these conditions, every proton that is coupled to the one being irradiated will be uncoupled. As the irradiation frequency approaches the frequency of the proton being decoupled, an interference pattern is generated (see Fig. 9.12).

In addition to the Bloch-Siegert shift, the spectrum of the 2-chloro-4-methylpyrimidine is simplified, and a singlet appears for the proton adjacent to the ring nitrogen when the other ring proton is irradiated (spectrum *a*). If the proton adjacent to the ring nitrogen is irradiated, the other ring proton becomes a singlet (spectrum *b*).

Another example of spin–spin decoupling, in this instance applied to an *ABX* system, is given in Figue 9.13. The various patterns are self-explanatory.

In the field sweep mode, homonuclear decoupling experiments are somewhat more complicated. In an *AB* system, for example, decoupling is observed only if the frequency difference between *A* and *B* accidently corresponds to the difference between the two resonance frequencies. Fortunately, most NMR instruments can be operated in either the frequency or field sweep mode. In fact, field sweep techniques are rarely ever used [19].

9.2.2.3 Selective Spin Decoupling

If the irradiation band width covers only part of a particular multiplet, incomplete decoupling is observed in the multiplet's *counterpart*. This "selective decoupling" technique can provide information leading to the establishment of the relative sign

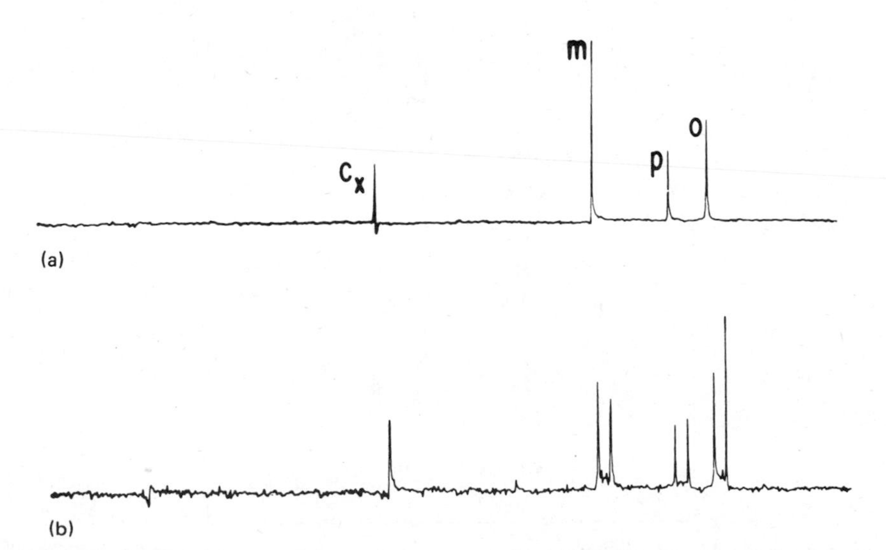

Figure 9.11. C^{13} phenol spectrum. (a) Wide-band H-decoupled. (b) Off-resonance H-decoupled. [Levy and Nelson, *Carbon-13 NMR for Organic Chemists*, John Wiley, 1972, p. 10. Reproduced with permission.]

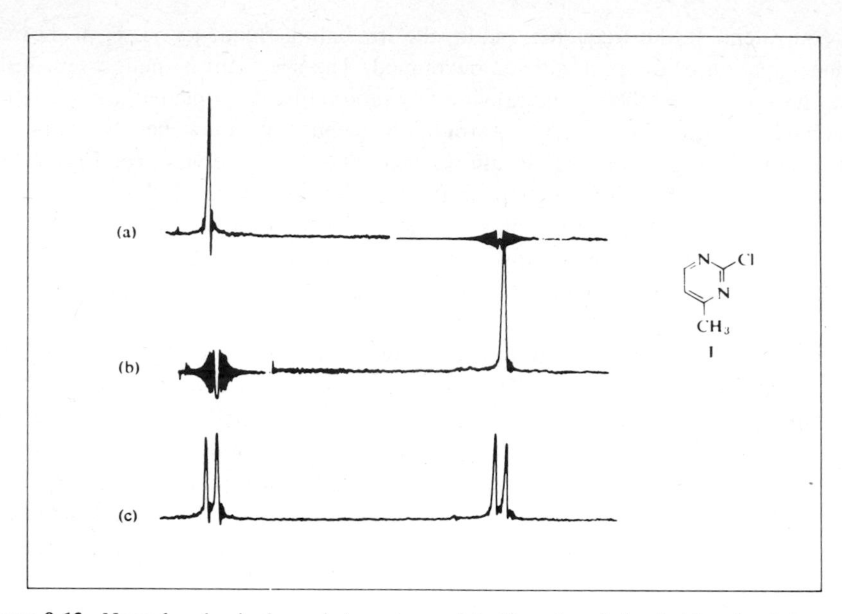

Figure 9.12. Normal and spin-decoupled spectrum of 2-chloro-4-methylpyrimidine (methyl protons not shown). (a) Irradiated at shielded doublet. (b) Irradiated at deshielded doublet. (c) Normal spectrum.

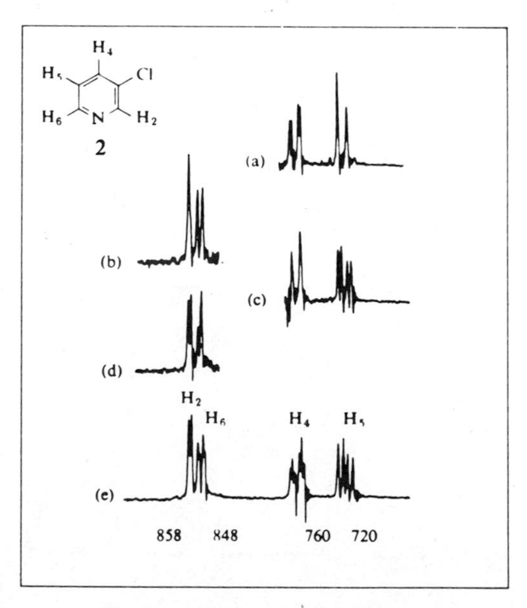

Figure 9.13. 100-MHz spectra of 3-chloropyridine **2** in CCl_4: (a) irradiated at H_6; (b) irradiated at H_4; (c) irradiated at H_2; (d) irradiated at H_5; (e) normal spectrum.

of a coupling constant and energy level distributions in multinuclear systems [20]. Some applications of this technique are described later in this chapter.

9.2.2.4 Spin Tickling

When the band width employed to provide selective decoupling is even more narrow or the intensity is very low, only a single transition may be affected. This can result in the observation of additional splitting patterns of other nuclei in the spectrum. This spin tickling process also provides relative signs of coupling constants [16]. Some applications are described later in this section.

9.2.2.5 Internuclear Double Resonance (INDOR)

This type of spectrum provides a record of the variations in intensity of a particular NMR resonance line detected with an RF field of intensity H_1 and a frequency ν_1, while sweeping another RF field of frequency ν_2 and intensity H_2 over the remaining resonance lines of the NMR spectrum. The experiment can be performed in either the homonuclear or heteronuclear sense [20–22].

One of the original heteronuclear–INDOR experiments demonstrates this technique. The P^{31} nucleus, with a spin quantum number of 1/2, in trimethylphosphite, $(CH_3O)_3P$, is adjacent to nine equivalent protons. Thus, the phosphorus atom will appear as a 10-line multiplet. The experimental P^{31} spectrum (Fig. 9.14) confirms this (see ref. 1 in Chapter 8 and ref. 6 in Chapter 7). The H NMR spectrum of

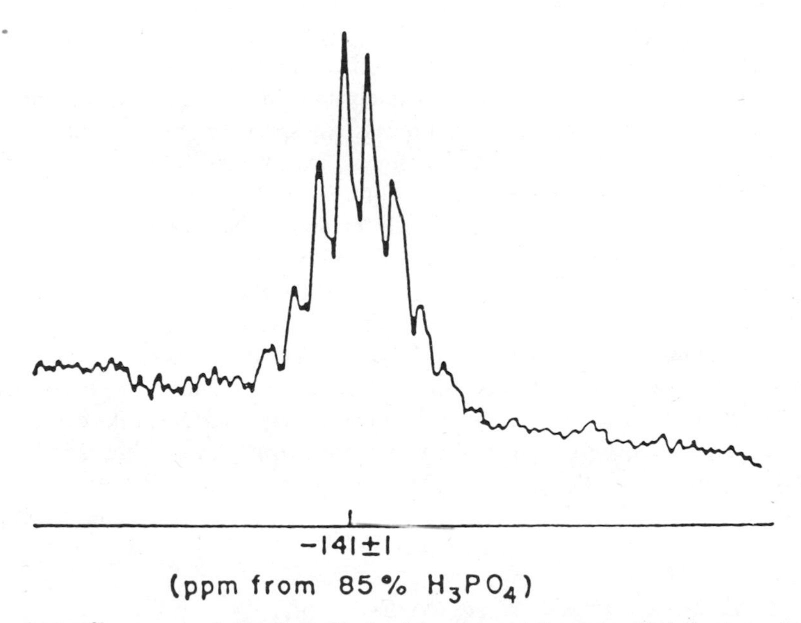

Figure 9.14. P^{31} spectrum of $(CH_3O)_3P$. [Pople, Schneider, and Bernstein, *High Resolution Nuclear Magnetic Resonance*, McGraw-Hill, New York, 1959, p. 352. Reproduced with permission.]

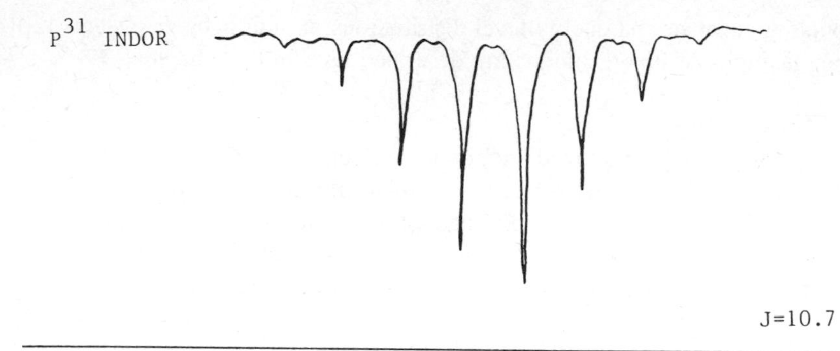

Figure 9.15. INDOR spectrum of $(CH_3O)_3P$. [From E. B. Baker, L. W. Burd, and C. V. Root, *Rev. Sci. Instrum.* **34,** 243 (1963).]

this compound is simply a doublet. The spacing between all of these multiplets, $J_{H,P}$, both proton and phosphorus, is 10.7 Hz.

The INDOR experiment is conducted by setting the proton NMR frequency at one of the methyl group peaks, in a 1.41-T magnetic field in this particular instance, while at the same time sweeping the sample with an RF from 24.29572 to 24.29582 MHz (this is the frequency required for phosphorus resonance to occur in a 1.41-T magnetic field) and recording the proton intensities of the methyl groups. When the irradiating frequency matches the resonance frequency of the P^{31} nucleus, the proton is decoupled from those P nuclei that resonate at that particular frequency. A negative peak will result if the intensity of the proton peak is decreased, and a positive one will result if the proton peak intensity is increased during the irradiation sequence. The result of this scanning process is a spectrum that is identical to that obtained from a normal P^{31} NMR experiment, except that the intensities are negative and the observation is made using a proton NMR frequency (see Fig. 9.15).

The INDOR method is only applicable to nuclei that are coupled to a high-sensitivity nucleus—for example, H and F. With the advent of the multinuclear spectrometers and associated FT technique, the INDOR process, applied to heteronuclear situations, has lost much of its earlier appeal. However, the application of INDOR to homonuclear analyses has increased, since it can potentially be applied to the various possible decoupling techniques. Contrary to heteronuclear INDOR, where all peaks are negative, in homonuclear INDOR peaks can be either positive or negative (although they can all become negative if the applied field intensity is high enough!).

The use of the INDOR method in the determination of the signs of coupling constant is delineated later in this chapter.

9.2.2.6 *Nuclear Overhauser Effect (NOE)*

There are occasions when the stereochemistry of a particular molecule is such that, although not bonded to each other, two nuclei are interacting with each other by mutual relaxation through interaction between their respective dipole moments. If one of these nuclei is irradiated while the other one is being examined in the normal fashion, the Boltzman distribution of the latter nucleus is altered, and the intensity of the resonance is perturbed [23, 24]. This phenomenon, an example of intramolecular spin–lattice relaxation (T_1) occurring intermolecularly, was first observed by Overhauser in an instance where the irradiated "nucleus" was actually a pair of electrons. The effect on the interacting nucleus was, however, still observed. Generally, it is of much broader import, however, that the situation prevails where both the spin systems are nuclei—hence the term "nuclear Overhauser effect."

To observe this effect to its experimentally most intense, the compound to be studied should be dissolved in a magnetically inert solvent, or at least in a proton- and fluorine-free one. In addition, the presence of paramagnetic species (such as molecular oxygen) must be kept to an absolute minimum.

The contribution to T_1 from this *intramolecular* interaction is given by:

$$\frac{1}{T_1^{AB}} = \frac{\hbar \gamma_A{}^2 \gamma_B{}^2 \tau^*}{d^6}. \tag{9.1}$$

In this expression [25], T_1^{AB} is the contribution to T_1 for nucleus A or nucleus B, d is the *internuclear* distance between A and B, and tau* is the correlation time for random molecular rotation.

The experimental "nuclear Overhauser enhancement factor" (NOEF) is given by the ratio of the peak intensity area with NOE present divided by the peak intensity area without NOE, minus 1.

The maximum attainable theoretical intensity increase depends on the ratio of the *gyromagnetic* values of nuclei A and B. If both of the nuclei are protons, a factor of 1.5 (50%) in intensity increase is obtained when one of the nuclei is irradiated. If nucleus A is a proton and nucleus B is a C^{13}, the factor is 2.99 (200%) when the proton is irradiated. In larger and relatively rigid organic molecules, virtually all carbons are relaxed by the *dipole–dipole* mechanism. Consequently, when proton decoupling is in process in a given experiment, most carbon peak intensities are augmented by the NOE.

When one of the interacting nuclides has a negative magnetogyric ratio (e.g., N^{15}, and the only main group metal nuclide, Sn^{119}) the existence of an NOE can be a serious problem. For example, if the relaxation of a nitrogen atom in a particular compound occurs to the extent of only 20% by the dipolar process, the NOEF is -1, and the nitrogen signal cannot be observed!

If the NOEF is in the range -2 to 0, the resonance intensities compared to those when the NOE is suppressed are more intense [5, 12, 26]. An example, drawn from H NMR studies, demonstrates the application of this technique: The chemical shifts of the methyl group protons of β,β'-dimethylacrylic acid dissolved in benzene-D_6 are 1.42 and 1.97 ppm, respectively.

$$\begin{array}{lll} \textbf{(a)} & CH_3 & H \\ & & C=C \\ \textbf{(b)} & CH_3 & CO_2H \end{array}$$

When the high-field methyl group is irradiated, the olefinic proton signal intensity is increased by about 20%. No such intensity change occurs to the olefinic proton signal when the low-field methyl group protons are irradiated. Thus, the methyl group that resonates at 1.42 ppm is the one that is *cis* to the olefinic proton.

9.2.2.7 Applications to Quadrupolar Nuclei

A nucleus with a quadrupole moment has an *ellipsoidal* charge distribution, in comparison to the *spherical* distribution in spin 1/2 nuclides. This type of nucleus can have three ways of aligning itself with an applied magnetic field (I = −1, 0, +1). These unsymmetrical electron clouds generate a fluctuating electric field, which causes a torque on the unsymmetrical nuclear charge and consequently alters its alignment in the applied magnetic field. This is the magnetic equivalent of changing the spin quantum number. The interaction between the quadrupolar nucleus and the unsymmetrical electron cloud can induce relaxation—in this instance *longitudinal relaxation*. This relaxation can occur only when both the nucleus and the electron cloud have unsymmetrical charge distributions. If the nucleus is not quadrupolar (e.g., H, C^{13}, and N^{15}), the fluctuating electric field cannot interact with it. When the electron cloud is tetrahedrally symmetrical (ammonium ion, borohydride ion), the electric field does not fluctuate, and no effect is noted.

Nuclei such as Cl^{35} are subject to strong quadrupolar relaxation when covalently bonded, and as a result line widths of 200–300 kHz are observed. Most of the N^{14} nuclei, on the other hand, have line widths of 10–60 Hz.

In the instance where relaxation of a N^{14} nucleus is extremely slow, a proton bonded to it will be split into a triplet. In a situation where the relaxation of the nitrogen nucleus is very rapid, a proton bonded to it appears as a singlet. However, the relaxation time of most N^{14}–H systems generally lies between these two extremes, and the proton will appear as a broadened singlet, often broad enough to become invisible over the instrumentation noise. This problem can be overcome by applying the heteronuclear decoupling technique. The normal and the N^{14} decoupled spectra of pyrrole illustrate the result of this type of decoupling technique (see Fig. 9.16).

9.2.3 Relative Signs of Coupling Constants

It is, perhaps, not at all intuitively obvious why spin–spin coupling constants should be either positive or negative. However, if the magnetic interaction between two nuclei A and B is expressed by $J_{AB}I_AI_B$, where the I_A and I_B are the nuclear magnetic moments (in units of $\hbar/2\pi$) and J_{AB} (in Hz) is the coupling constant, so that $E = hJ_{ab}\vec{I}_a\vec{I}_b$ (in ergs), the coupling constant can be described as positive when the nuclear spins are antiparallel (a more stable arrangement) or negative when they are parallel (a less stable arrangement) [27].

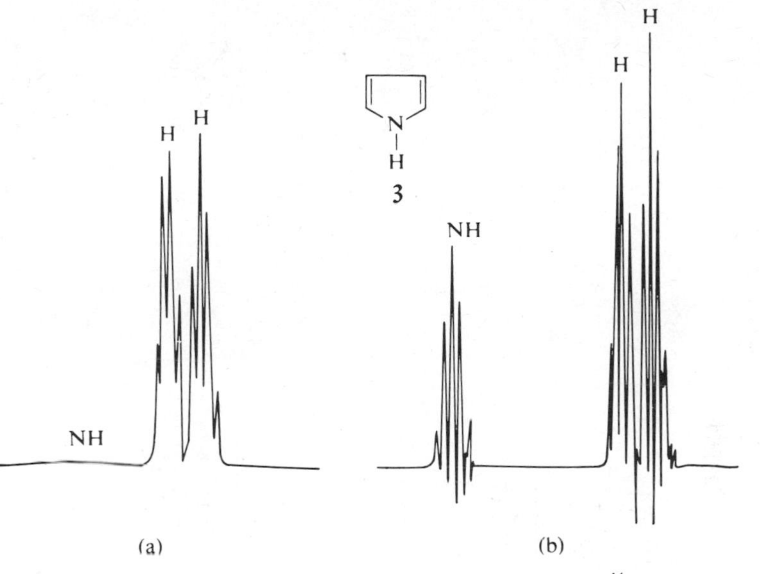

Figure 9.16. Proton NMR spectra of pyrrole 3. (a) Normal spectrum. (b) N^{14}-decoupled spectrum. [Spectrum supplied by Jeolco (U.S.A.), Inc. and reproduced with their permission.]

The application of double-resonance experiments with narrow-band irradiation (i.e., selective decoupling and spin tickling) provide experimental information about the relative signs of coupling constants.

9.2.3.1 Theoretical Considerations

The relative signs of coupling constants of many non-first-order spin systems have a significant effect on the relative line intensities of these spectra. If a spectrum is a non-first-order one and the pattern is computer-simulated, the experimental and theoretical spectra will only match if the relative signs of the coupling constants are the same in the theoretical and experimental spectra. Figures 9.17 and 9.18 provide the experimental and theoretical spectra of 2,3-dibromopropionic acid obtained at 60 MHz, where the spectrum is a first-order one, and consequently coupling constant changes do not alter the line intensities. The spectrum of the same compound obtained at 15 MHz, where the pattern is not first order, shows the difference between the line intensities with changes in the relative signs of the coupling constants employed in generating the theoretical spectra. Clearly, the theoretical spectrum computed on the assumption of unlike signs of the coupling constants matches the experimental spectrum very adequately. This quasi-experimental technique can be applied only with difficulty to complex systems and is, in fact, rarely used.

9.2.3.2 Selective Decoupling

This technique, already briefly described, is one of the two preferred ones for determining the relative signs of the coupling constants. The determination of

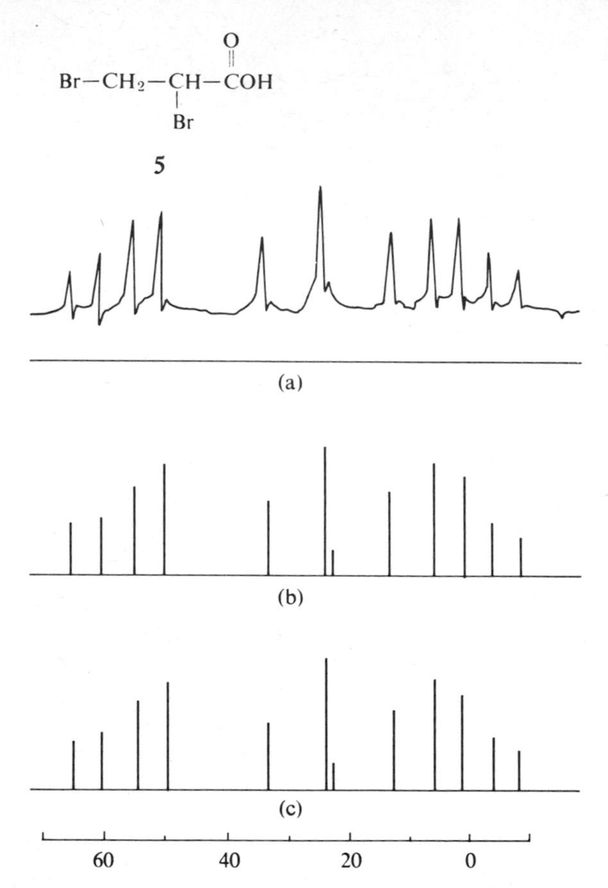

Figure 9.17. 60-MHz spectra of 2,3-dibromopropionic acid **5**. (a) Experimental spectrum. (b) Calculated spectrum with all the coupling constants of the same sign. (c) Calculated spectrum with the two vicinal coupling constants of opposite sign to the geminal one. [From R. Freeman, K. A. McLaughlin, J. J. Musher, and K. L. R. Pachler, *J. Mol. Phys.*, **5** 321 (1962). Used with permission of Taylor and Francis, Ltd.]

the relative signs of the coupling constants of the thallium diethyl cation $[(CH_3CH_2)_2Tl^+]$ will serve to demonstrate an application of this technique [19]. When the most deshielded triplet, caused by Tl^{205}–CH_3 coupling, was irradiated, the most shielded quartet, owing to Tl^{205}–CH_2 coupling, collapsed to a singlet (see Fig. 9.19). Thus, these two signals are associated with the same Tl^{205} spin state (either + or $-\frac{1}{2}$). Since these two affected multiplets are on opposite sides in the proton spectrum, it follows that the signs of J_{Tl-CH_3} and J_{Tl-CH_2} are opposite.

The relative coupling constant signs of three different interacting nuclei are similarly determined. However, since such a spectrum has at least 12 lines (if first order) and a very large number of possible spin state combinations, the analysis of such a spectrum is somewhat of a logistical problem.

In developing an energy level diagram for a three-nucleus system, the assump-

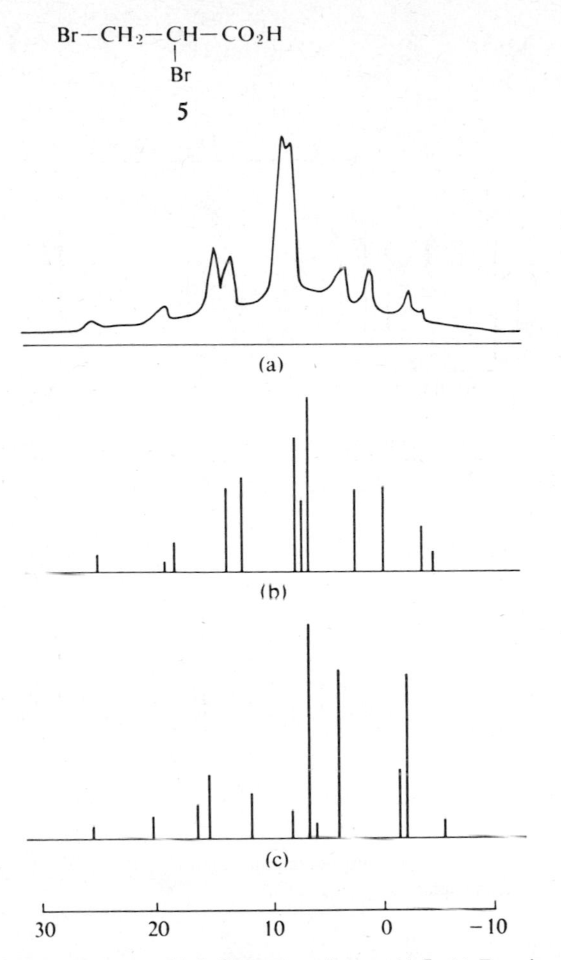

Figure 9.18. 15.086-MHz spectrum of 2,3-dibromopropionic acid **5**. (a) Experimental 15.086-MHz spectrum. (b) Spectrum calculated for unlike signs of geminal and vicinal coupling constants. (c) Spectrum calculated for like signs. [From R. Freeman, K. A. McLaughlin, J. J. Musher, and K. L. R. Pachler, *Mol. Phys.*, **5** 321 (1962). Used with permission of Taylor and Francis, Ltd.]

tion that $J_{BC} > J_{AC} > J_{AB}$ and that $\nu_A > \nu_B > \nu_C$ will be made. This produces the *line-number*-labeled part of the energy level diagram given in Figure 9.20.

On the assumption that all of the coupling constants are of equal sign, lines *f*, *d*, and *b* must be in the same spin states (either + or −). Consequently, if these lines represent positive spin states, then lines *e*, *c*, and *a* must represent negative spin states. This is indicated in row *C* columns *A* and *B* and row *B* column *C*, respectively, in the diagram. For example, lines 11 and 12 are composed of the same *C* nucleus spin state (+) as are lines 3 and 4. The smaller coupling constant, J_{AB}, causes splitting of lines *f*, *e*, *d*, and *c* into doublets by the + and − spin states of the *B* and *A* nuclei, respectively (row *B*, column *A*; and row *A*, column *B*). The

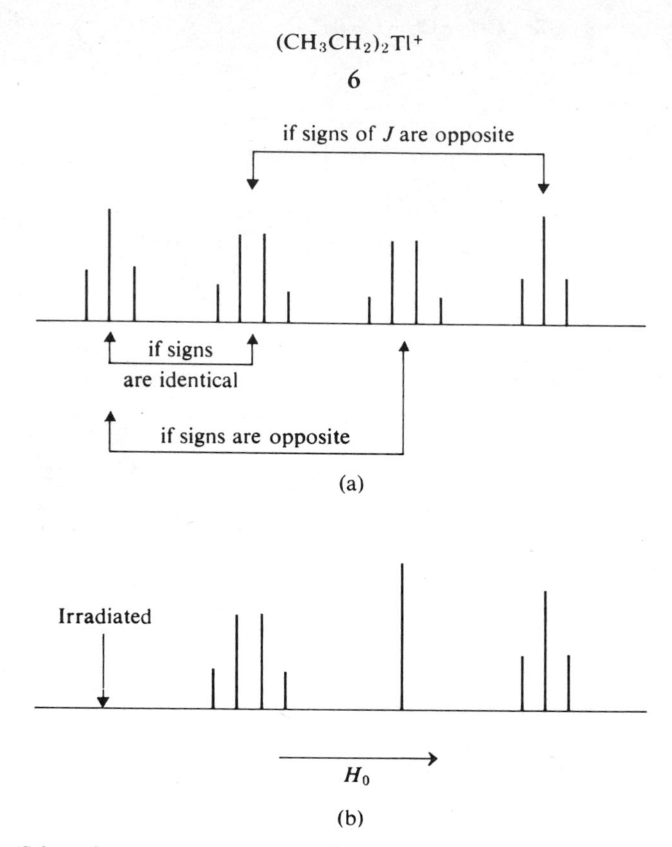

Figure 9.19. Schematic proton spectrum of thallium diethyl cation (VI). (a) Normal spectrum. (b) Irradiated at the deshielded triplet.

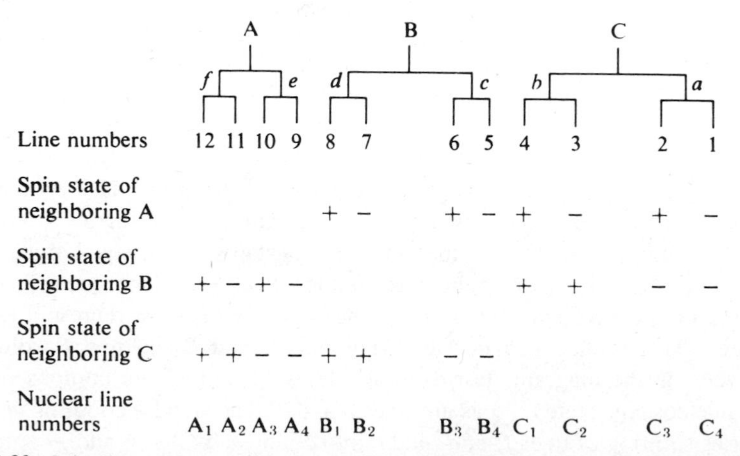

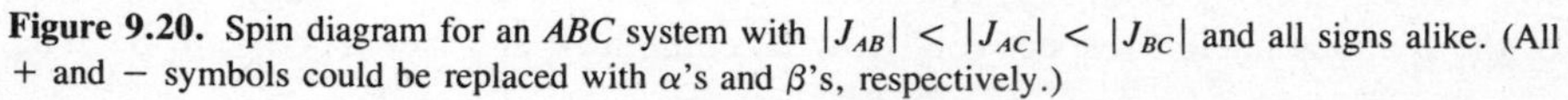

Figure 9.20. Spin diagram for an *ABC* system with $|J_{AB}| < |J_{AC}| < |J_{BC}|$ and all signs alike. (All + and − symbols could be replaced with α's and β's, respectively.)

spin table can now be completed by considering the remaining spin and its effect on lines a and b.

An examination of the diagram indicates that irradiation at a frequency intermediate between lines 5 and 6 would cause lines 9 and 10 to collapse into a singlet, since these two doublets have the same spin state of nucleus C in common. This is tantamount to decoupling those A and B nuclei from the C nucleus that have a spin quantum number of $-\frac{1}{2}$ and would demonstrate, in an experimental spectrum, that J_{AC} and J_{BC} are of equal sign.

Similarly, if a position between lines 1 and 2 is irradiated, lines 9 and 11 will coalesce into a singlet, since it is these two transitions that have the same spin quantum number in common with lines 1 and 2. If this occurred in an experimental spectrum, the conclusion that J_{BC} and J_{AC} have the same sign could be drawn.

A theoretical diagram reflecting a system where J_{AC} and J_{BC} have opposite signs and J_{AB} and J_{BC} have the same sign would be generated when either the signs in row C of column A (J_{AC}) or those in row C of column B are reversed. Since the symbols α and β are used interchangeably with + and − in describing spin states, the relative signs for this energy diagram are presented using this sign convention (Fig. 9.21).

When this spin sign combination exists in a particular compound, irradiation at a position intermediate between lines 5 and 6 will cause lines 11 and 12 to coalesce into a singlet, thus establishing that in this particular compound J_{BC} and J_{AC} have opposite signs.

A number of selectively decoupled spectra of 1,8-naphthyridine are given in Figure 9.22. The coupling constants are readily determined from an examination of trace a in Figure 9.14 ($J_{2,3} = 4.2$, $J_{3,4} = 8.0$, $J_{2,4} = 2.0$). The a trace results when irradiation between lines 1 and 2 is done, and since lines 9 and 11 coalesce into a singlet (line 10 is hidden by this singlet), $J_{2,4}$ and $J_{3,3}$ are of equal sign. Spectrum d shows that $J_{3,4}$ and $J_{2,3}$ are of equal sign. The remaining trace can be similarly used to conclude that all three of the coupling constants in this compound have the same sign.

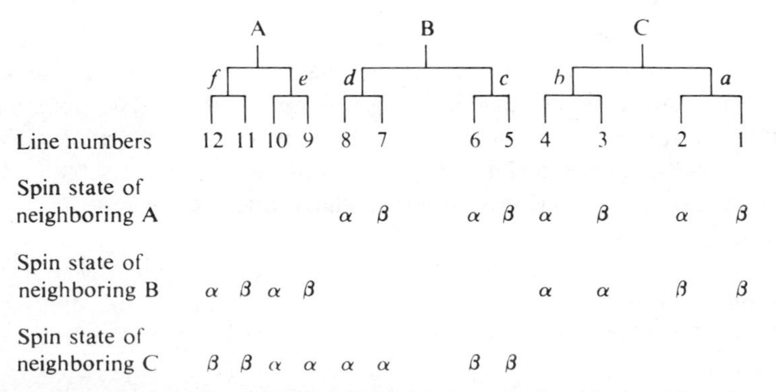

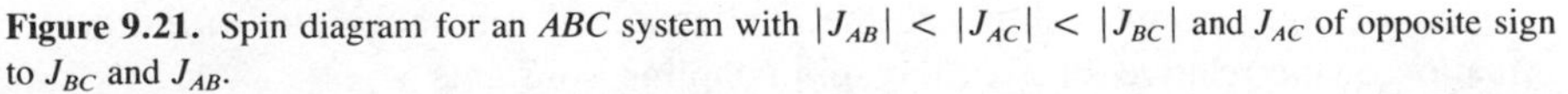

Figure 9.21. Spin diagram for an ABC system with $|J_{AB}| < |J_{AC}| < |J_{BC}|$ and J_{AC} of opposite sign to J_{BC} and J_{AB}.

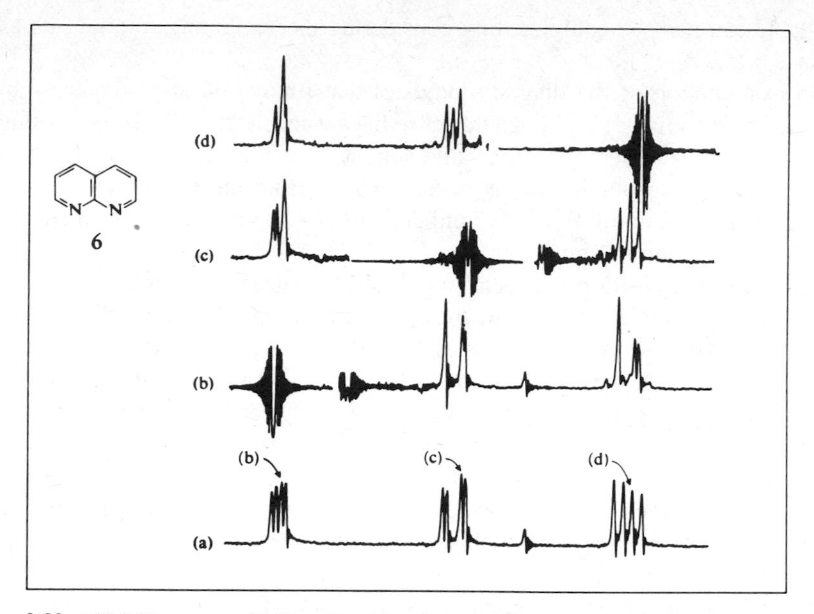

Figure 9.22. 100-MHz spectra of 1,8-naphthyridine **6**. (a) Normal spectrum. (b–d) Irradiated at position indicated.

9.2.3. Spin Tickling

Since it is often difficult if not impossible to partially decouple weakly coupled nuclei without interfering with other spectral lines, the selective decoupling technique is only applicable to strongly coupled nuclear systems. When a weak RF field is applied to a specific transition frequency, while scanning the spectrum in the normal frequency sweep mode, the already outlined spin tickling process is in operation [16].

This technique is applicable to weakly as well as strongly coupled spin systems and allows the interpretation of the various transitions in terms of the energy level diagram of the system.

The following rules apply when this technique is being used. *Rule 1:* A transition split into a doublet by the perturbing RF field has an energy level in common with the nondegenerate transition that is being radiated. *Rule 2:* The common level determined by rule 1 is one of three involved in the two transitions. The spin quantum numbers of the remaining two levels must either be the same ($\Delta M = 0$) or differ by two units ($\Delta M = 2$). When the former is the case (a regressive transition), the observed doublets are well resolved, but if the latter situation exists (a progressive transition), the doublets are poorly resolved. *Rule 3:* The magnitude of the splitting in the well-resolved doublets is proportional to the strength of the applied RF field as well as to the square root of the intensity of the line perturbed by this field.

Two general examples illustrate the application of this technique to the determination of the relative sizes of the spin coupling constants.

TWO-SPIN SYSTEM. Another way of representing the energy level listing (Table 3.1) for an *AB* system is the pictorial representation:

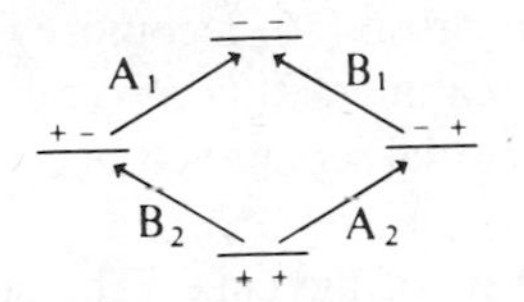

This figure graphically represents the various permitted transitions in an *AB* spin system. As the data in Table 3.1 have already demonstrated, the pictograph also shows that neither the two *A* nor the two *B* transitions have any energy levels in common. If a very weak irradiation is applied at the B_1 frequency, the $--$ and $-+$ states are rapidly interconverted. Thus, two new states are formed, and two new transitions are now allowed. The overall result is an increase in the total number of possible *A* lines to four by splitting the A_1 and A_2 lines into two lines each.

Figure 9.23 is an *AB* system subjected to spin tickling. Since the A_1 transition is split into a clean doublet when B_1 is irradiated, its terminal magnetic quantum number is the same as that of the S1 transition (i.e., $M = 0$), as predicted by the energy level diagram. On the other hand, the A_2 transition is split into a poorly split doublet, which, according to the energy level diagram, is expected since the transition spans an interval $M = 2$ ($+1$ to -1). The spectrum resulting from spin tickling at B_2 is equally readily interpreted in terms of the transition interactions between B_2 and A_1 and A_2.

THREE-SPIN SYSTEMS. A spin system composed of three different spin $+/-\frac{1}{2}$ nuclei can have 2^3, or 8, possible spin state combinations. These eight states give rise to 12 strong transitions composed of four sets of four lines each. The flipping

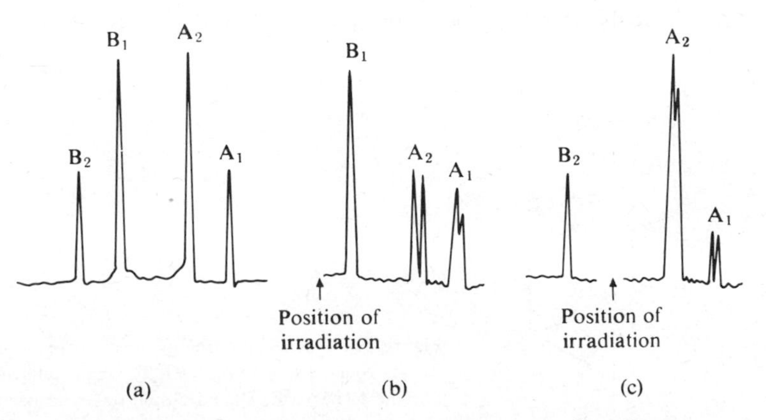

Figure 9.23. (a) Normal and (b, c) spin-tickling spectra of an *AB* system.

of three spins causes the generation of two additional resonance lines under some circumstances. The tabular representation of the various interactions of an *ABX* systems (Fig. 9.20, for example) is rather cumbersome to develop and interpret. Fortunately, Freeman and Anderson [16] recognized the existence of a topological relationship between the 12 spin interaction possibilities of an *ABX* system and the 12 edges of a cube. In this pictorial representation, the eight different energy levels (+ + +, + + −, + − +, − + +, + − −, − − +, − + −, − − −) of an *ABX* system are placed at the corners of the cube. The cube is oriented with one body diagonal vertical so that the energy levels are grouped (1, 3, 3, 1) corresponding to the states $M = +\frac{3}{2}$ (1 state), $+\frac{1}{2}$ (3 states), $-\frac{1}{2}$ (3 states), and $-\frac{3}{2}$ (1 state), respectively.

The three sets of parallel edges are taken to represent the quartets of the *A*, *B*, and *C* nuclei so that the diagram in Figure 9.24 can be drawn. A vertical body diagonal corresponds to the one triple quantum transition (often not observed) that occurs at the center of the spectrum. The other three body diagonals are assigned to the weak combination lines, which are generally indistinguishable from the other 12 lines of the spectrum. When all three coupling constants are of equal sign, the various transitions can be assigned to the cube pictorial (Fig. 9.24). The most deshielded transitions in each of the quartets involve the + + + (*ABC*) site and a transition to − + + + for the A_1 line, to + − + for the B_1 line, and to + + − for the C_1 transition line. Similarly, the most shielded lines of each of the quartets originate at the − − − site involve transitions to + − −, − + −, and − − +, respectively, for the A_4, B_4, and C_4 transition lines. The remaining transitions, the second and third lines of each quartet, can then be assigned as shown in Figure 9.24 for the instance where all coupling constants are either + or −.

Three other combinations of the three coupling constants in ABX systems are possible: (1) the sign of J_{AB} is opposite to those of J_{AC} and J_{BC}; (2) the sign of J_{BC} is opposite to those of J_{AB} and J_{AC}; (3) the sign of J_{AC} is opposite to those of J_{AB} and J_{BC}. The schematic energy diagrams for these situations are given in Figure 9.25. The application of the cubes to the analysis of a particular *ABX* pattern serves to illustrate the method.

Figure 9.26 is the H NMR spectrum of 1,8-naphthyridine (1,8-diazanaphthal-

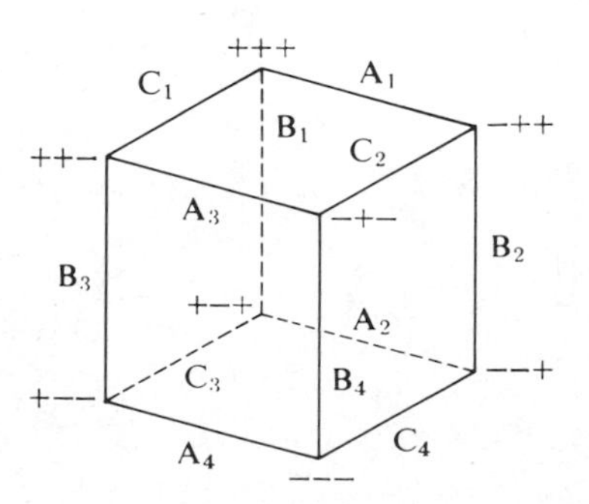

Figure 9.24. Schematic energy level diagram for three weakly coupled spin 1/2 nuclei. The signs refer, from left to right, to nuclei *A*, *B*, and *C* spin states, respectively.

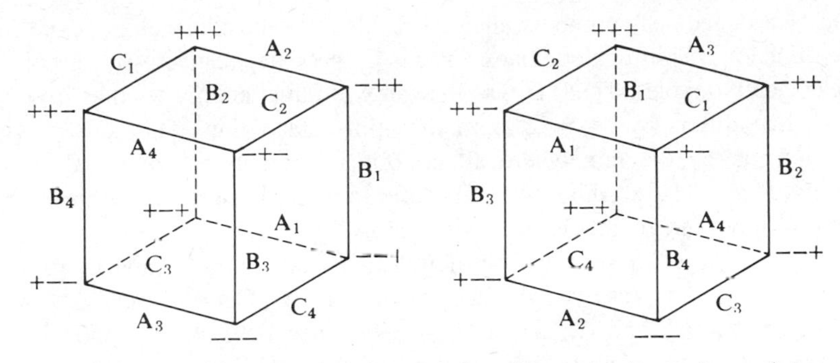

J_{AB} of opposite sign to J_{AC} and J_{BC} J_{AC} of opposite sign to J_{AB} and J_{BC}

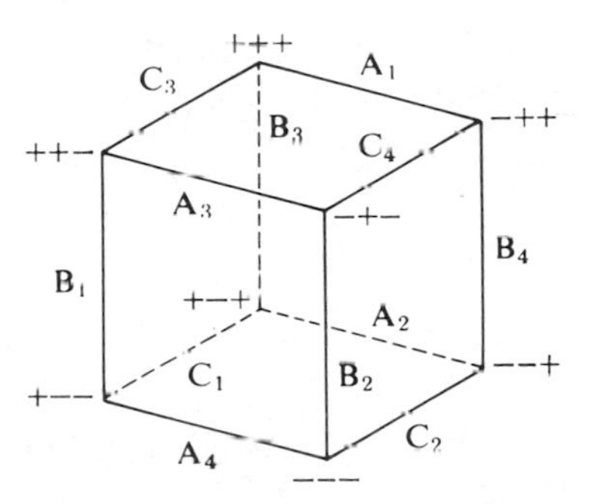

J_{BC} of opposite sign to J_{AB} and J_{AC}

Figure 9.25. Schematic energy level diagram for three weakly coupled spin 1/2 nuclei.

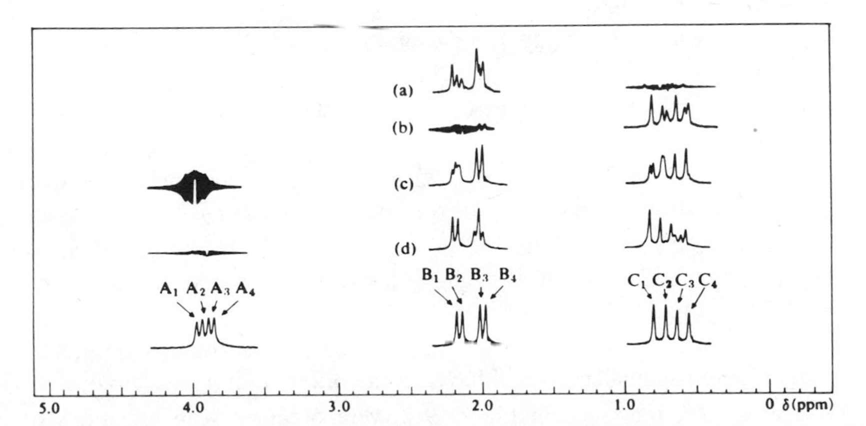

Figure 9.26. Spin-tickling spectra of an *ABX* system. (a, b) The reader should identify the lines that were irradiated to give these two patterns; assume all *J*'s have equal signs. (c) Irradiated at A_1. (d) Irradiated at A_4.

ene), its assigned peak numbers, and the results of some spin-tickling experiments. When line A_4 is spin tickled, lines C_3 and C_4 become, respectively, a poorly and a clearly split doublet. Thus, C_4 has the same terminal energy level as line A_4, and the C_3 line differs by $M = 2$ from the spin-tickled line. This agrees with the transition energy diagram where all coupling constants are of equal sign. The splitting of B_1 and B_2, while not analyzable in detail because of the peak overlap, does not, however, contradict this conclusion.

Irradiation of A_1 causes collapse of B_1 and B_2 as well as C_1 and C_2 into clearly resolved and poorly resolved doublets, respectively. Consequently $\Delta M = 0$ between lines A_1, B_1, and C_1, and $\Delta M = 2$ between lines A_1, B_2, and C_1. Again, this agrees with the diagram in which all of the coupling constants are of equal sign.

Modifications of this technique, exemplified by transitory selective irradiation have also been developed and are available [28–30].

9.3 PROBLEMS

1. Analyze the NMR spectrum of 3-chloropyridine (Fig. 9.13) in detail (without taking recourse to the analysis in the book!).
2. Determine the sizes of the various pyrrole coupling constants from the spectrum given in the text.
3. Predict the relative chemical shifts of the C^{13} nuclei in quinoline (verify your answer with the spectrum given in this chapter).
4. Why can coupling constants be either positive or negative?
5. Develop an energy level diagram for an ABX_3 system assuming (1) that J_{AB} and J_{AX} have the same sign as J_{BX}, and (2) that all three coupling constants have the same sign.
6. Draw a schematic energy level diagram for an ABX system where J_{AC} is zero.
7. Could the relative signs of the coupling constants in problem 6 be evaluated? If so, how could this be done?
8. Analyze the spectra in Figures 9.22 and 9.26 in detail, including the determination of the relative signs of the coupling constants.
9. What would be the significance if the pattern of the spectrum in Figure 9.12 were reversed? What, if any, changes would be noticed in the *a*, *b*, and *c* spectra in this figure?
10. Predict the effect, if any, that irradiation of the *tert.*-butyl group in 1,4-di-*tert.*-butylnaphthalene would have on the different aromatic protons.
11. The spectra presented in Figure 9.2 were obtained with variations of the pulse angle. What is the molecular significance of the peak intensity variations?

12. Describe the concepts leading to the use of *refocusing pulses*.

13. What is meant by "polarization transfer?"

14. What is the pulse sequence in the INEPT technique? What is the technique's major application?

15. Figure 9.5 represents the spectra of an inversion recovery experiment on the Cu^{63} ion. What causes the reversal from a negative to a positive signal?

16. Why is there a significant difference in the spectral band width of the Cu^{63} and the Pt^{195} ions?

17. Describe the technique of noise decoupling.

REFERENCES

1. A. Allerhand and D. W. Cochran, *J. Am. Chem. Soc.* **92,** 4482 (1970).
2. D. Shaw, *FT-NMR Spectroscopy*, Elsevier Scientific, Amsterdam, 1976.
3. R. R. Shoup, E. D. Becker, and T. C. Farrar, *J. Magn. Res.* 458 (1972).
4. E. Fukushima and S. B. Roeder, *J. Magn. Res.* 199 (1979).
5. J. J. Dechter, *Prog. Inorg. Chem.* **29,** 293 (1982).
6. W. McFarlane, *Ann. Rep. NMR Spectrosc.* **5A,** 353 (1972).
7. T. C. Farrar and E. D. Becker, *Pulse and FT-NMR, Introduction to Theory and Methods*, Academic Press, New York, 1971.
8. R. Freeman and H. D. W. Hill, *Chem. Phys. Lett.*, 3367 (1971).
9. R. D. Bertrand, W. B. Monic, A. N. Garroway, and G. C. Chingas, *J. Am. Chem. Soc.* **100,** 5227 (1978).
10. G. C. Chingas, A. N. Garroway, W. B. Moniz, and R. D. Bertrand, *J. Am. Chem. Soc.* **102,** 2526 (1980).
11. G. A. Morris and R. Freeman, *J. Am. Chem. Soc.* **101,** 760 (1979).
12. G. C. Levy, R. L. Lichter, and G. L. Nelson, *Carbon-13 NMR Spectroscopy*, 2d ed., Wiley, New York, 1980.
13. M. Witanowski and G. A. Webb (eds.), *Nitrogen NMR*, Plenum, London, 1973, and references therein.
14. F. Bloch, *Physical Rev.* **102,** 104 (1956).
15. A. L. Bloom and J. N. Shoolery, *Physical Rev.* **97,** 1261 (1955).
16. R. Freeman and W. A. Anderson, *J. Chem. Phys.* **97,** 2053 (1962).
17. J. D. Baldeschwieler, *J. Chem. Phys.* **36,** 152 (1962).
18. D. H. Whiffen and R. Freeman, *Proc. Phys. Soc.* **79,** 794 (1962).
19. J. R. Maher and D. F. Evans, *Proc. Chem. Soc.* 208 (1961).
20. E. B. Baker, L. W. Burt, and C. V. Root, *Rev. Sci. Instrum.* **34,** 243 (1963).
21. R. Kaiser, *J. Chem. Phys.* **39,** 2435 (1963).
22. M. Czekalski, M. E. deMilov, and V. J. Kowalewski, *J. Magn. Res.* **41,** 61 (1980).

23. R. Kaiser, *J. Chem. Phys.* **42,** 1838 (1965).
24. F. A. L. Anet and A. J. R. Bourn, *J. Chem. Soc.* **87,** 5250 (1965).
25. N. Bloemberger, E. M. Purcell, and R. V. Pound, *Phys. Rev.* **73,** 679 (1948).
26. J. S. Noggle and R. E. Schirmer, *The Nuclear Overhauser Effect*, Academic Press, New York, 1971.
27. H. M. McConnell, *J. Chem. Phys.*, **23,** 2454 (1955).
28. R. A. Hoffman and B. Gestbloom, *J. Mol. Spectr.* **13,** 221 (1964).
29. J. A. Ferretti and R. Freeman, *J. Chem. Phys.* **44,** 2054 (1966).
30. K. Kuhlmann and J. D. Baldeschweiler, *J. Am. Chem. Soc.* **85,** 1010 (1963).

10

SPECIAL APPLICATIONS

10.1 TWO-DIMENSIONAL FT NMR SPECTROSCOPY

10.1.1 General Considerations

The assignment of chemical shifts and coupling constants in complex systems has traditionally been simplified by obtaining NMR spectra, generally either H^1 or C^{13}, at higher magnetic fields. However, the complexity of the NMR spectra of many high-molecular-weight compounds and polymers can still not be sufficiently simplified by this technique, and overalapping of multiplets remains a problem.

The concept of 2-D NMR spectroscopy addresses itself to the resolution of these problems. In its most general description, as originally proposed by Jeener and implemented by Ernst [1], [2], two-dimensional NMR spectroscopy is based on the concept of plotting NMR spectral data obtained in one time domain against data obtained in another.

In one-dimensional spectroscopy, a signal is presented as a function $S(\omega)$ of a single frequency variable [3]. It is, fundamentally, an expression of the number of nuclei that experience a particular magnetic field (a peak intensity-related expression). Similarly, in a 2-D spectrum, two independent variables, $S(\omega_1)$ and $S(\omega_2)$ or equally $S(\omega_{1,2})$, are employed to represent the spectral data. By separating two different interactions in two dimensions, 2-D NMR spectroscopy greatly facilitates analyses of complex nuclear interactions. The technique permits, among other applications, the resolution of overlapping multiplets in heteronuclear systems such as are generated by the C^{13}–H coupling patterns and the C^{13} chemical shifts and the separation of homonuclear coupling multiplets and their corresponding chemical shifts. The technique is being applied to liquid and solid ("magic angle spinning" is required) samples.

The data obtained through a 2-D experiment can be displayed in a number of different ways, all of which generate a correlation map. These maps show interactions among nuclei of one species or between two different nuclides in a molecule. The former type of map is referred to as an autocorrelation, and the latter as a cross-correlation one.

There are three important applications of 2-D correlated NMR spectroscopy:

1. *2-D-autocorrelation spectroscopy.* Major application: Proton network analyses, often in protein structure determination.
2. *2-D-cross-correlation mapping utilizing chemical shift data.* Major application: Determination of H–C connections in organic compounds.
3. *C^{13}–C^{13} natural abundance coupling analyses.* Major application: Elucidation of carbon networks.

Generally, 2-D-NMR experiments involve four separate time intervals: (1) preparation; (2) evolution; (3) mixing; and (4) detection.

The preparation time is set so that the sample reaches a steady-state condition in the applied magnetic field. It can also involve the deliberate saturation of some spin states (to cause decoupling of an A–X hetero- or homonuclear system, for example). Generally, this step ends with the application of an RF pulse to generate transverse magnetization.

During the *evolution time*, the magnetization vector processes at a rate based on the spin and environmental interactions of the nuclear spin sample. This step can, and often does, include the application of a refocusing pulse.

The *mixing time* depends on the magnetization that must be redistributed and can vary from a one-pulse duration time to a number of seconds. If, for example, Z magnetization is redistributed between various frequencies as a result of a chemical exchange process in the molecule being examined, the mixing time can be extraordinarily long.

The *detection period*, earlier defined as t_2, is used to record the FID of the observed nucleus; it is a constant for each value of t_1 (the evolution time). G. A. Gray of Varian Associates describes this in terms of a running time axis from 0 to t_2 (a maximum), just as t_1 runs from 0 to some maximum value (it is usual to collect full FIDs (0 to t_2 (max)) for each value of t_1 (evolution time)).

One of the initial experiments in this area was done by Johnson and Hurd (cited in Ref. 4) who used an RF pulse to tip either C^{13} or H^1 spin vectors by 90°. After an evolution period, the sample was pulsed again, and the resulting FID was subjected to FT analyses. Repetition of this experiment under identical conditions, except with a different evolution period, provides spectra in the second dimension.

When one such spectrum is plotted in one time domain versus another in a different time domain, the signals, expressed either as density contours (Fig. 10.1b) or as two-dimensional Gaussian (Fig. 10.1a) representations, that lie on the diagonal correspond to the normal NMR spectrum of the compound.

The additional signals that result from mutual relaxation of spin-coupled nuclei appear at positions away from the diagonal and form a square with those on the diagonal. Figure 10.2 is a 2-D NMR spectrum obtained on a Varian model XL spectrometer. The normal H spectrum is shown on the top of the contour graph. The contour map itself is a result of 24 repetitions based on an evolution delay of 0.285s and an acquisition time of 0.285s. Each cross peak on this map represents a particular H–H coupling interaction.

The concepts embedded in this technique have been and continue to be expanded to include applications of ever greater complexity. In fact, even during the initial developmental phases of this technique, Ferretti and Balaban (cited in Ref. 4) placed an anesthetized rat into a sample chamber and followed the conversion of

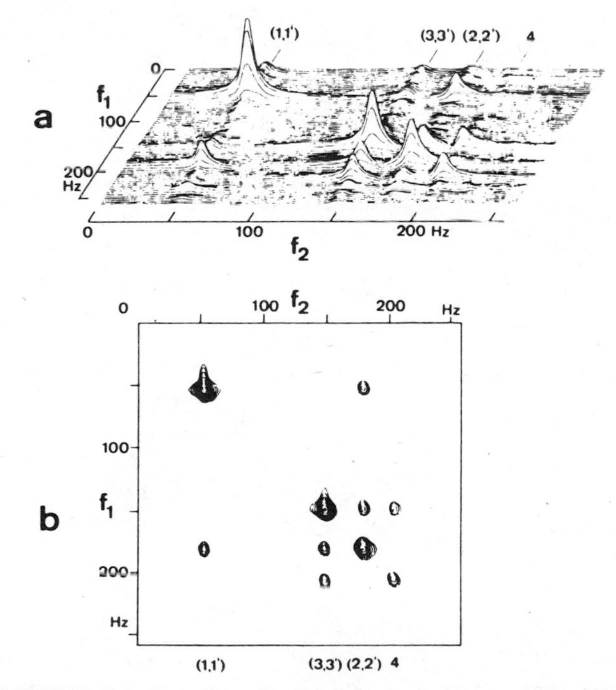

Figure 10.1. 2-D-NMR plots. Two-dimensional and density contour representations. [Reprinted with permission from J. Jeener, B. H. Meier, P. Bachmann, and R. R. Ernst. *J. Chem. Phys.* **71,** 4546 (1979). Copyright (1979) American Chemical Society.]

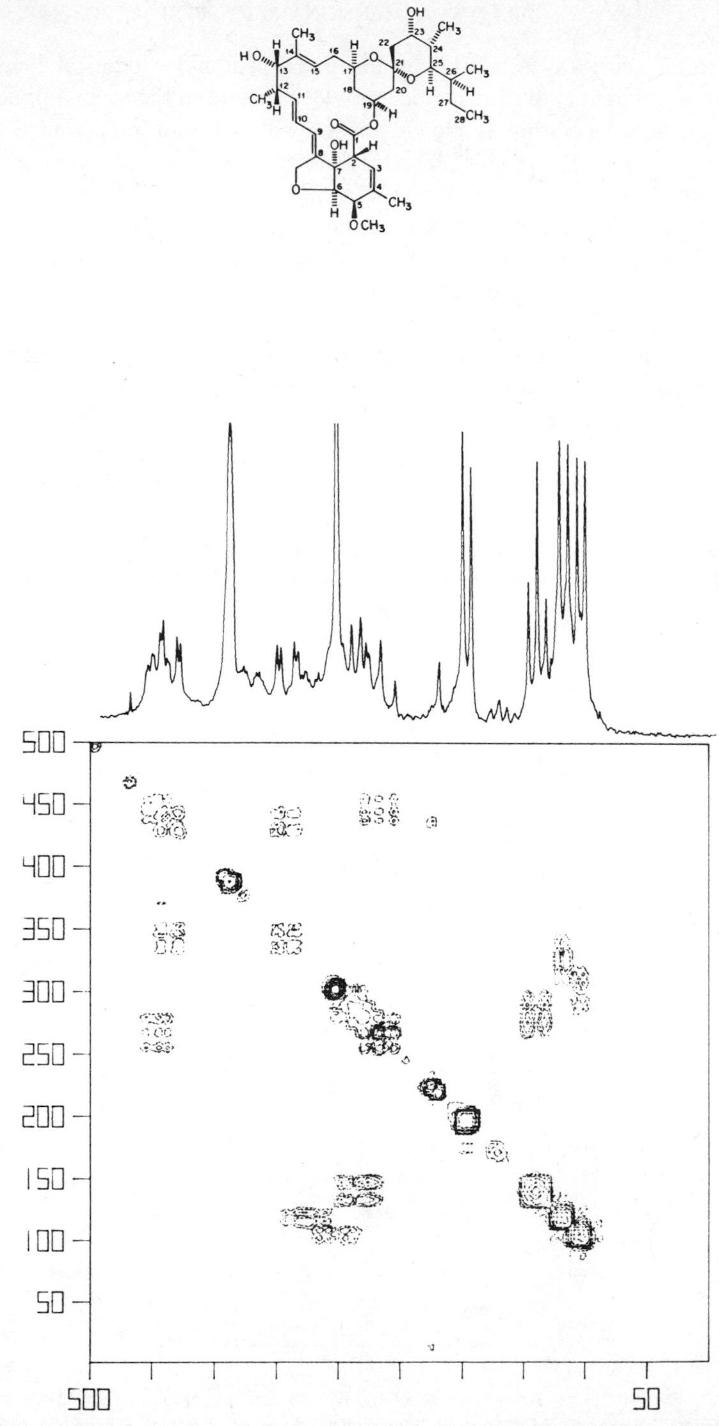

Figure 10.2. 2-D NMR spectrum of avermectin. [*Varian XL-300 Spectra Catalog No. 1.* Reproduced with permission.]

adenosine triphosphate to the diphosphate and the monophosphate by means of the 2-D NMR technique.

10.1.2 *J*-Resolved 2-D NMR Spectroscopy

This technique, designed to facilitate analyses of very complex spectra, is based on the concept of controlling the evolution time in pulsed NMR spectroscopy in a manner so that J couplings will modulate the signal of the observe nucleus during t_2 (detection time). This is accomplished by exciting the observe nucleus with a strong 90° pulse followed by a reversal pulse of 180° halfway through the evolution time. The latter will refocus the precessing nuclei at the end of the evolution time.

In heteronuclear 2-D J NMR spectroscopy it is only necessary to turn off the decoupler frequency for half of the evolution time in order to obtain J modulation as a function of t_1 and to eliminate the chemical shift component. This "gated decoupled" technique is, in fact, preferentially used in heteronuclear 2-D J NMR spectroscopy.

An example of a gated decoupled 2-D NMR spectrum, provided by G. A. Gray of Varian Associates, is given in Figure 10.3. The quasi X axis (frequently labeled F_1) presents the C^{13}–H coupling constants, while the H chemical shifts are given on the quasi Y axis (frequently labeled F_2).

10.1.3 2-D Correlation Spectroscopy (COSY)

The original 2-D experiment as proposed by Jenner has become the method of choice for analyzing proton coupling networks, particularly in the determination of protons belonging to the same amino acid residue. The time and pulse sequences used in this process must, of course, be done on non-H–H decoupled spectra, or no coupling information can be obtained.

In this, as in all of the 2-D NMR experiments, it is necessary to define the evolution time. During this time interval any existing homonuclear coupling interactions cause spreading of the mangetization vectors. This modulation can be transferred by application of a 90° pulse just prior to reaching t_2. As already described, this type of pulse has a different effect on different magnetization vectors depending on their location in the X–Y plane.

Finally, the application of another 90° pulse along the X axis will flip all vectors along the Y axis, without altering the spin components in the X–Y plane. Since the orientation of a spin vector in the X–Y plane is determined by its chemical shift and its coupling interactions, these pulses will modulate the detected signal and cause the *diagonal* intensities in the 2-D spectrum. An important, additional effect of the second 90° pulse is a mixing of the spin states within the spin system. Any X–Y magnetization that is the result of one particular proton can be transferred to another coupled one by a pulse applied along the axis in the X–Y plane. Any magnetization that has been *coded* in t_1 is then detected, at another chemical shift, in t_2. The graphic result of this is that the intensity contour that has the chemical

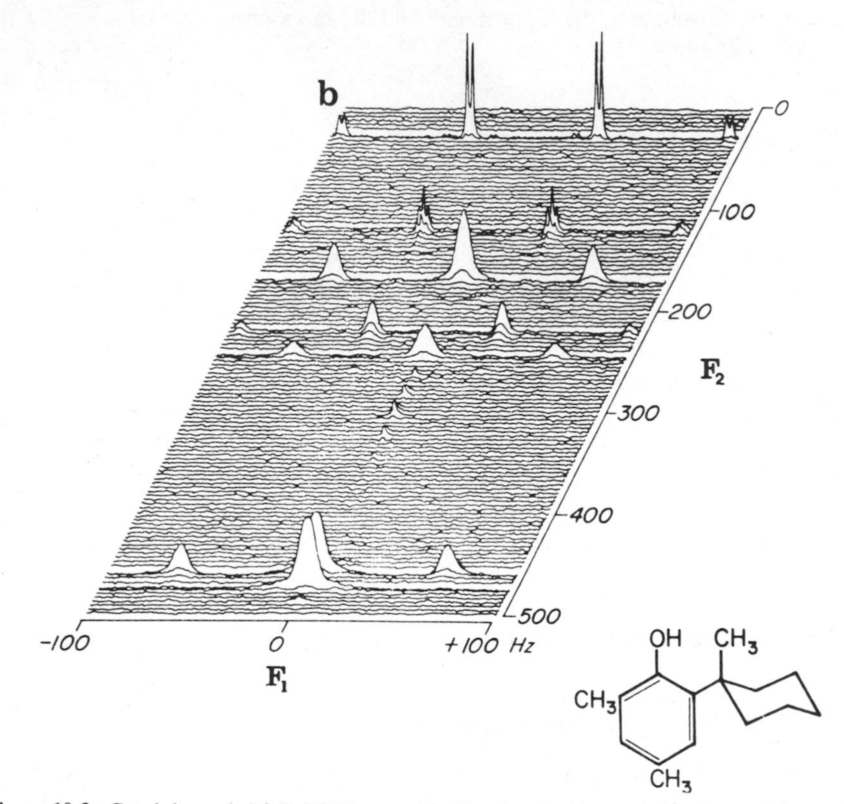

Figure 10.3. Gated-decoupled 2-D NMR spectrum. [Reprinted with permission from *NMR in Chemistry and Biology, Echoes and New Focuses*, Science Symposia Series, ACS, Two Dimensional Nuclear Magnetic Resonance, slide 32. Copyright American Chemical Society.]

shift of the first proton will appear at a coordinate of F_1 (the quasi X axis), and the coordinate of the F_2 (the quasi Y axis) value will correspond to the chemical shift of the detected nucleus. Thus the intensity contour for this particular situation will be *off-diagonal*, and constitutes proof of H–H coupling between the two protons.

The data are generally accumulated with anywhere from 128 to 1024 FIDs, and the results are given in the form of a contour map of a 2-D representation. A one-dimensional spectrum of the compound is also normally shown either alongside or above the contour map (see Fig. 10.4).

10.1.4 2-D Nuclear Overhauser Enhancement Spectroscopy (NOESY)

The normal NOE, described in Chapter 9, is caused by *cross-relaxation* between spins of interacting nuclei. The measurement of this effect, originally largely em-

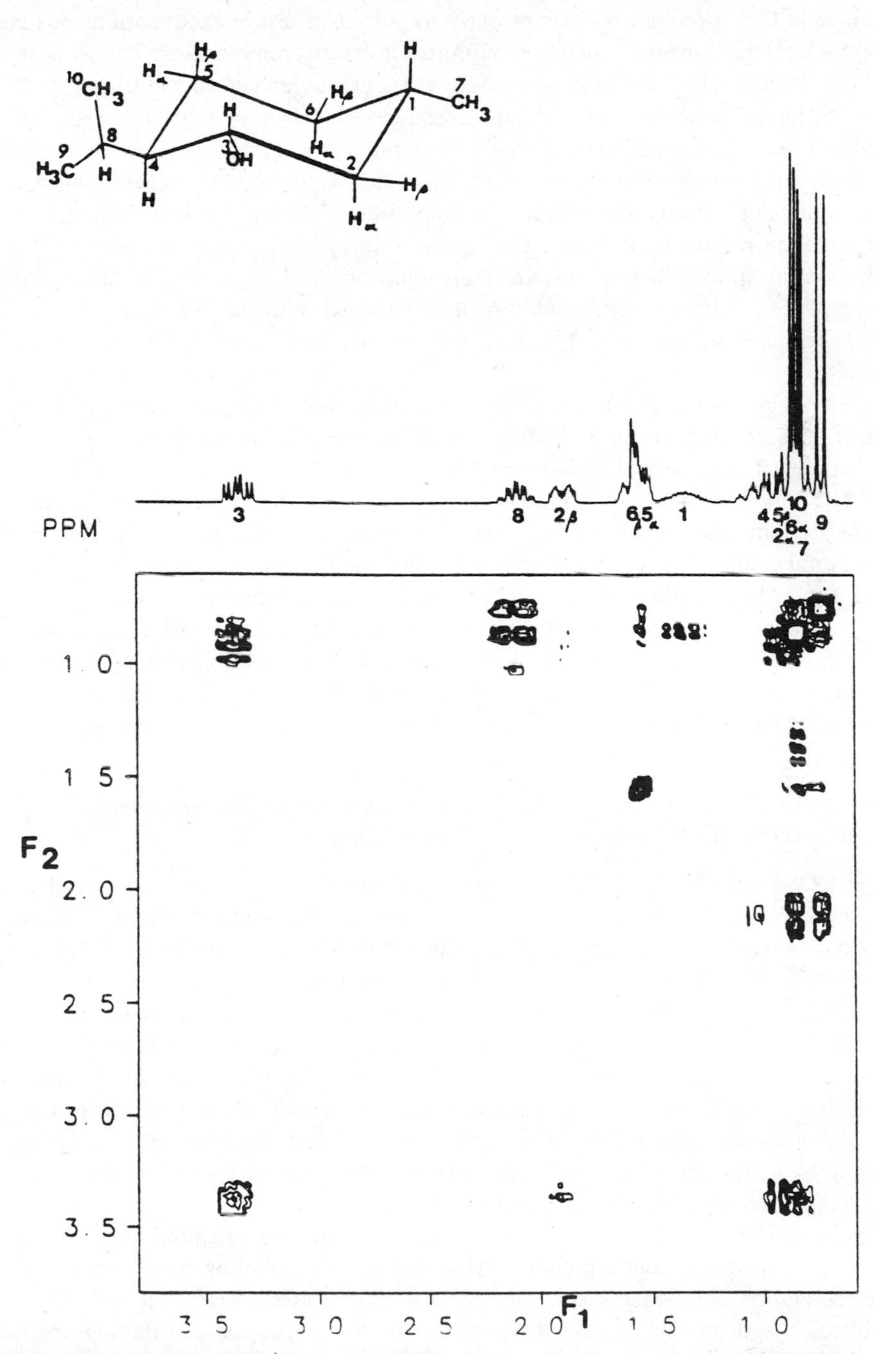

Figure 10.4. Homonuclear 2-D chemical shift correlation. [Reprinted with permission from G. A. Gray, Varian Associates, Palo Alto, CA, preprint: *Methods of Two-Dimensional NMR*.]

ployed in C^{13} spectroscopy, has recently found much use in examinations dealing with the establishment of nuclear proximities in biomacromolecules. The extension of this approach to 2-D NMR spectroscopy has greatly enhanced its utility [5], [6]. The technique utilizes three 90° pulse sequences. The introduction of the "evolution time" is accomplished after the first pulse and labels the spins with their Larmor frequencies. The second pulse turns the magnetization vector along the y direction parallel to the z direction. The amplitude of the longitudinal magnetization depends on t_1 (evolution time). During the mixing time, after the second pulse, the nuclear spins will relax toward their equilibrium distribution. In large molecules, the equilibration has, in addition to the normal spin–lattice relaxation, a large cross-relaxation component. This is the process that is responsible for the 2-D NOE effects.

Haasnoot [7] has reported the assignment of the H^1 NMR peaks of the antibiotic bleomycin. A part of the 500-MHz 2-D NOE spectrum resulting from this Herculean effort is presented in Figure 10.5.

The connectivities between the resonances in the 6–9 ppm range and the rest of the spectrum are shown on this graph. The abbreviations represent the following structural components: S, γ-aminopropyldimethylsulfonium; V, methylvalerate; B, bithiazole; H, β-hydroxyhistidine; T, threonine; Py, pyrimidinyl group.

It is appropriate to point out that data accumulation for NOESY experiments can be very time-consuming. It is not unusual for studies on large molecules to require 12–70 h! The application of COSY and SECSY in their 2-D modes provides the assignments for the remaining protons in this very complex molecule.

10.1.5 Combined Correlated and Nuclear Overhauser Enhancement Spectroscopy (COCONSY)

The COSY and NOESY 2-D techniques are the most commonly employed 2-D methods to date. The combination of these two methods is a powerful extension for resolving and assigning individual resonances in highly complex H NMR spectra. This process allows the identification of the nuclei belonging to a particular coupling network (via the COSY approach), and through-space connectivities between hydrogen atoms located at short distances from each other (about 4.5 Å) [8]. One of the problems of combining the two techniques is caused by the requirement that the two sets of spectra must be run under identical conditions.

The normal COSY sequence, already described, ends after two 90° pulses separated by t_1 (the evolution time). The resulting free induction decay is immediately recorded after the second pulse, and the data can be stored prior to Fourier transformation. The mixing pulse also creates longitudinal magnetization with the components having different amplitudes as a result of the labeling by t_1. Since these magnetization components will mix because of the cross-relaxation process, the induced magnetization transfer will correspond to the normal distribution created in a NOESY experiment. Consequently, the normal COSY sequence can be extended with an extra 90° pulse, and the resulting data can be separately stored.

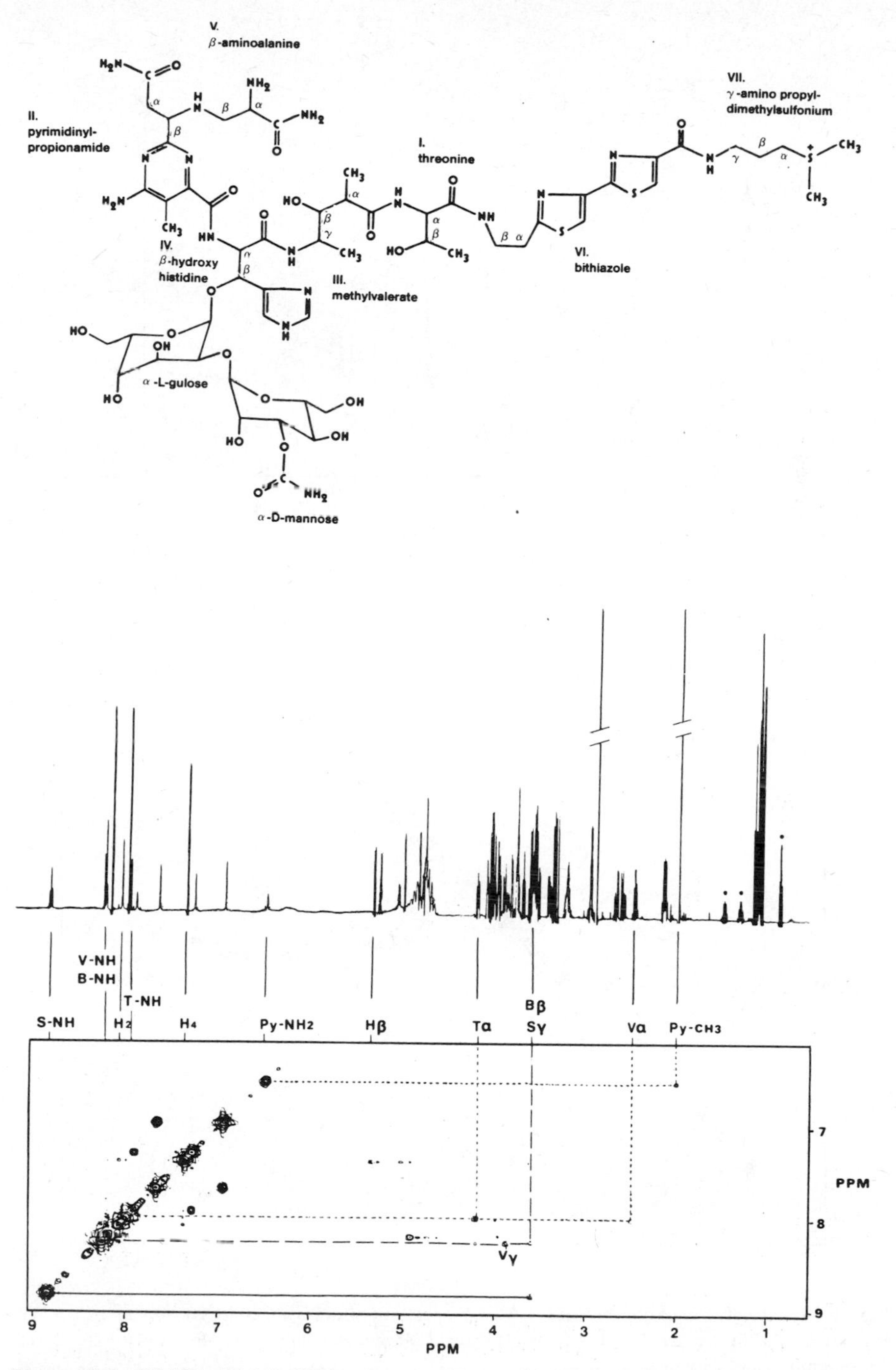

Figure 10.5. 500-MHz H[1] 2-D NOE spectrum of bleomycin. [C. A. H. Haasnoot, U. K. Pandit, C. Krut, and C. W. Hilbers, *J. Biomol. Struct. Dynamics* **2,** 449 (1984). Reproduced with permission of Adenine Press.]

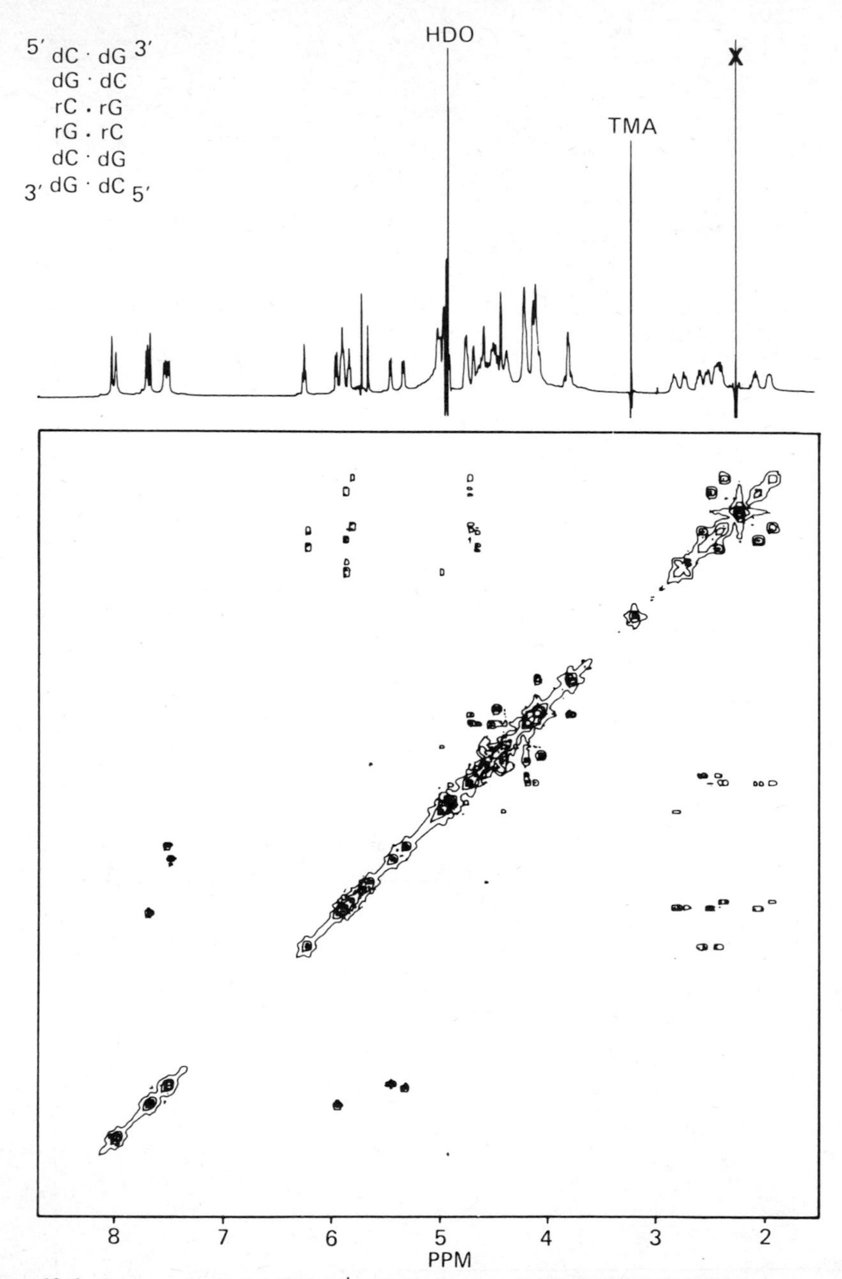

Figure 10.6. Contour plot of 500-MHz H^1 COSY spectrum of d(CG)r(CG)d(CG). [Source: C. A. G. Haasnoot, F. J. M. van de Ven, and C. W. Hilbers, *J. Magn. Resonance* **56,** 346 (1984). Reproduced with permission of Academic Press.]

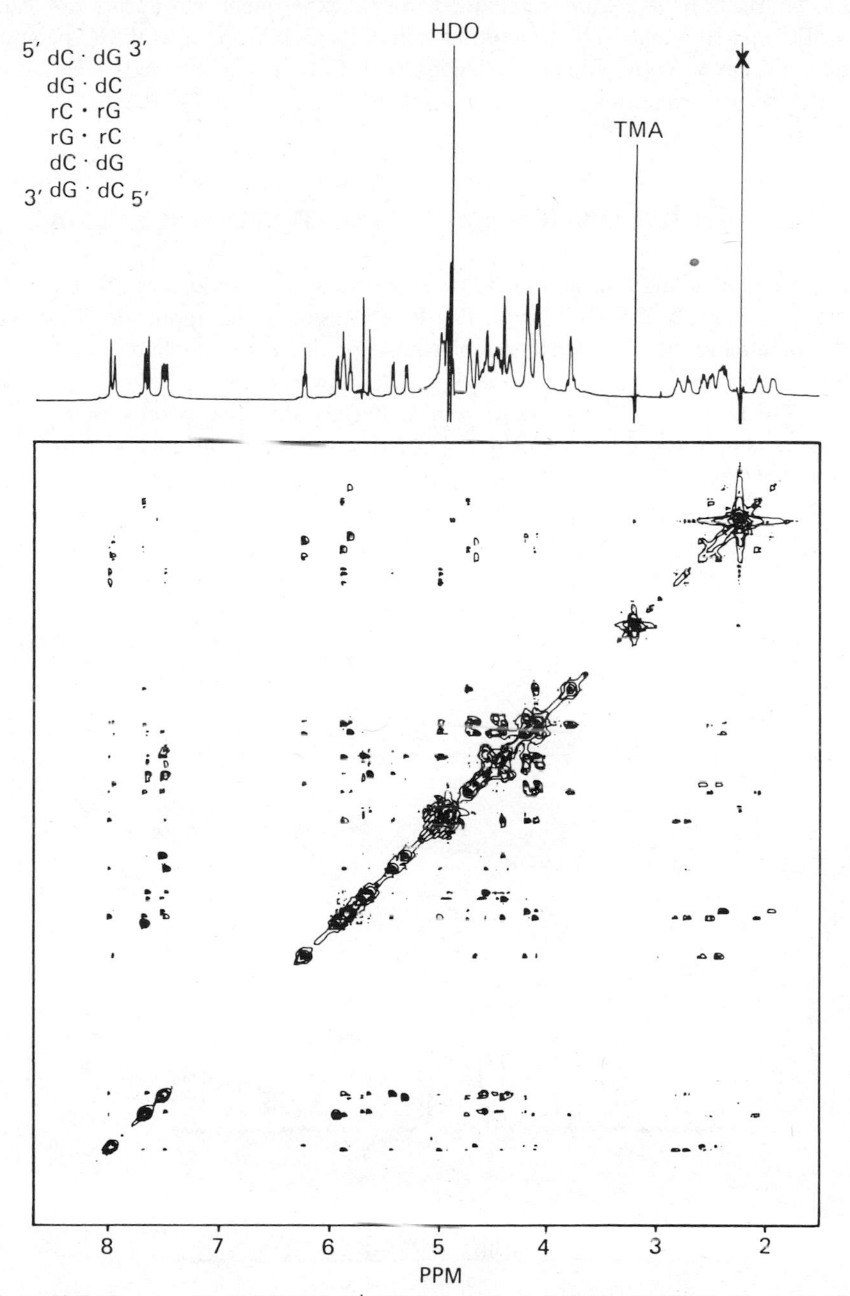

Figure 10.7. Contour plot of 500-MHz H[1] NOESY spectrum of d(CG)r(CG)d(CG). [C. A. G. Haasnoot, F. J. M. van de Ven, and C. W. Hilbers, *J. Magn. Resonance* **56,** 347 (1984). Reproduced with permission of Academic Press.]

Thus, both sets of data can be recorded in one experiment and during the same time. Figures 10.6 and 10.7 are reproductions of a COSY and a NOESY spectrum of a hybrid DNA–RNA oligonucleotide d(CG) r(CG) d(CG) obtained by Haasnoot under these new conditions.

10.2 CROSS-POLARIZATION MAGIC ANGLE SPINNING (CPMAS)

The technique of high-resolution NMR spectroscopy of solids was first experimentally realized by Schaeffer et al. [9]. It is a result of the application of magic angle spinning to the cross-polarization acquisition process as applied to multipulse NMR spectroscopy.

CPMAS produces C^{13} spectra of solid materials with line widths as low as 2 Hz. An example the C^{13} spectra of a benzil derivative obtained under different physical and experimental conditions is given in Figure 10.8.

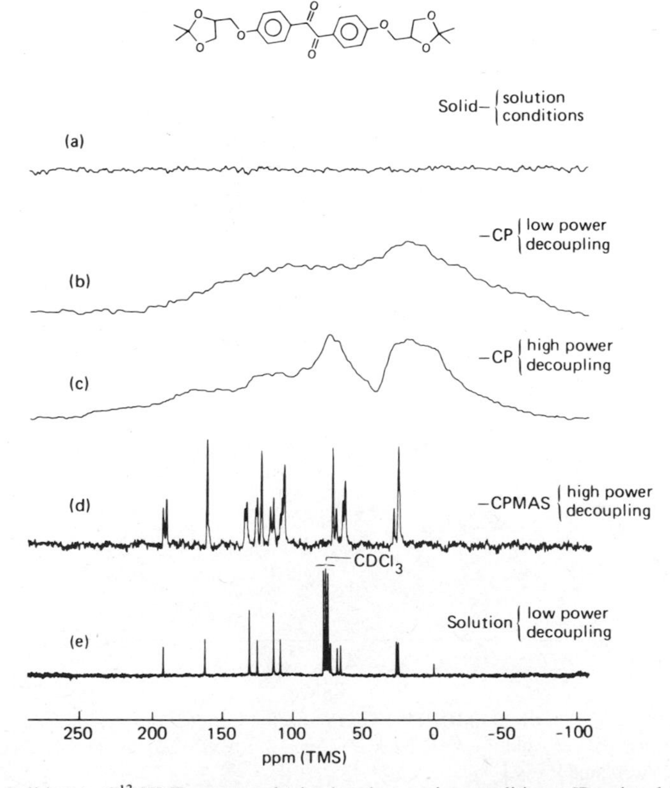

Figure 10.8. Solid-state C^{13} NMR spectra obtained under varying conditions. [Reprinted with permission from C. S. Yannoni, *ACC Chem. Res.* **15,** 202 (1982). Copyright American Chemical Society.]

Spectrum *a*, obtained on a solid sample using the conventional 60° carbon pulse with a 10-s delay between scans and with 6-kHz proton decoupling, clearly does not produce absorption peaks. The application of low- (spectrum *b*) and high- (spectrum *c*) power decoupling improves the spectrum somewhat. Finally, when low-power decoupling (proton–carbon cross-polarization with 1 s delay between scans) and magic angle spinning are combined, spectrum *d* is obtained. This solid-state CPMAS spectrum (obtained on 34 mg of material!) is clearly almost as good as the solution spectrum shown in Figure 10.8e. These superb examples, reported by one of the pioneers of this technique, C. S. Yannoni, clearly demonstrate the power of this technique.

Although most of the work employing CPMAS has been done on C^{13} spectra, other spin $\frac{1}{2}$ nuclei with weak mutual coupling (N^{15}, Si^{29}, P^{31}, Hg^{199}, etc.) can also be analyzed by this technique. Generally, the single-contact spin-lock cross-polarization CP RF pulse sequence, which invoves a pi/2 pulse, a 90° phase shift, a spin lock/decouple sequence with time intervals of 10^{-6}, 10^{-3}, 10^{-1}, and 1 s (proton polarization time) is employed in these experiments [10–13].

The technical limitations of this superb technique rest with the orientation of the spinning axis which must be within 0.5° of the magic-angle in order to decrease the chemical shift anisotropy (CSA) effect. The line widths obtained (2 ppm) can usually be reduced further by careful adjustments of the spinning angle and modifications of the spinning speed.

Thus, CPMAS uses high-power decoupling to remove the dipolar broadening of carbon spectra, the magic angle spinning to decrease CSA effects, and proton–carbon cross-relaxation to enhance the C^{13} sensitivities.

Yannoni and co-workers and others have extended this technique to variable temperature (VT) studies on polymers and many different classes of organic solids [11, 12, 14, 15]. The amount and importance of the information that can be obtained from these studies on solid samples are truly impressive.

One of the major applications of VTCPMAS is the study of proton-transfer reactions in the solid state by means of this DNMR (dynamic NMR) technique. For example, the VTCPMAS spectrum of the naphthoquinone

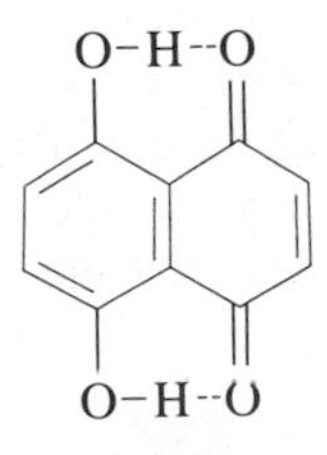

provides three C^{13} peaks due to, respectively, C's$_{1,4,5\,\mathrm{and}\,8}$, C's$_{2,3,6\,\mathrm{and}\,7}$, and C's$_{9\,\mathrm{and}\,10}$, at room temperature. At −160°, six peaks with very approximate area ratios of 1 : 1 : 2 : 1 : 2 : 1 are observed. It is thought that the spectra are caused by a second-order phase transition in the solid state between a low-temperature phase without proton exchange (H–O–H) and a high-temperature one where proton exchange is possible.

A study of N^{15}-enriched *meso*-tetraarylporphines [14] by the VTCPMAS process revealed two N^{15} peaks at 210 K, which coalesce to a broad singlet at 295 K. Thus, at the low temperature there exist two unequally populated tautomers that average out at higher temperatures.

The VTCPMAS spectrum of cyclooctatetraene diironpentacarbonyl shows a singlet for the ring carbons when observed at 100 K. At 32 K this singlet has separated into two peaks, with the downfield peak of one third the intensity of the upfield peak. This represents the first example for the slowing of a chemical exchange process in this compound!

10.3 CHEMICALLY INDUCED DYNAMIC NUCLEAR POLARIZATION (CIDNP)

NMR spectra of samples in which free radical reactions are occurring frequently show strongly affected intensities for the lines that belong to the reaction products. This observation has developed into the method of choice for the study reactions that involve short-lived radical intermediates [16].

For example, when benzoylperoxide ($C_6H_5CO{-}COC_6H_5$) dissolved in cyclohexane is heated at 110°C, the initial spectrum, typical of that expected for this compound (an AB_2C_2 pattern), changes with time. After 4 min, the intensity of the pattern has decreased significantly, and a strong benezene peak appears with a negative absorption. After 8 min, this peak has disappeared, and the intensity of the benzoylperoxide peaks has further decreased. Finally, after 12 min, the benzene peak has reappeared and is now in the normal positive intensity mode. The reverse intensity peak has been shown to be due to reverse spin polarization during the radical reaction.

Spin polarization in a diamagnetic product represents evidence that the nuclear spins were, at some point in time during reaction, in a radical pair. Although ESR is the traditional technique used to study free radical reactions, CIDNP is more sensitive and lends itself somewhat more readily to structural identification which may ascertain that a reaction path radical is being observed in a given reaction rather than a radical produced in a minor by-reaction. It is of significance to keep in mind that polarization occurs only in molecules that recombine within the initially formed radical pair. Any radicals that are not retained within the solvent cage do not contribute to the induced polarization and consequently go undetected by this technique [17].

Experimentally, photo-CIDNP experiments are done by placing the sample solution, often containing an initiator, into the magnet cavity and irradiating with a laser (argon ion lasers are frequently used) prior to application of any RF pulse and acquisition of the FID if pulsed NMR spectroscopy is the technique of choice. FID's obtained under alternating dark and irradiated conditions are analyzed, and, after subtraction of the dark spectra, the photo-CIDNP difference spectra are recorded. In the case of flavin-induced spectra of tyrosine, the 3 and 5 protons on the phenyl ring appear as negative-intensity peaks. Thus, these species are involved in the spin polarization process [18].

10.4 MAGNETIC RESONANCE IMAGING

Perhaps the most unique application of the NMR concept was initiated with a short paper by P. C. Lauterbur [19] entitled "Image formation by induced local interactions: Examples employing nuclear magnetic resonance." This almost nonexciting title has led to the most significant diagnostic tool development in medicine since the discovery of x-rays by Roentgen!

A direct quote taken from Lauterbur's note is appropriate to initiate the discussion of this technique:

> An image of an object may be defined as a graphical representation of the spatial distribution of one or more of its properties. Image formation usually requires that the object interact with a matter or radiation field characterized by a wavelength comparable to or smaller than the smallest features to be distinguished, so that the region of interaction may be restricted and a resolved image generated.

The initial experiment was performed with a 60-MHz RF field and a magnetic field equivalent to cause proton resonances at about 700 Hz. The sample was a 4.2 mm inside-diameter glass tube filled with D_2O. This tube, in turn, contained two 1 mm inside-diameter capillaries filled with H_2O. If a uniform signal strength across the region within the transmitter–receiver coil is assumed, the proton signal represents a one-dimensional projection of the H_2O content of the object, integrated over different planes (slices), which are perpendicular to the direction of the gradient. A 2-D image can be obtained by combining a number of different projections resulting from rotation of the sample about an axis perpendicular to the gradient direction. This results, after appropriate mathematical computations, in a 2-D image of a slice of the water in the capillary tubes in this particular sample.

At low RF power, the two capillary samples provided essentially identical images in the spectral representation (the term *zeugmatography* from the Greek "that which is used for joining" was proposed by Lauterbur [19] as an appropriate descriptor for this technique). At higher power levels, pure water samples gave much more saturated signals, which disappeared at even greater power levels than samples whose spin–lattice relaxation time had been shortened by the addition of the paramagnetic Mn^{+2} ion. Thus, samples with long relaxation times can be differentiated from those with short ones by the expedient of obtaining data at different power levels.

A most interesting application of this signal difference depending on relaxation times involves the study of malignancies, since tumors have been shown to have much longer proton relaxation times than does healthy tissue [20]. The powerful extensions of this initial experimentation will become clear in the forthcoming paragraphs.

The concept of field-focusing NMR spectroscopy described by Damadian and co-workers [21] resulted from an earlier idea [22] suggesting that whole-body NMR spectroscopy might be feasible. In normal NMR experiments, the signal is detected by pickup coils surrounding the sample without any differentiation of the nuclide concentration across the sample container (in fact, much effort is spent to assure proper averaging of the signal intensity in normal NMR spectroscopy). The FONAR (field-focusing NMR) technique permits direct focusing of the NMR signal within the interior of a sample.

Fundamentally, the process can be explained by realizing that for any one selection choice of the RF field there can be only one value of the magnetic field (H_0) that causes a resonance signal for a particular nuclide. In reality, the magnetic field is linear over a narrow sample width, and the resonance condition prevails over a small volume, referred to as the "resonance aperture." The position of the resonance aperture in any particular sample can be altered by adjusting either the magnetic field or the RF frequency in a manner that will move the resonance focus to another resonance aperture within the sample located in the magnet gap. The recognized method utilizes a shaped DC magnetic field whose resonance aperture can be varied by appropriate adjustments of the DC-generated field.

The initial experiments that achieved the first human NMR scan utilized a superconducting magnet composed of two Helmholtz pairs with a 135-cm inner diameter. Each half of the magnet contained a sweep coil and z-gradient coil in addition to the magnet wiring. Although this particular magnet was theoretically capable of generating a magnetic field of 5 T, field strengths of 0.5 and 1 T were used in these experiments. As is true for all *supercons*, the magnetic field stability of this magnet was superb. The RF pulses were generated using an appropriate-size single-coil probe surrounding the sample (a portion of a human being, in this instance). The coil operated at 2.18 MHz and delivered 10 W over 60 μs. Ninety-degree pulses were repeated with a period of 800 μs. The spectral images of the maximum P–P amplitude of a constant 5-kc off-resonance beat pattern of the phase-detected proton signal are appropriately stored. Although this initial experiment required 4.5 h to generate an intensity contour map of a vertical slice (the reference is a human body lying on the back) of a human chest, it nevertheless provided excellent proton intensity contours of this part of the chest.

Modern instrumentation, operating on similar principles, provides a great variety of options for the accumulation and analysis of data obtained by this process. It is the current purchasing cost (about $1 million!) of these superb instruments, generally referred to as "magnetic imaging spectrometers" that has prevented their widespread use. However, it is encouraging to note that many of the larger hospitals in the country do own a magnetic resonance imaging instrument.

The General Electric 1.5-T Signa system, [23] as the name implies, operates at 1.5 T. It utilizes a 6-megabyte computer to facilitate instrumentation control (location of the resonance aperture) and analysis of the data. In this instrument the location of each resonance aperture (also referred to as "voxel") is determined

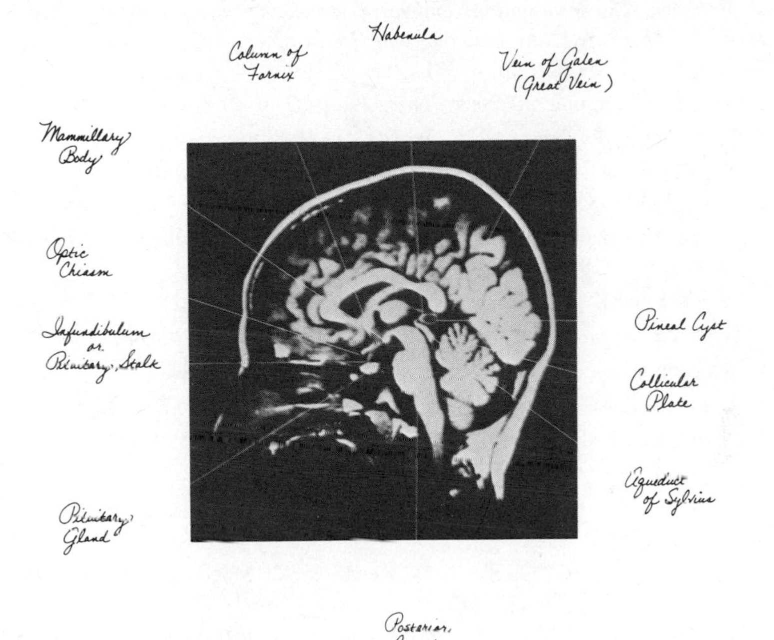

Magnetic image "slice" of a human brain.

Figure 10.9. Magnetic imaging picture.

with gradient coils by altering the magnetic field in carefully controlled steps. Application of FT analysis to each of the voxels obtained at several thousand different X and Y coordinates provides, ultimately, an intensity contour map representing one plane. Movement of the patient within the magnetic cavity generates contour maps of additional slices (or planes), and, ultimately, a total three-dimensional intensity map could be generated. Computer manipulation also permits recomputation of the data so that if a sufficient number of vertical slices (see Fig. 10.9) are available, some of the data can be used to generate a horizontal slice.

Finally, intensity differences, as already noted by Lauterbur, due to variations in relaxation times of protons in different cellular regions provide additional information for the diagnostic analysis of magnetic imaging-generated information [23].

REFERENCES

1. J. Jenner, *Ampere International Summer School*, Basko Polje, Yugoslavia, 1971, and J. Jenner, B. H. Meier, P. Bachman, and R. R. Ernst, *J. Am. Chem. Soc.* **101,** 6441 (1979).
2. L. Mueller, A. Kumar, and R. R. Ernst, *J. Chem. Phys.* **63,** 5490 (1975).

3. R. R. Ernst, *The Information Content of Two-Dimensional Fourier Spectroscopy in NMR Spectroscopy: New Methods and Applications*, ACS Symposium series 191, Washington, DC, 1982, pp. 47–61.
4. *Chemical and Engineering News*, Nov. 29, 1982, p. 21 and references therein.
5. S. Mora and R. R. Ernst, *Mol. Phys.* **41,** 95 (1980).
6. C. W. Hilbers, A. Heershap, C. A. Haasnot, and J. A. L. Walters, *J. Biomol. Struct. Dynam.*, 183 (1983).
7. C. A. H. Haasnot, U. K. Pandit, C. Kruk, and C. W. Hilbers, *J. Biomol. Struct. Dynam.* **2,** 449 (1984).
8. G. Wagner and K. Wuethrich, *J. Mol. Biol.* **155,** 347 (1982).
9. J. Schaeffer, E. O. Stejskal, and R. Buchdahl, *Macromolecules* **8,** 291 (1975).
10. C. S. Yannoni, *Acct. Chem. Res.* **15,** 201 (1982).
11. J. R. Lyerla, C. S. Yannoni, and C. A. Fyda, *Acct. Chem. Res.* **15,** 208 (1982).
12. V. Mach, R. Kendrick, and C. S. Yannoni, *J. Magn. Res.* **52,** 450 (1983).
13. Ref. 3, p. 219.
14. H. H. Limback, J. Henning, R. Kendrick, and C. S. Yannoni, *J. Am. Chem. Soc.* **106,** 4059 (1984).
15. C. S. Yannoni, T. C. Clark, R. D. Kendrick, V. Macho, and R. D. Miller, *Mol. Cryst. Liquid Cryst.* **96,** 305 (1983).
16. L. T. Muus, P. W. Atkins, K. A. McClauchton, and J. B. Pederson (eds.), *Chemically Induced Magnetic Polarization*, D. Reidel, Dordrecht, The Netherlands, 1977.
17. J. Bargon, H. Fischer, and V. Johnsen, *Z. Naturforsch.* **A22,** 551 (1967).
18. Ref. 3, p. 308.
19. P. C. Lauterbur, *Nature* **242,** 190 (1973).
20. I. D. Weisman, L. H. Bennett, L. R. Maxwell Sr., *Science* **178,** 1288 (1972).
21. R. Damadian, L. Minkoff, M. Goldsmith, and J. A. Koutcher, *Naturwissenschaften* **65,** 250 (1978).
22. R. Damadian, *Science* **171,** 1151 (1971).
23. General Electric Company, *Magnetic Resonance System: SIGNA*, General Electric Co, Milwaukee, WI. 1985.

APPENDIX

SUPPLEMENTARY PROBLEMS

Establish the structures and determine the various NMR parameters for the following. The descriptions correspond to the numbers shown on the spectra. Spectra were obtained on a Varian HA-100 instrument.

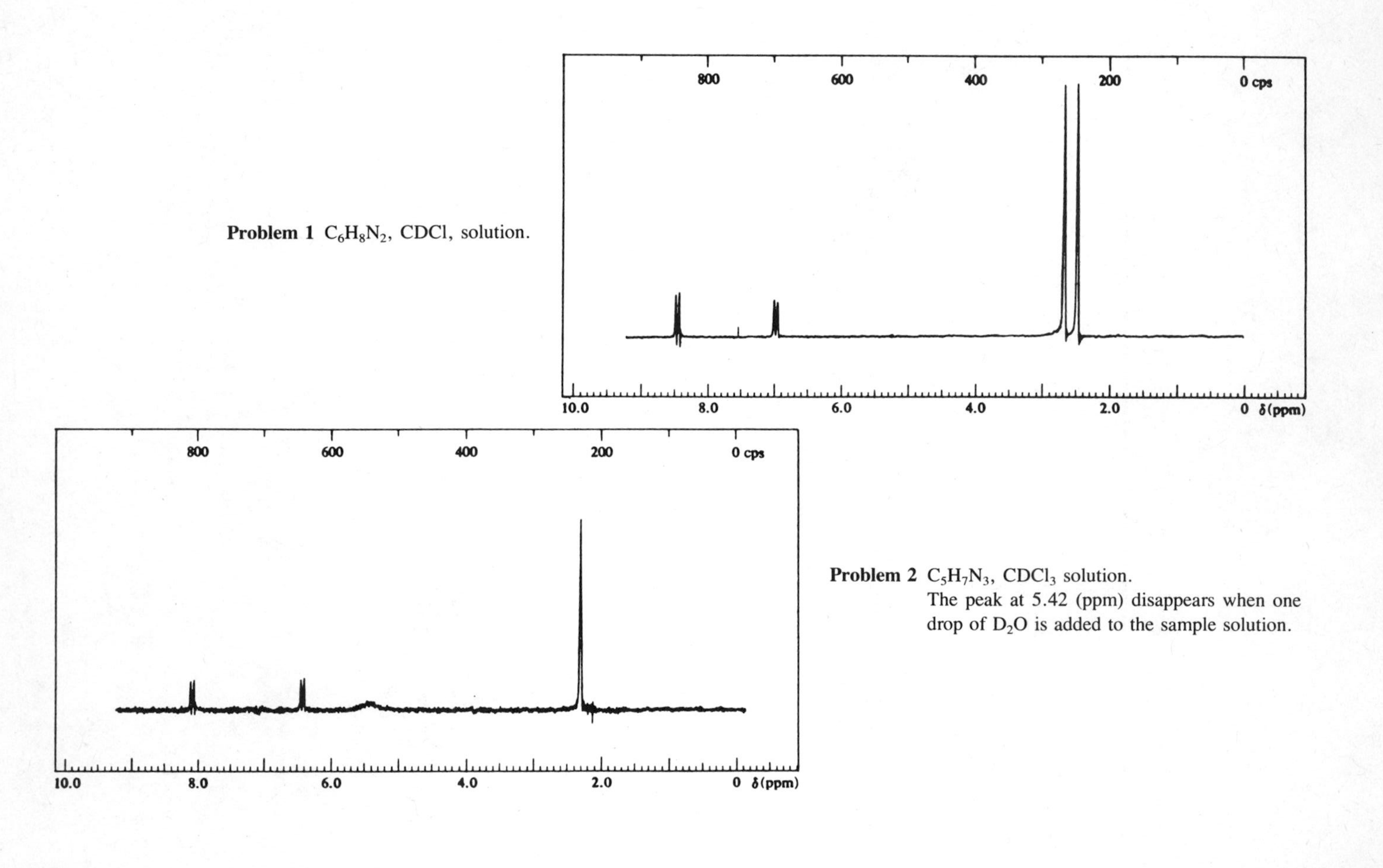

Problem 1 $C_6H_8N_2$, CDCl, solution.

Problem 2 $C_5H_7N_3$, $CDCl_3$ solution.
The peak at 5.42 (ppm) disappears when one drop of D_2O is added to the sample solution.

Problem 3 $C_8H_{11}N$, $CDCl_3$ solution.
In addition to the peaks shown on the chart there also are three 3-proton singlets.

Problem 4 $C_{13}H_{18}N_2$, $CDCl_3$ solution.
The compound does not form a salt with HCl.

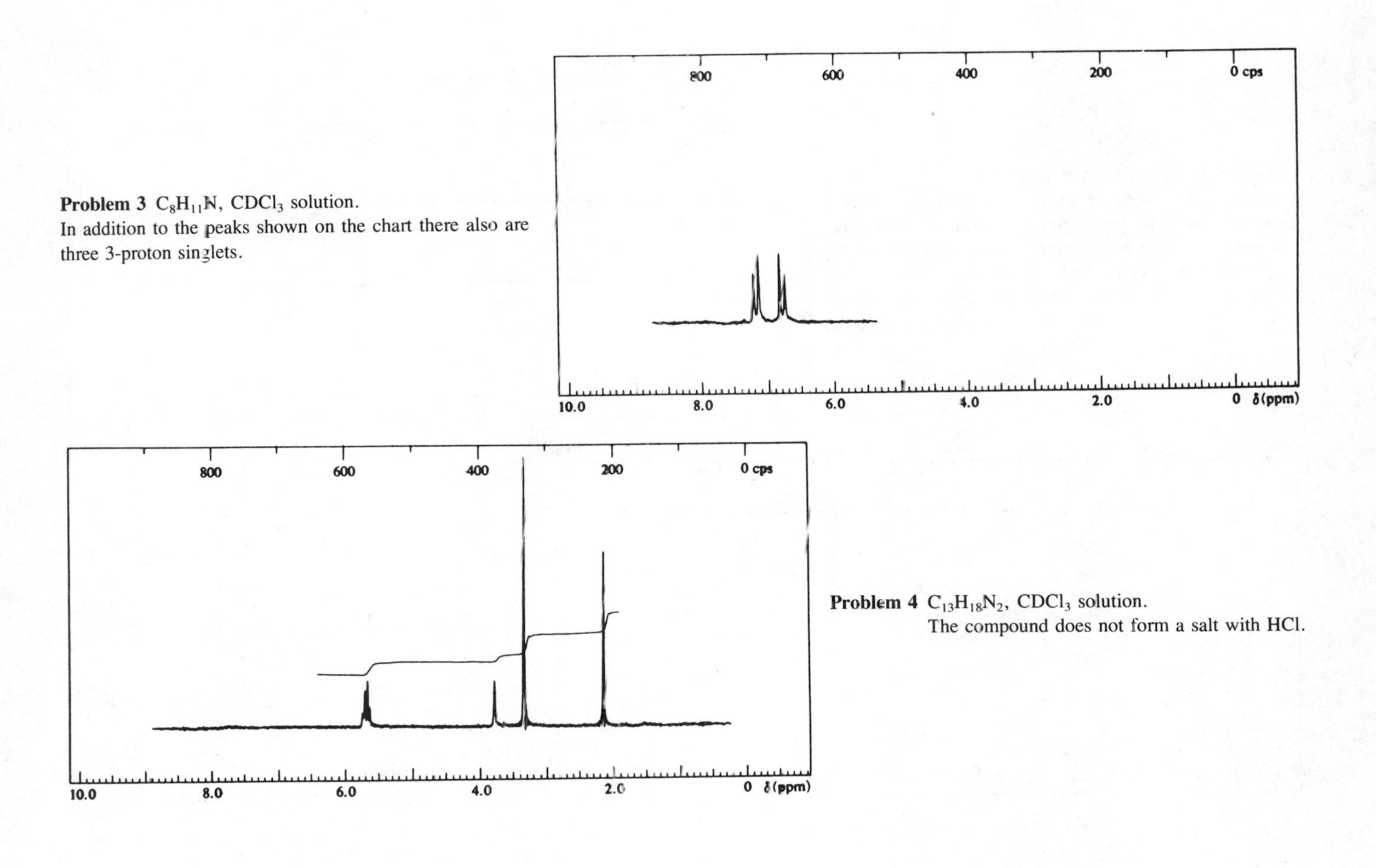

Problem 5 $C_{12}H_{12}O_2$, $CDCl_3$ solution.

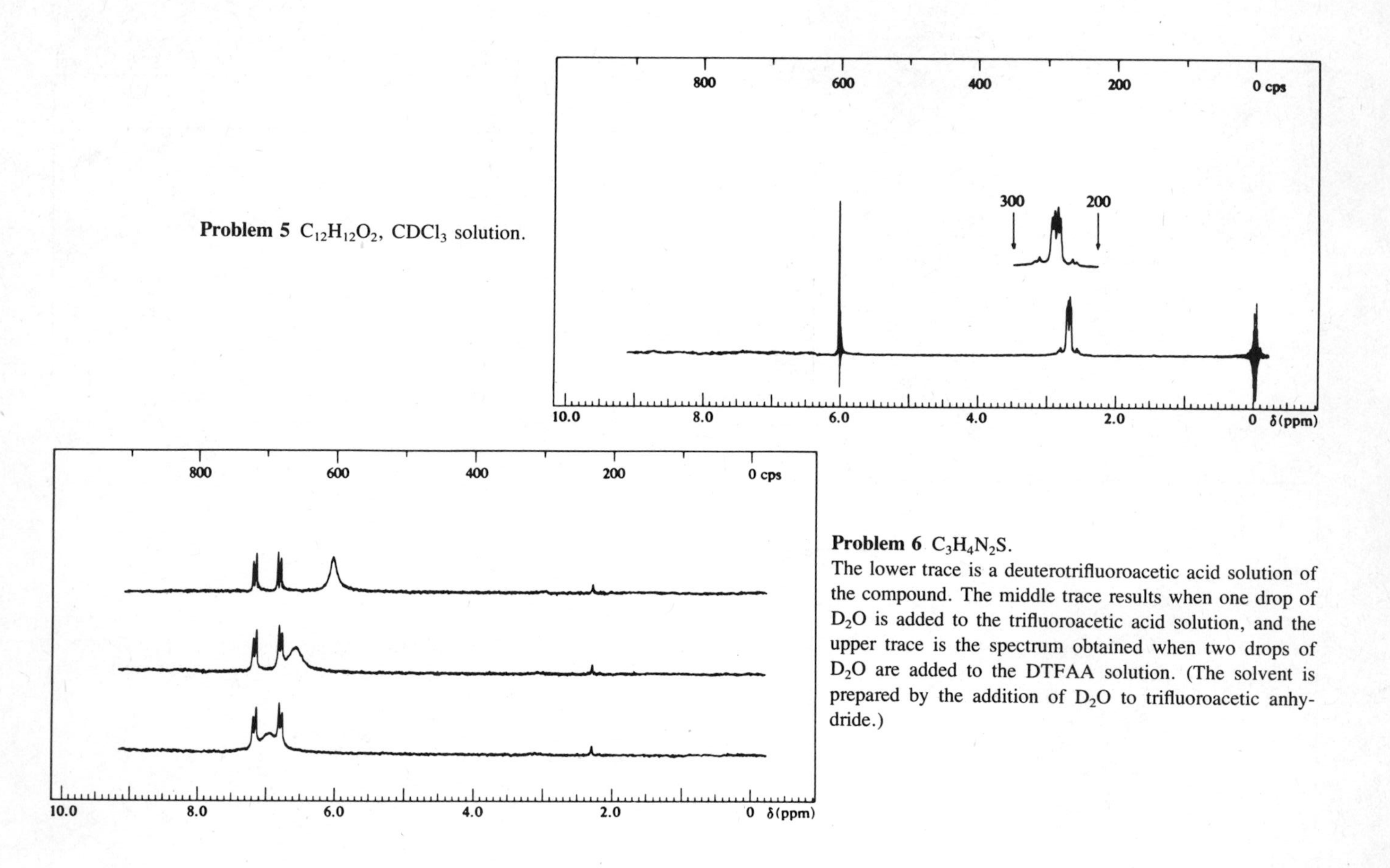

Problem 6 $C_3H_4N_2S$.
The lower trace is a deuterotrifluoroacetic acid solution of the compound. The middle trace results when one drop of D_2O is added to the trifluoroacetic acid solution, and the upper trace is the spectrum obtained when two drops of D_2O are added to the DTFAA solution. (The solvent is prepared by the addition of D_2O to trifluoroacetic anhydride.)

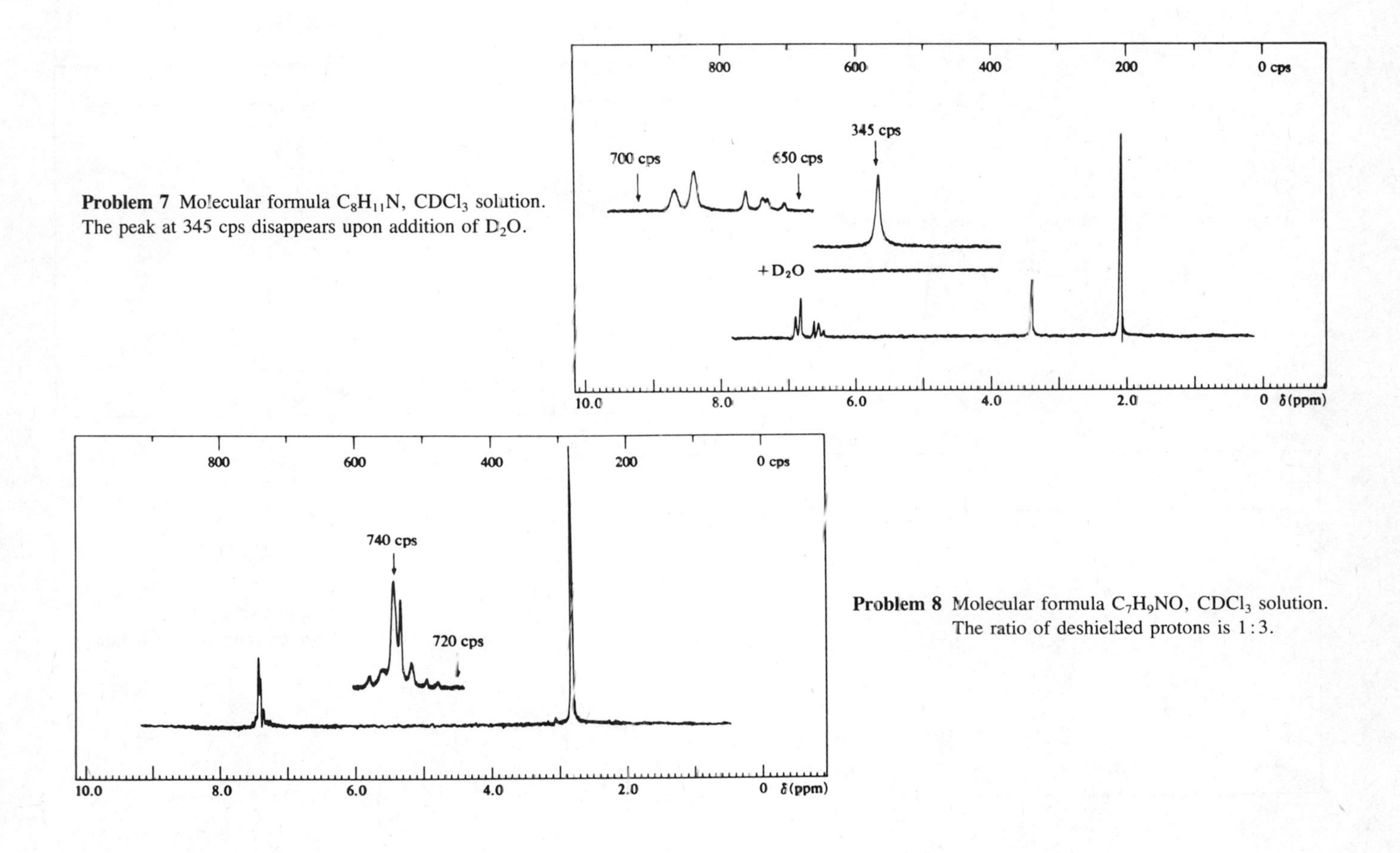

Problem 7 Molecular formula $C_8H_{11}N$, $CDCl_3$ solution. The peak at 345 cps disappears upon addition of D_2O.

Problem 8 Molecular formula C_7H_9NO, $CDCl_3$ solution. The ratio of deshielded protons is 1:3.

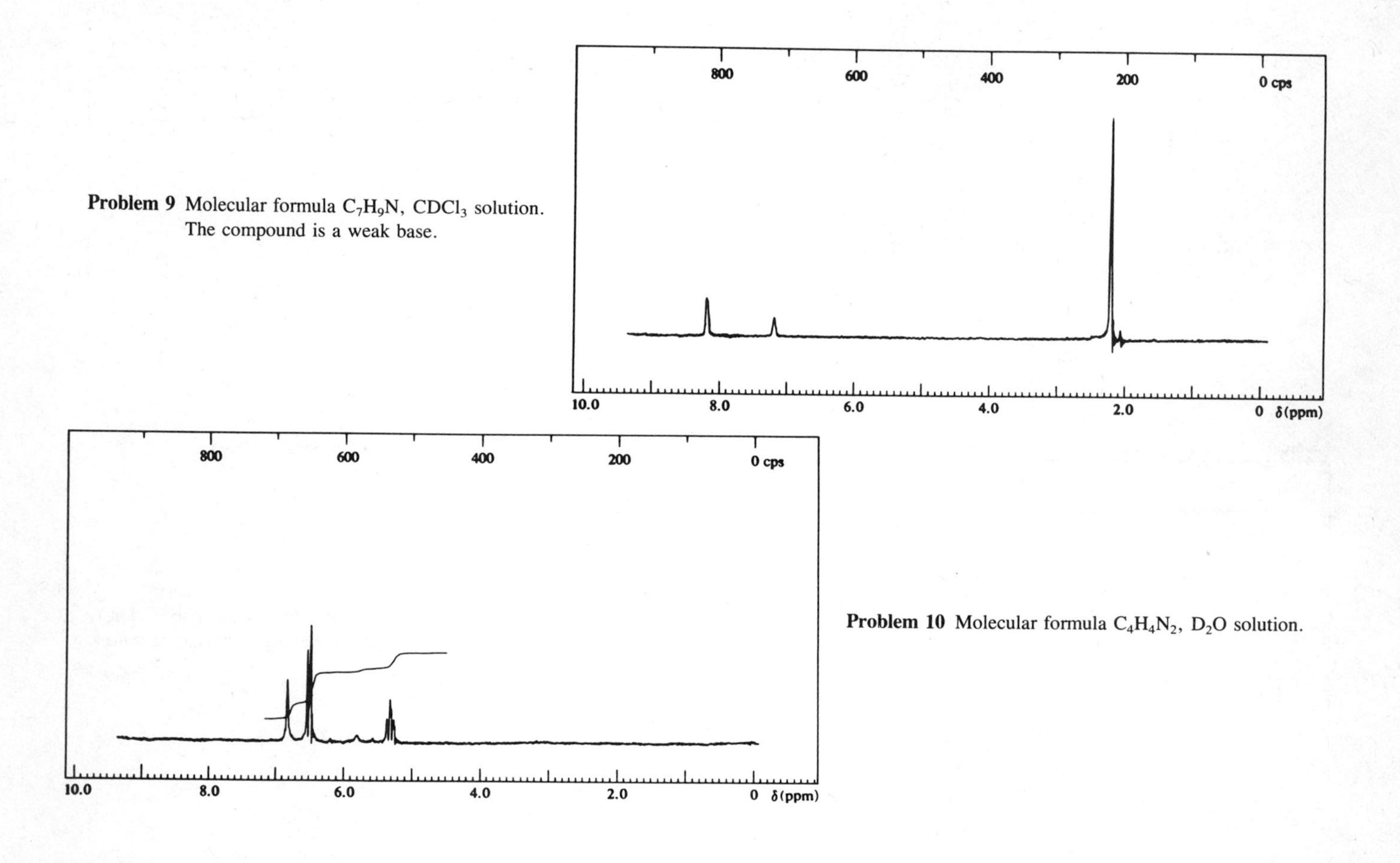

Problem 9 Molecular formula C_7H_9N, $CDCl_3$ solution. The compound is a weak base.

Problem 10 Molecular formula $C_4H_4N_2$, D_2O solution.

Problem 11 Molecular formula $C_6H_8N_2$, $CDCl_3$ solution.
The peak at 600 cps disappears upon addition of D_2O.

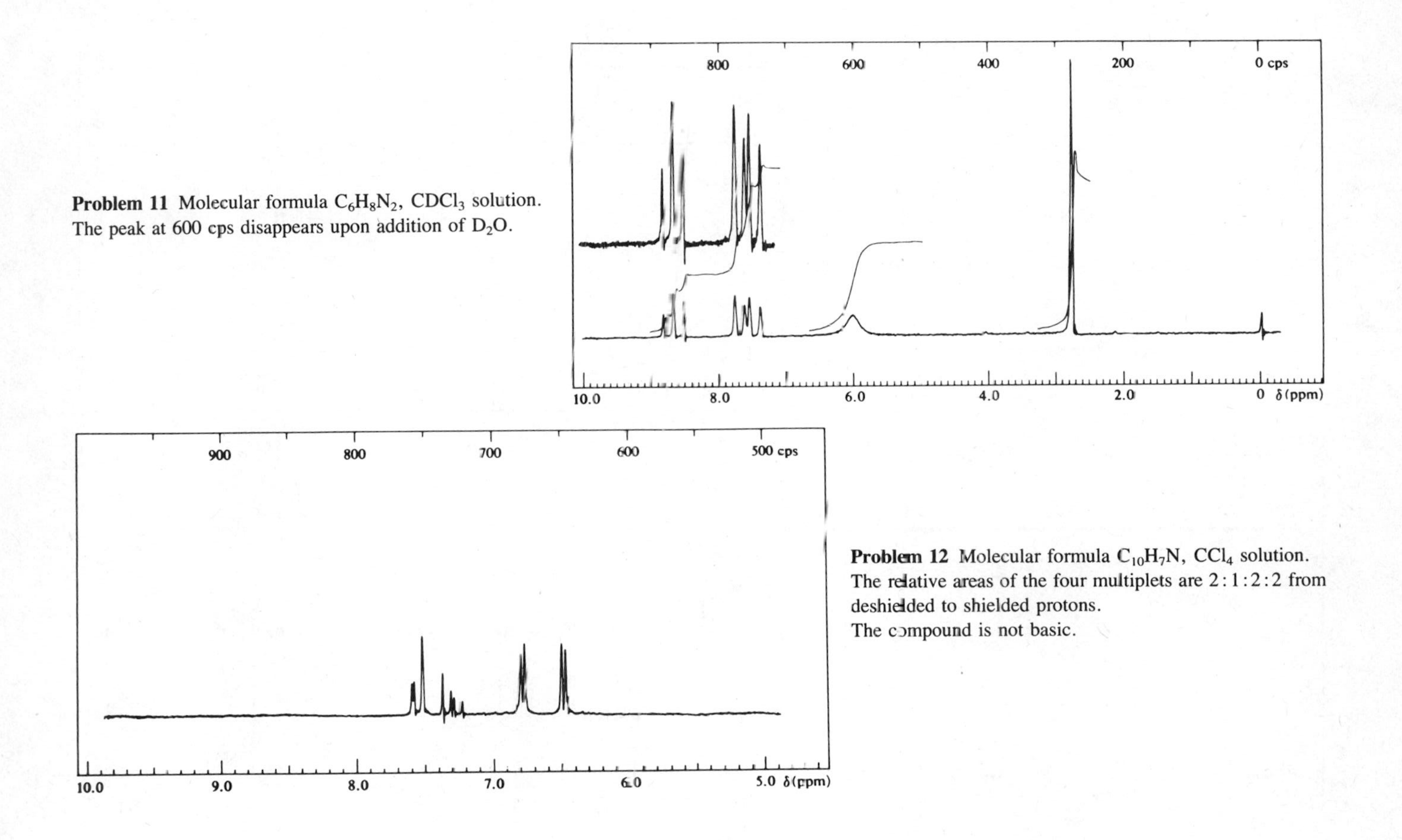

Problem 12 Molecular formula $C_{10}H_7N$, CCl_4 solution.
The relative areas of the four multiplets are 2 : 1 : 2 : 2 from deshielded to shielded protons.
The compound is not basic.

Problem 13 Molecular formula $C_8H_4O_3$, $CDCl_3$ solution.

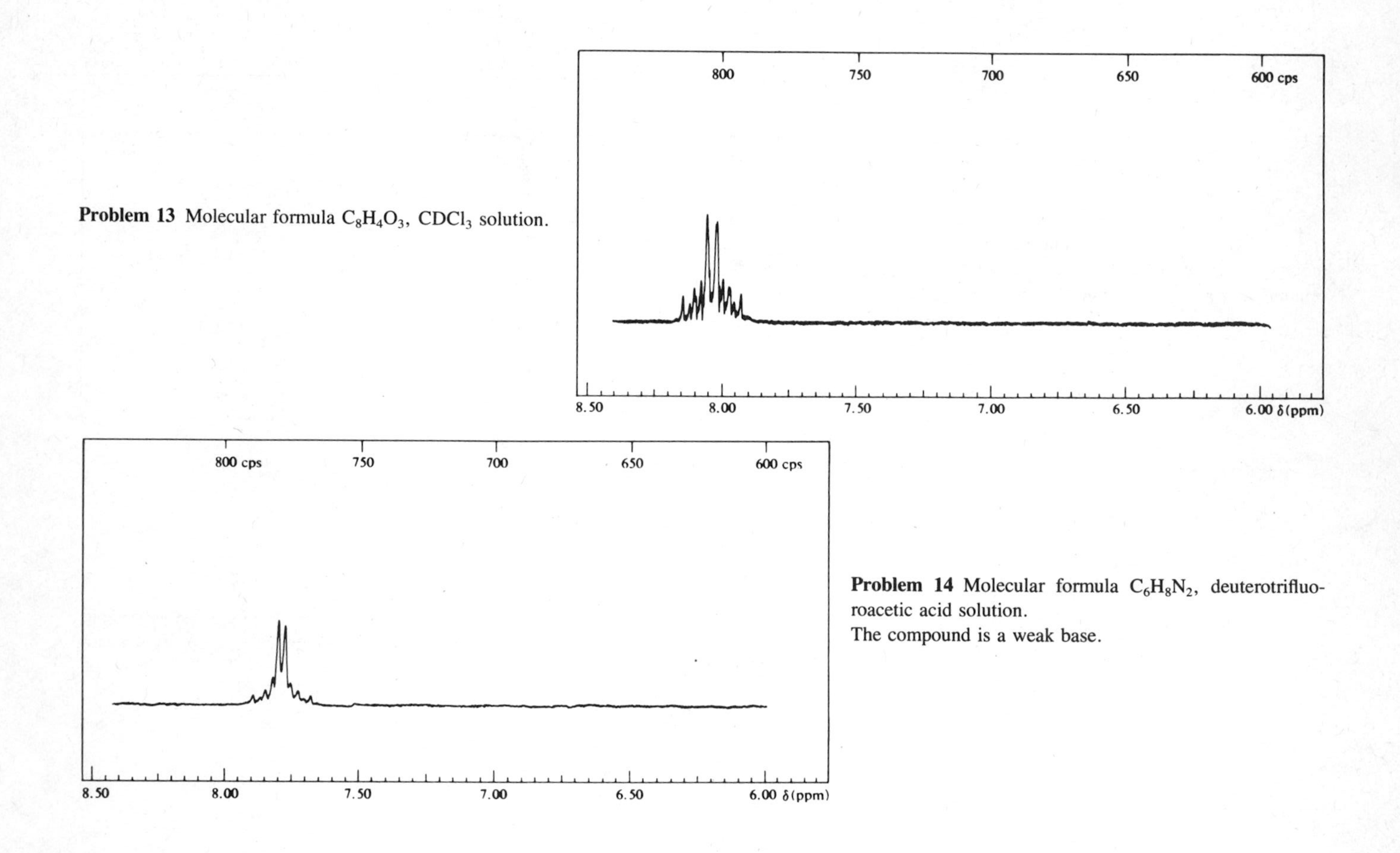

Problem 14 Molecular formula $C_6H_8N_2$, deuterotrifluoroacetic acid solution.
The compound is a weak base.

Problem 15 Molecular formula $C_6H_5N_3$, $CDCl_3$ solution. One single proton does not appear in the spectrum. The compound is a weak acid.

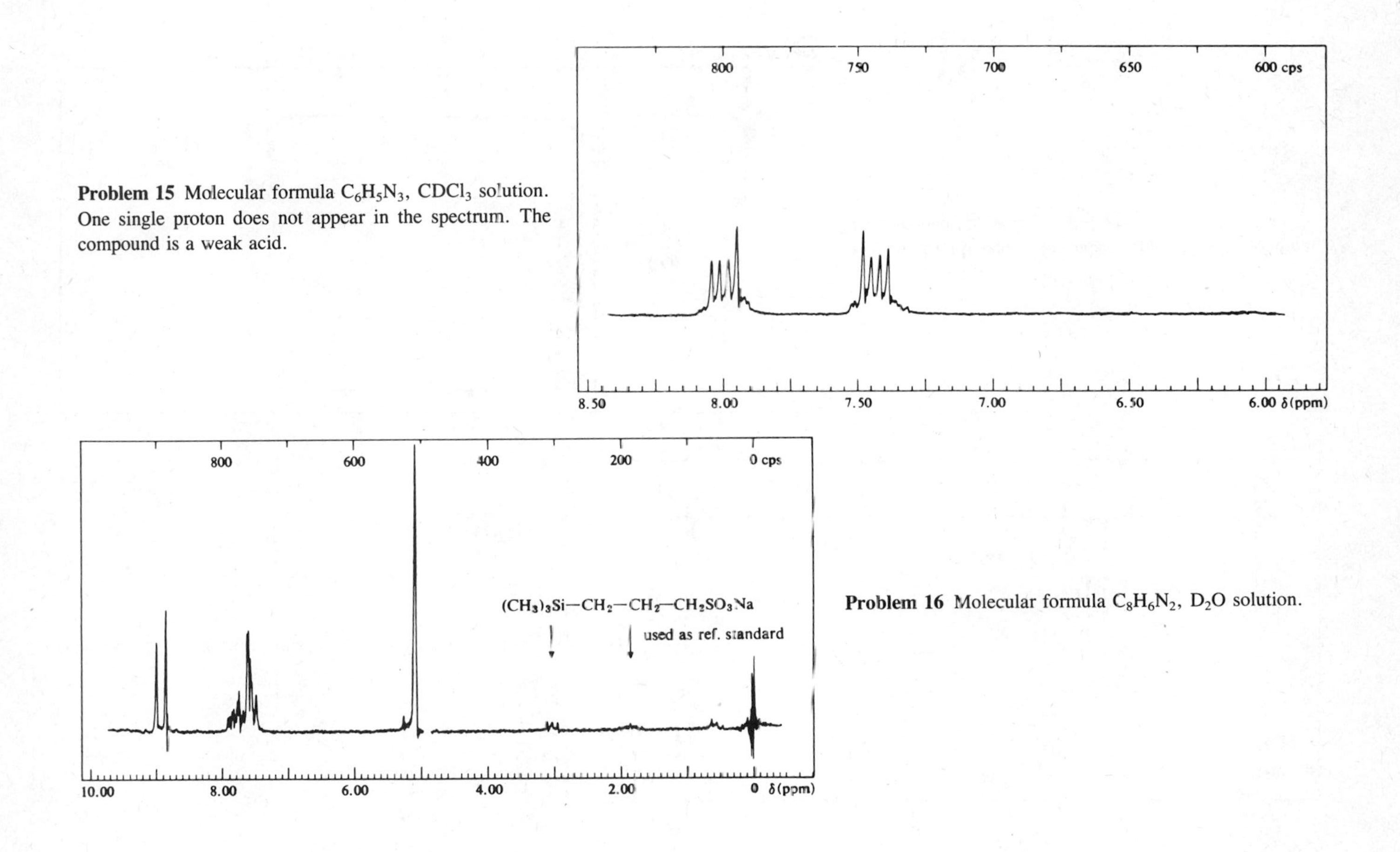

Problem 16 Molecular formula $C_8H_6N_2$, D_2O solution.

Problem 17 Molecular formula C_7H_9N, $CDCl_3$ solution. The compound is basic.

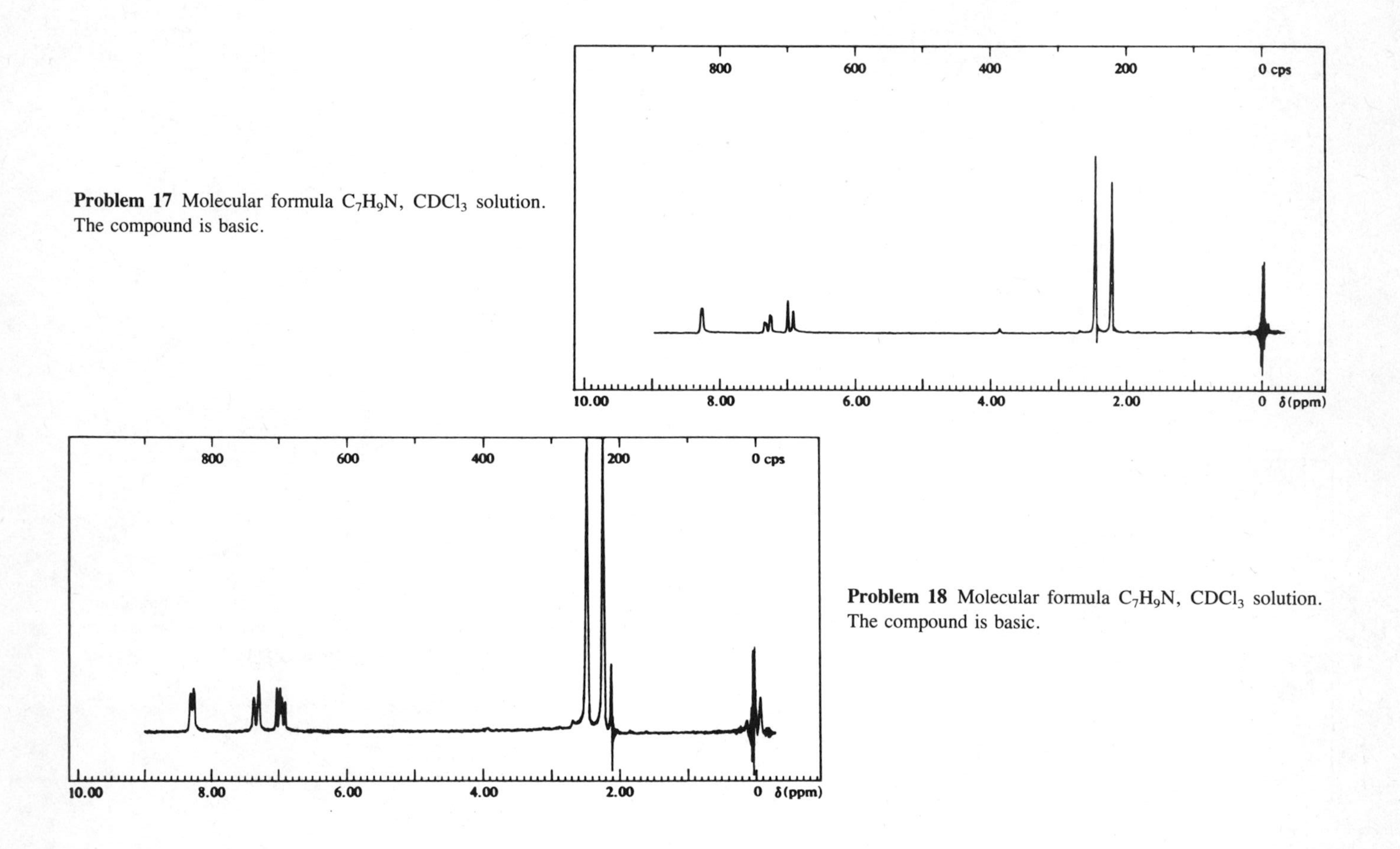

Problem 18 Molecular formula C_7H_9N, $CDCl_3$ solution. The compound is basic.

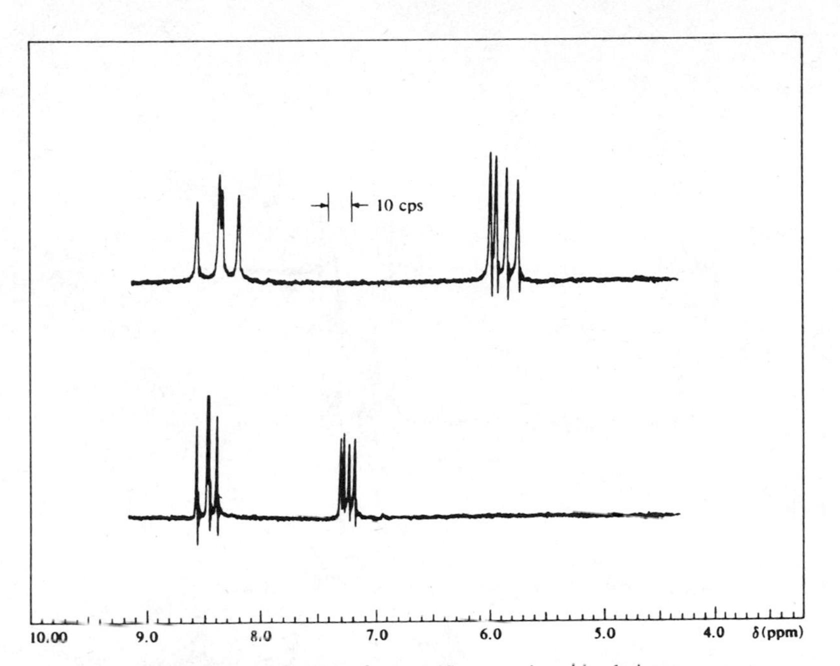

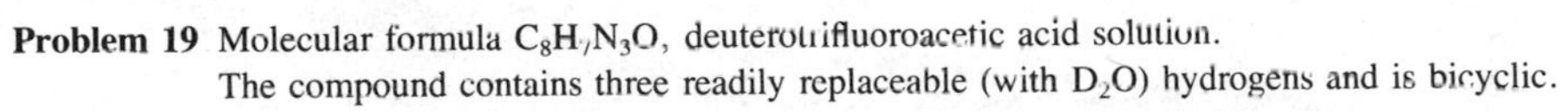

Problem 19 Molecular formula $C_8H_7N_3O$, deuterotrifluoroacetic acid solution.
The compound contains three readily replaceable (with D_2O) hydrogens and is bicyclic.

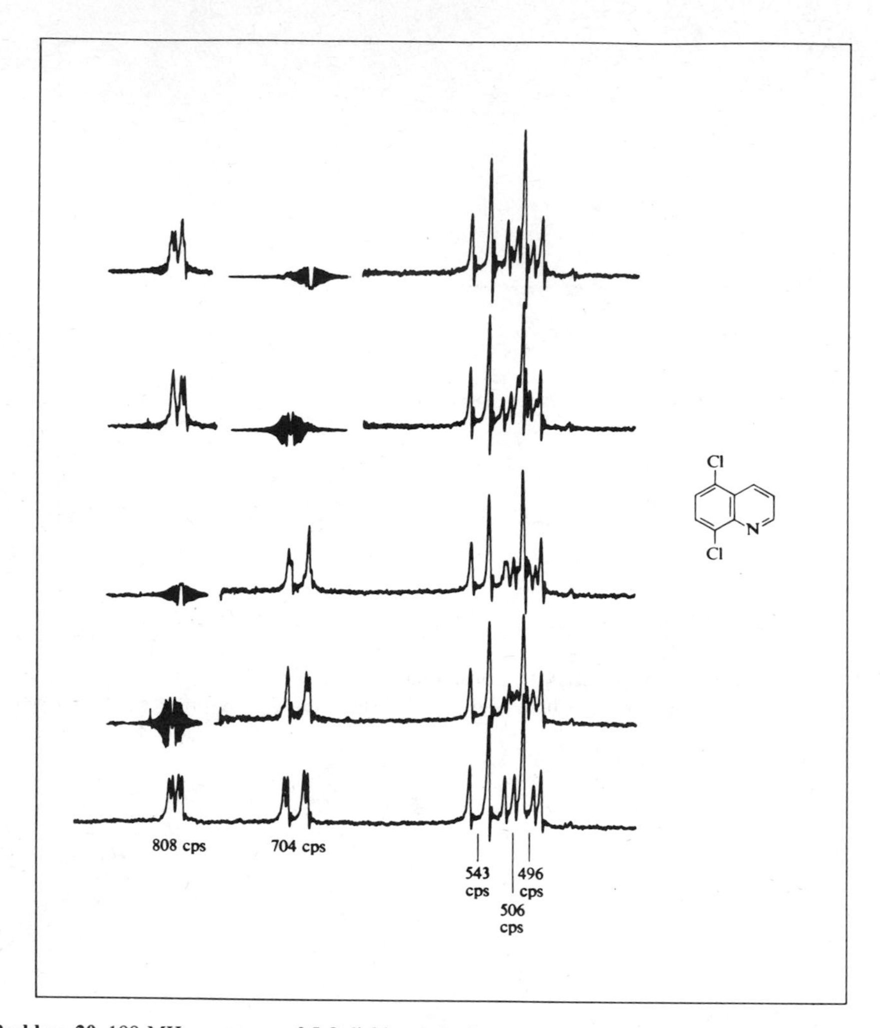

Problem 20 100-MHz spectrum of 5,8-dichloroquinoline.
Analyze this spectrum in detail, assigning chemical shifts and coupling constants from the data given.

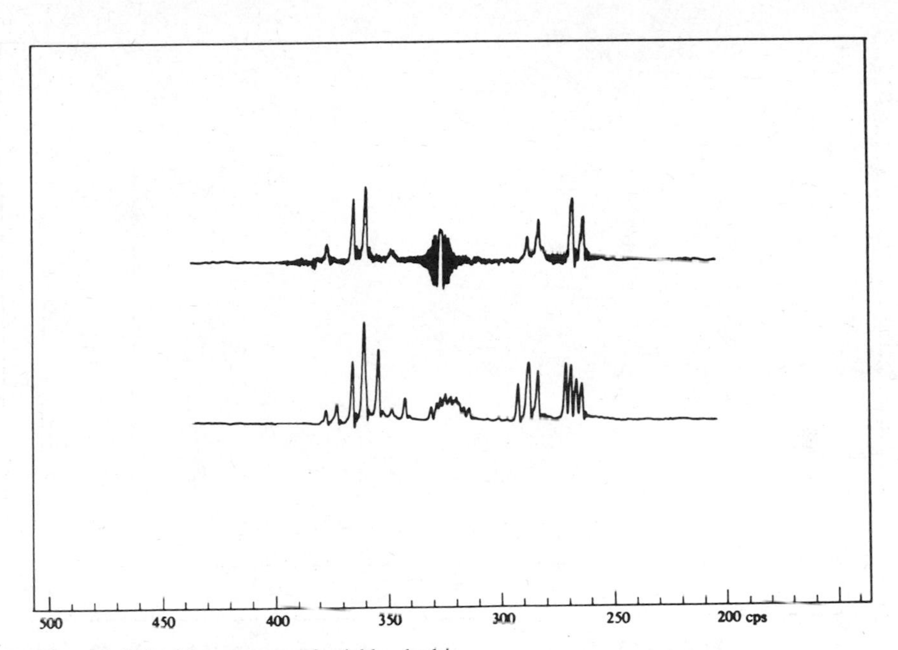

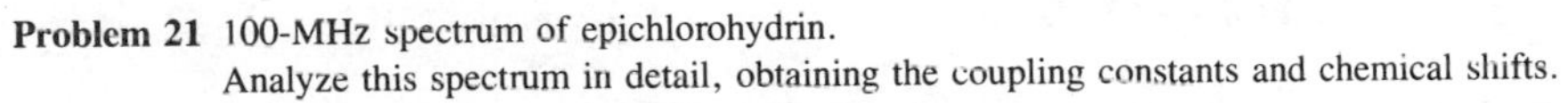

Problem 21 100-MHz spectrum of epichlorohydrin.
Analyze this spectrum in detail, obtaining the coupling constants and chemical shifts.

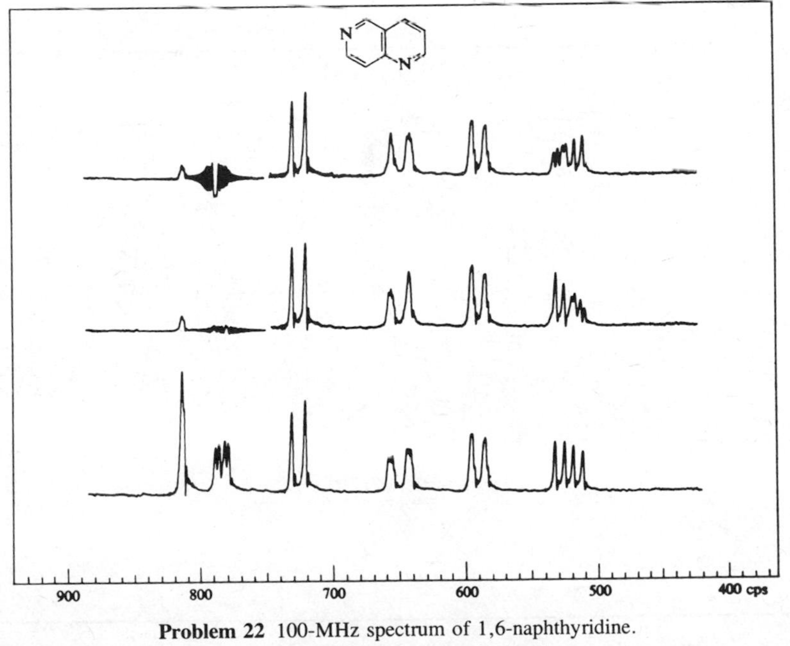

Problem 22 100-MHz spectrum of 1,6-naphthyridine.
Analyze this spectrum in detail.

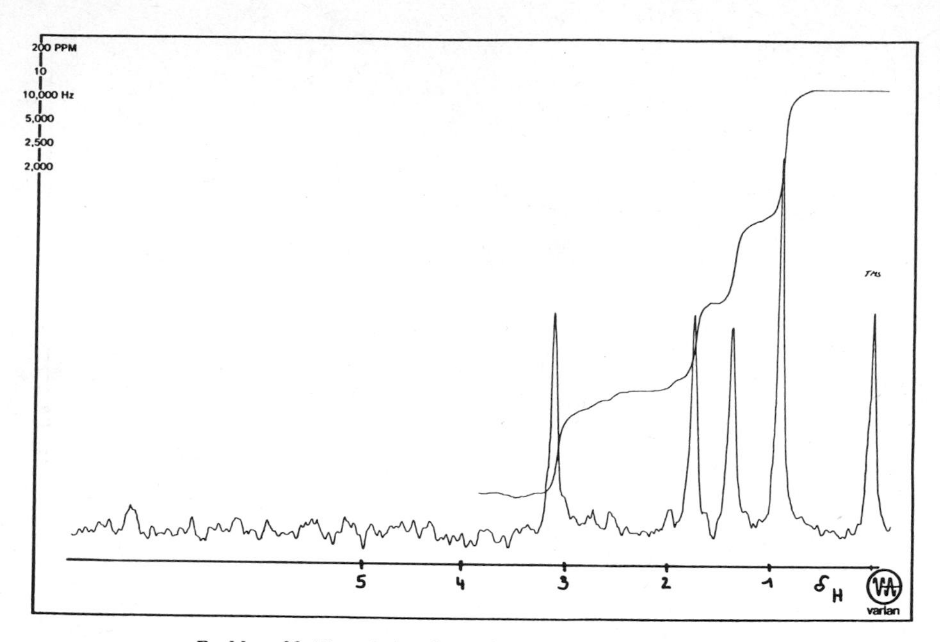

Problem 23 Natural abundance deutrium spectrum of C_4H_9I.

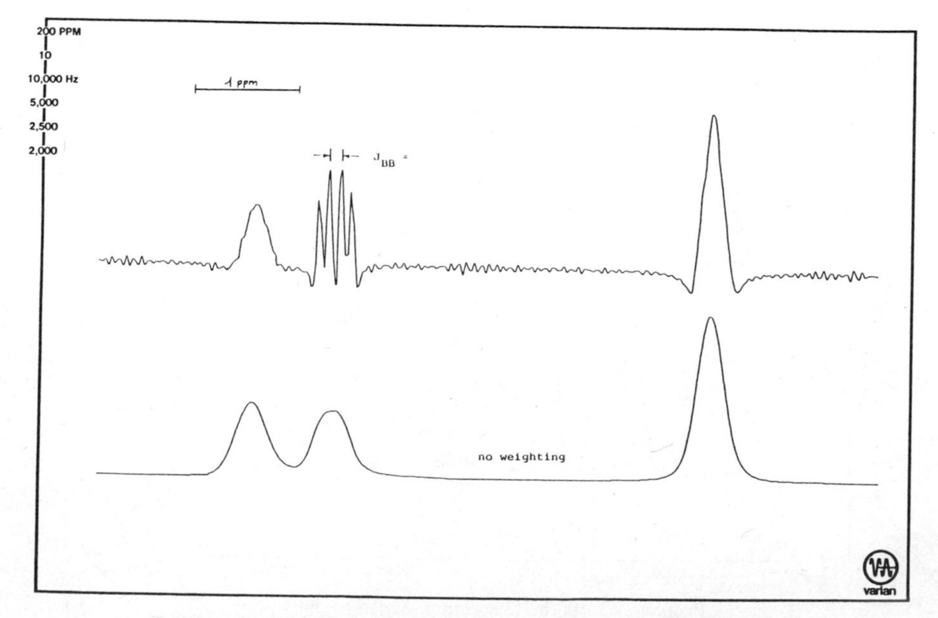

Problem 24 Explain the boron pattern of this compound (decarborane).

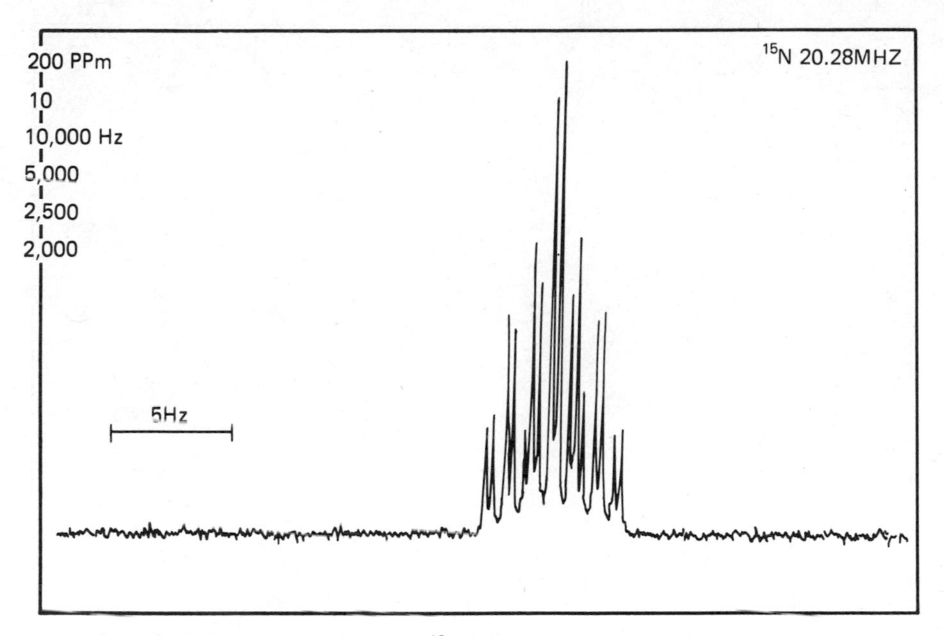

Problem 25 Account for the N^{15} pattern of this compound (nitrobenzene).

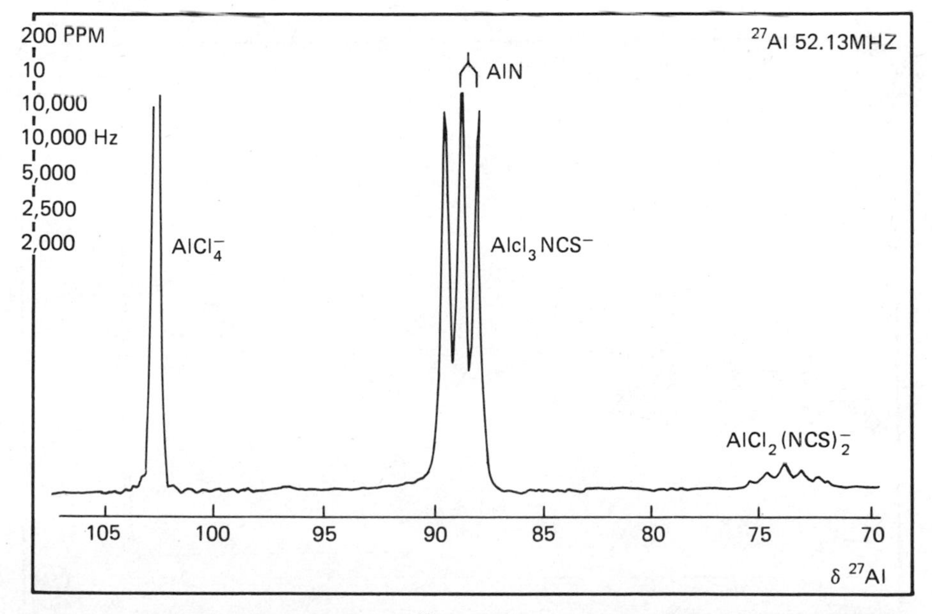

Problem 26 Rationalize the pattern of this Al^{27} spectrum generated from a mixture of $AlCl_3$ and KNCS dissolved in acetonitrile.

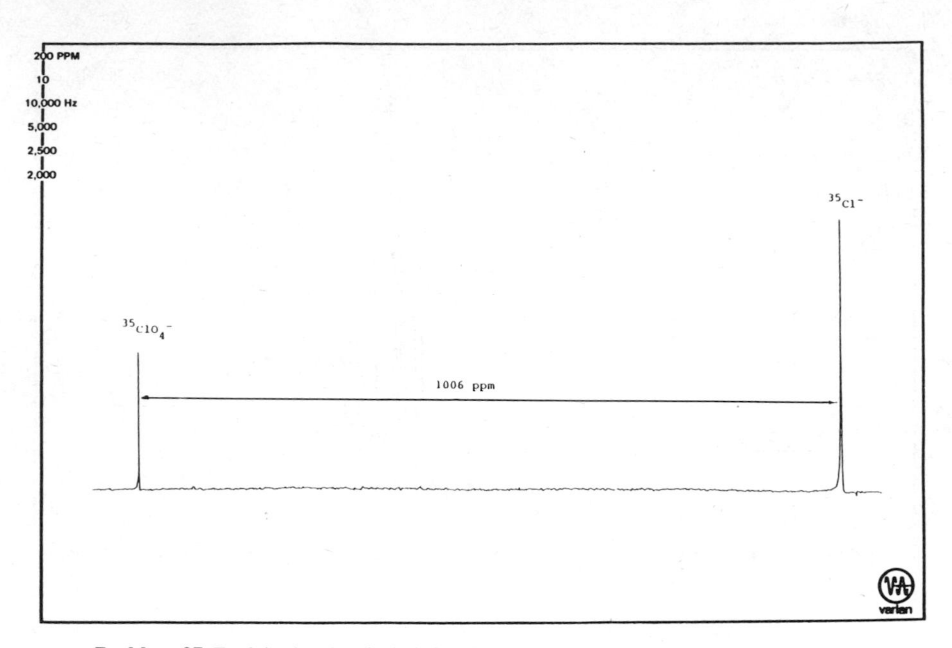

Problem 27 Explain the chemical shift difference between these two chlorine species.

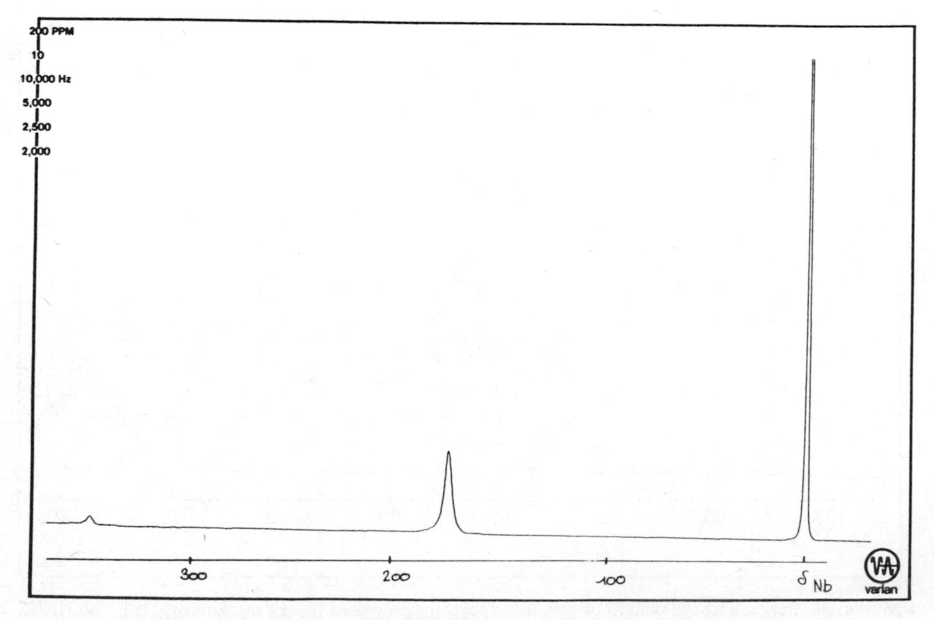

Problem 28 Explain the chemical shift difference between these Nb species.

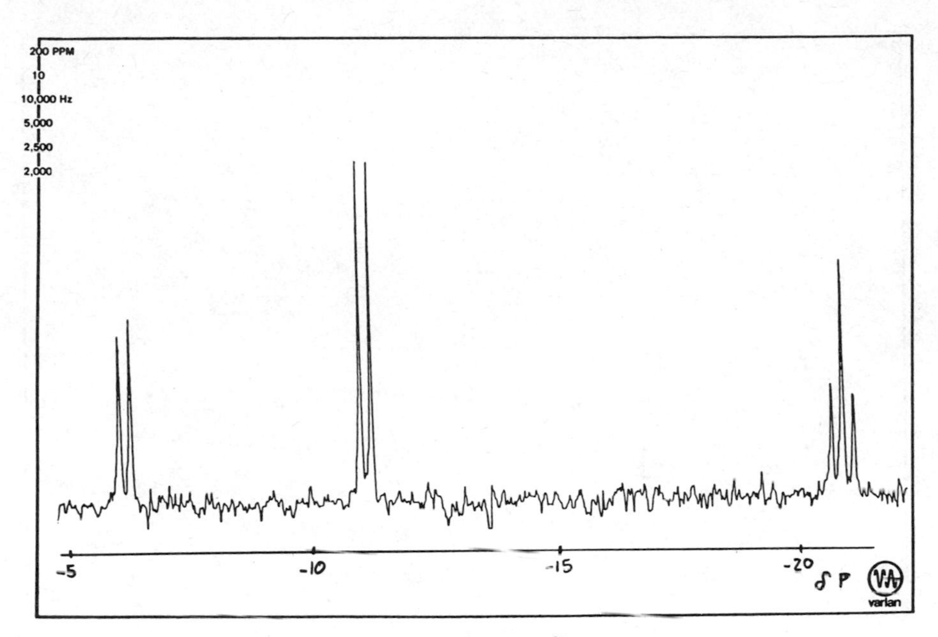

Problem 29 Explain the different P^{31} chemical shifts.

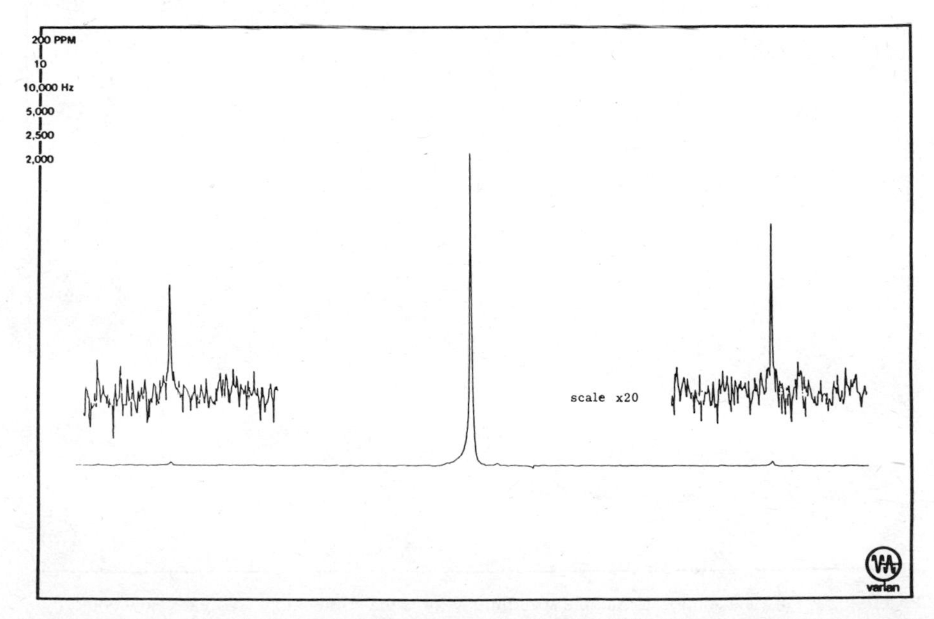

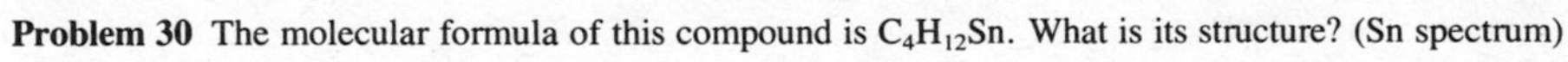

Problem 30 The molecular formula of this compound is $C_4H_{12}Sn$. What is its structure? (Sn spectrum)

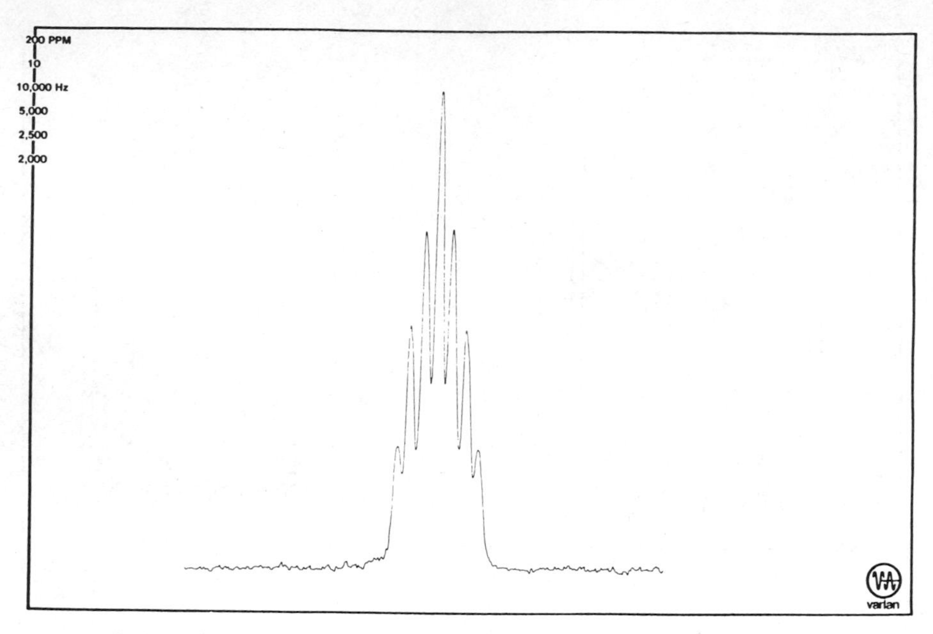

Problem 31 Rationalize the multiplicity of the Te absorption in this compound.

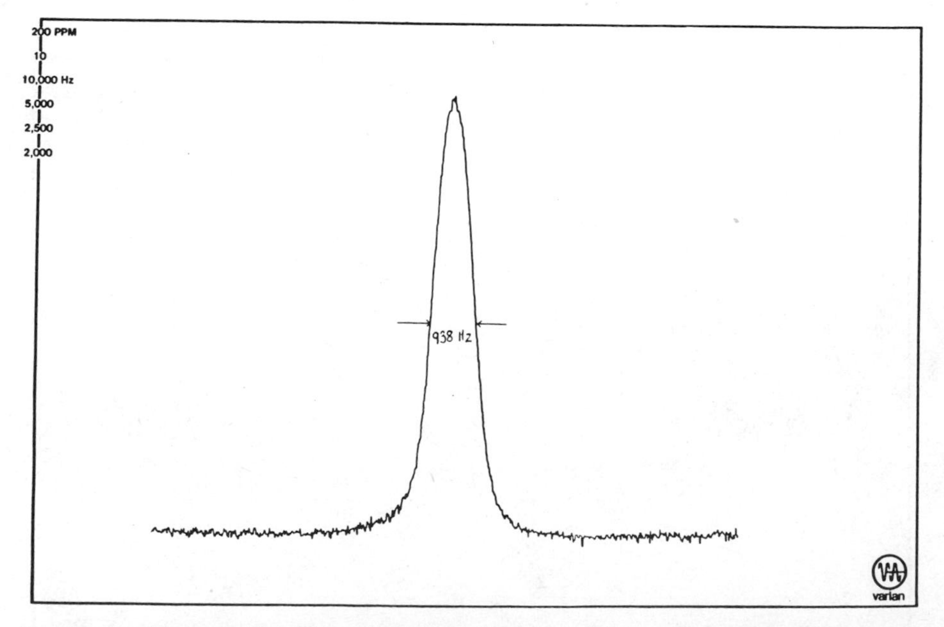

Problem 32 This is a solid CsBr sample spectrum. Why is the $\frac{1}{2}$ line width so great? How could it be improved?

INDEX